Textbook of

MEAT HYGIENE

THE AUTHOR

Dr. Chandra Shekhar is currently Professor, Department of Veterinary Public Health and Epidemiology, College of Veterinary Science and Animal Husbandry, Acharya Narendra Deva University of Agriculture and Technology, Kumarganj, Ayodhya (U.P.), India. He obtained his B. V. Sc. and A. H. and M. V. Sc. (Veterinary Public Health) degrees from College of Veterinary Science and Animal Husbandry, Mathura, C. S. A. U. A. and T., Kanpur. He earned his Ph.D. from College of Veterinary and Animal Sciences, G. B. P. U. A. and T., Pantnagar. He has reported first *Salmonella* Drogana serovar from human source in India. He has to his credit more than 60 scientific papers in various journals of repute. He has published a book on *"Zoonoses: Infectious Diseases of Animals Transmissible to Humans"* in 2018 through Daya Publishing House® (A Division of Astral International Pvt. Ltd.), New Delhi. Dr. Chandra Shekhar has also attended and presented papers in various national and international conferences/seminars. He is a life member of *Indian Association of Veterinary Public Health Specialists* (IAVPHS) and member of the editorial board of *"The Asian Journal of Animal Science"*, *"Veterinary Science Research Journal"* and *"Research Journal of Animal Husbandry and Dairy Science"*. He is the reviewer of the journals *"International Journal of Basic and Applied Agricultural Research"*, *"South Asian Journal of Research in Microbiology"* and *"Indian Journal of Animal Sciences"*. He has been conferred the awards "Fellow" by *Hind Agri-Horticultural Society (FHAS)* Muzaffarnagar (U.P.) India, "Certificate of Excellence in Reviewing" in recognition of an outstanding contribution to the quality of the journal *"South Asian Journal of Research in Microbiology"* and "Best Paper Presentation Award" in the National Seminar. Besides this he has been the recipient of many honors, awards and distinctions from various scientific bodies.

Textbook of MEAT HYGIENE

Chandra Shekhar
M.V. Sc., Ph. D.

Professor
Department of Veterinary Public Health & Epidemiology
College of Veterinary Science & Animal Husbandry
A.N.D. University of Agriculture & Technology
Kumarganj – 224 229, Ayodhya, U.P.

2022

Daya Publishing House®
A Division of
Astral International Pvt. Ltd.
New Delhi – 110 002

ISBN: 978-93-5222-139-4 (HB)

Published by : **Daya Publishing House®**
A Division of
Astral International Pvt. Ltd.
– ISO 9001:2015 Certified Company –
4736/23, Ansari Road, Darya Ganj,
New Delhi-110 002
Ph. 011-43549197, 23278134
E-mail: info@astralint.com
Website: www.astralint.com

Digitally Printed at : **Replika Press Pvt. Ltd.**

Dedication

THIS BOOK IS DEDICATED TO STUDENTS/READERS

Preface

In developing countries, a large number of animals are slaughtered in rural areas under very primitive conditions where even minimal technical and hygienic requirements do not meet. Animals are slaughtered in all kinds of places such as converted buildings or rooms, under the shade of trees, and on open, bare ground. Meat produced under such conditions has high microbial loads which may lead to rapid spoilage and spread of meat-borne diseases in the consumers. Moreover, unsatisfactory slaughtering techniques used by untrained butchers also cause unnecessary losses of meat and valuable by-products.

Meat acts as a vehicle for many microorganisms which causes transmission of meat-borne diseases in the consumers. Although the spectrum of meat-borne diseases of public health importance has changed with changing production and processing systems, continuation of the problem has been well illustrated in recent years by human surveillance studies of specific meat-borne pathogens such as *Escherichia coli* O157:H7, *Salmonella* spp., *Campylobacter* spp. and *Yersinia enterocolitica*. In addition to existing physical, chemical and biological hazards, new hazards are also appearing *e.g.*, the agent of bovine spongiform encephalopathy (BSE).

A contemporary risk-based approach to meat hygiene requires that hygienic measures should be applied at those points in the food chain where they will be of greatest value in reducing food-borne risks to consumers. This should be reflected in application of specific measures based on science and risk assessment, with a greater emphasis on prevention and control of contamination during all aspects of production of meat and it's further processing. Application of HACCP principles is an essential element. The measure of success of contemporary programmes is an objective demonstration of levels of hazard control in food that are correlated with required levels of consumer protection, rather than by concentrating on detailed and prescriptive measures that give an unknown outcome.

The activities of the Competent Authority having jurisdiction at the slaughterhouse (usually Veterinary Administrations) very often serve animal health as well as public health objectives. This is particularly the case in relation to ante-mortem and post-mortem inspection where the slaughterhouse is a key point in animal health surveillance, including zoonoses. Regardless of jurisdictional arrangements, it is important that this duality of functions is recognized and relevant public health and animal health activities are integrated.

Chandra Shekhar

Contents

Abbreviations

α: Alpha
β: Beta
μg: Microgram
AGMARK: Agricultural Produce (Grading and Marking)
APEDA: Agricultural and Processed Food Products Export Development Authority
BIS: Bureau of Indian Standards
BOD: Biochemical Oxygen Demand
CAC: Codex Alimentarius Commission
CCPs: Critical Control Points
cm: Centimetre
DFD: Dark, Firm and Dry
DMI: Directorate of Marketing and Inspection
DNA: Deoxyribonucleic acid
e.g.: Exempli gratia (for example)
EDTA: Ethylene diamine tetra acetic acid
EEC: European Economic Community
EOU: Export Oriented Units
FAO: Food and Agriculture Organization of the United Nations
FDA: Food and Drug Administration
FPO: Fruit Product Order

FSMS: Food Safety Management Systems
FSSA: Food Safety and Standards Act
FSSAI: Food Safety and Standards Authority of India
g: gram
GHPs: Good Hygienic Practices
GMPs: Good Manufacturing Practices
h: hour
HACCP: Hazard Analysis Critical Control Points
HAS: Hygiene Assessment System
HMS: Hygiene Management System
ICMR: Indian Council of Medical Research
ICMSF: International Commission of Microbiological Specification for Foods
INFOSAN: International Network of Food Safety Authorities
IS: Indian Standards
ISI: Indian Standards Institution
ISO: International Organization for Standardization
Kg: Kilogram
kGy: Kilogray
L: Litre
LWK: Live Weight Kill
m: Metre
mA: Milli Ampere
MFPO: Meat Food Products Order
mg: Milligram
ml: Millilitre
MMPO: Milk and Milk Product Order
MT: Metric Ton
nm: Nanometer
NSAI: National Standards Authority of India
°C: Degree Celsius
OIE: Office International des Epizooties (World Organization for Animal Health)
PCR: Polymerase chain reaction
PFA: Prevention of Food Adulteration
PPE: Personal Protective Equipment
ppm: Part Per Million

PSE: Pale, Soft and Exudative
QMSCS: Quality Management System Certification Scheme
RNA: Ribonucleic acid
spp.: Species
SPS: Sanitary and Phytosanitary
SSOPs: Sanitation Standard Operating Procedures
TBT: Technical Barriers to Trade
USDA: United States Department of Agriculture
Viz.: Videlicet (namely)
WHO: World Health Organization
WTO: World Trade Organization

1
Introduction

Meat is an excellent source of protein that can fulfill the protein scarcity in the human diet. In developing countries, where diets are composed of a few staple foods, meat and meat products are especially important in preventing malnutrition and contributing to food security. Paleontological evidence suggests that meat constituted a substantial proportion of the diet of even the earliest humans. Early hunter-gatherers depended on the organized hunting of large animals such as bison and deer. The domestication of animals allowed the systematic production of meat and the breeding of animals with a view to improve meat production. The animals which are now the principal source of meat were domesticated in conjunction with the development of early civilizations. The advent of civilization allowed the domestication of animals such as cattle, sheep, pigs, chickens *etc.* This eventually led to their use in meat production on an industrial scale with the aid of modern slaughterhouses.

The rate of meat production in the world has increased by 280 per cent in the last four decades. At present the demand of meat and meat products has also increased due to growing urban population with changing diets and sufficient income to increase animal products in their diets. In India, there are more than 30 export oriented modern abattoirs established on the guidelines of the Agricultural and Processed Food Products Export Development Authority (APEDA) and they are following Sanitary and Phytosanitary (SPS) measures, mandatory requirements of Hazard Analysis Critical Control Points (HACCP) and International Organization for Standardization (ISO) certification. India is the largest producer and exporter of buffalo meat in the world. In India, there are more than 4000 slaughterhouses registered with local bodies and more than 25000 unregistered premises where animals are slaughtered.

Meat

Meat is the flesh of an animal, typically a mammal or bird which is used as a food. The word *meat* often does not include the flesh of fish or seafood. Most often meat includes the skeletal muscles and associated fat and other tissues along with edible organs and offal. Meat and meat products have a great significance in human nutrition and thus for maintenance of consumer health. Meat is very rich source of proteins, containing all the essential amino acids and a good source of iron, phosphorus, zinc, selenium, riboflavin, niacin, vitamin B6, vitamin B12, choline *etc.* The term meat has been derived from the Old English word "mete", which referred to food in general.

Meat can be broadly classified as "red" or "white" depending on the concentration of myoglobin in muscle fibre. When myoglobin is exposed to oxygen, it is converted into oxymyoglobin which imparts red colour to the meat. The redness of meat varies according to animal species, age, and type of muscle fibre. Red meat contains more narrow muscle fibres that tend to operate over long periods without rest, while white meat contains more broad fibres that tend to work in short fast bursts. Generally, the meat of adult mammals such as cows, sheep, goats, pig and horses is considered red, while chicken and turkey breast meat is considered white. Red meat contains many essential nutrients necessary for healthy growth and development in children. Red meat is one of the best sources of iron and zinc which is well absorbed by the body.

Table 1.1. Technical Term of Meat of Different Animal Species

Animal Species	*Technical Term of Meat*	*Animal Species*	*Technical Term of Meat*
Buffalo	Carabeef	Pig	Pork
Calf	Veal	Pigeon	Squab
Camel	Camel	Poultry	Chicken or poultry meat
Cattle	Beef	Rabbit	Rabbit meat
Deer	Venison	Sheep	Mutton
Duck	Duck	Turkey	Turkey
Goat	Chevon	Wild animal	Bushmeat
Horse	Horsemeat		

Hygiene

Hygiene is the conditions and practices that help to maintain health and prevent the spread of diseases. Hygiene is very important in maintaining the safety and quality of food and protecting the health of animals and humans.

Meat Hygiene

Meat hygiene may be defined as "expert supervision of all meat products with the object to provide wholesome meat for human consumption and to prevent danger to public health". Meat hygiene may also be defined as "all conditions and measures necessary to ensure the safety and suitability of meat at all stages of the food chain". Adequate meat hygiene measures span the time of the live animals enter abattoir until the final meat product is consumed. Meat hygiene involves not only the steps taken during slaughter of the animal for meat production but also several important steps/activities as follows:

1. **Rearing of Animals:** The rearing of animals for meat production purpose is the beginning step in the meat hygiene activities. Obviously the rearing of animals under hygienic condition is an important step in the production of quality meat. Animals reared under hygienic conditions are less subjected to acquiring infectious diseases. Therefore, it is an important step in the production of good quality of raw material (animal) and thereby a quality meat production.
2. **Transportation of Animals:** The animals should be transported under comfortable conditions. Transportation stress may cause poor quality meat production. Preferably the animals should be transported as per the national/international regulation.
3. **Maintaining the Hygiene in the Slaughterhouse Environment:** Overcrowding of animals in the lairage and prevailing poor sanitary condition in it may lead to spread of pathogens among animals. Therefore, cleaning and sanitation of the slaughterhouse environment is important in production of hygienic quality of meat.
4. **Pre-Slaughter Care of Animals:** Pre-slaughter care of animals includes resting, feeding, and watering, and fasting (for a certain time period before slaughter). Pre-slaughter care of animals is important in reducing the microbial contamination of meat.
5. **Ante-Mortem Inspection:** Clinical examination of live animals is important in minimizing the microbial load in meat. Efficient ante-mortem inspection can eliminate the diseased animals. In this way the food production chain can be prevented from contamination from diseased animals.
6. **Hygienic Slaughter of Animals:** Hygienic conditions of slaughterhall, equipment, instrument, and hygienic practices followed by the butchers and meat handlers including their personal hygiene are important in production of hygienic quality of meat.
7. **Post-Mortem Inspection:** Efficient post-mortem inspection of carcase, offal and other parts of the slaughtered animals are important in reducing the microbial load in them by eliminating the diseased or abnormal carcase, offal and other parts of the slaughtered animals.
8. **Microbiological Examination:** Microbiological examination of meat is essential to determine the microbial load and type of organisms present in meat, so that the appropriate action can be taken to improve the hygienic quality of meat by preventing or minimizing its contamination.
9. **Hygienic Processing and Handling of Meat:** Cutting, packaging and handling of meat should be carried out in hygienic way to maintain the safety and quality of meat.

10. **Hygienic Distribution of Meat:** Maintenance of hygiene from production of meat to its transportation and distribution to meat markets and retail meat shops are essential to maintain the hygienic quality of meat and safeguard the health of meat consumers. Butcher's shops need license to operate and must have sufficient working space. The walls and floor should be constructed of durable material and be smooth, impervious, easily cleanable and light-coloured. There should be the provision of disposal of meat wastes.

Meat Inspection

Meat inspection is the supervision of the food animals and the meat derived from them. Meat inspection has two major components which include ante-mortem inspection of the food animals and post-mortem inspection of slaughtered animals that is carcase, offal and other parts. Meat inspection also includes veterinary supervision of municipal abattoirs, export oriented abattoirs, retail meat markets and butcher shops.

Principles of Meat Inspection

The basic principles of meat inspection are as follows:

1. Ensure the supply of clean, wholesome and nutritious meat from healthy animals.
2. Carry out ante-mortem inspection of animals and post-mortem inspection of slaughtered animals that is carcase, offal and other parts.
3. Supervise the transport and distribution of carcase/meat from slaughterhouses to the meat shops and their handling and storage.
4. Protect the health of the slaughterhouse and meat processing plant personnel.

Objectives of Meat Inspection

The main objectives of meat inspection are as follows:

1. To ensure the supply of clean, wholesome, disease-free and cheap meat to the consumers.
2. To ensure that all meat intended for human consumption is to be prepared, stored, transported and marketed under hygienic conditions.
3. To prevent the adulteration (substitution or mixing) of meat, *e.g.*, the carcase of a goat for that of a sheep, the carcase of a horse for that of an ox and the carcase of a foetal calf for that of a marketable calf.
4. To detect the outbreak of animal diseases and assists in the location and control of diseased flocks and herds.
5. To enforce the use of humane method of slaughter of food animals and keep food animals awaiting slaughter under proper hygienic condition.
6. To protect animals and human beings indirectly by interrupting the life cycle of certain parasites, *e.g.*, by excluding dogs from slaughterhouses and

by ensuring the destruction of larval tapeworms of hydatid cysts (caused by *Echinococcus* spp.) infesting the organs and tissues of slaughtered animals.

7. To assists in establishing and fostering meat export trade by ensuring that meat exported is clean and wholesome and that the regulation of the importing countries are observed.
8. To protect the honest butchers from unfair competition, *e.g.*, some butchers may involve selling of adulterated and diseased meat in absence of meat inspection, while honest butchers do not involve in such activities.
9. To prevent nuisance to neighbours and the public in general.
10. To ensure the proper disposal of condemned meat.

Meat Production and Consumption

In most of the developing countries, the agricultural progress has increased the production of staple crops and the introduction of industrial crops. Improved farming techniques, use of new varieties of seeds, greater use of fertilisers, irrigation and chemical control of pests have resulted in considerable increases in production. On the other hand developments in livestock production have lagged far behind. Although there has been an increase over the years in the amount of meat available in developing countries the quantities are small. Globally the meat production between 1980 and 1990 has been increased by 29 per cent ; 15.6 per cent in developed countries and 56 per cent in developing countries. The daily per caput availability of protein from meat increased by 24 per cent but this was an increase from only 4.9 g to 6.1 g in contrast to increase in developed countries by 8 per cent, from 27.4 to 33.9 g per day.

The problems of meat production are complex and include multiple biological, economic and social factors. The practices of the small-holder system of livestock production need to be gradually developed so as to fit local conditions and meet increasing demands. Modern scientific practices developed in industrialized countries are rarely directly transferable to developing countries and, if they are to be transferred usually need to be adapted to local conditions.

Animal products play important role in the socio- economic condition of India. Exports of animal products represent an important and significant contribution to the Indian agriculture sector. The export of animal products includes buffalo meat, sheep and goat meat, poultry products, animal casings, milk and milk products, honey *etc.* India's exports of animal products was Rs. 29,532.65 Crores in 2016-17, which include buffalo meat (Rs. 26,307.93 Crores), sheep/goat meat (Rs. 8, 71.08 Crores), poultry products (Rs. 531.65 Crores), animal casing (Rs. 13.84 Crores) and processed meat (Rs. 4.58 Crores). The demand for Indian buffalo meat in international market has sparked a sudden increase in the meat exports. Buffalo meat dominated the exports with a contribution of over 89.08 per cent in total animal products export from India. The main markets for Indian buffalo meat and other animal products are Malaysia, Indonesia, Vietnam, Arab Republic, Egypt and UAE.

Livestock Resources in India

Livestock sector plays an important role in India's economy in terms of income, equity and foreign exchange earnings. The demand for livestock product has increased tremendously due to the sustained economic growth and rising of domestic income. This has encouraged the rapid expansion of livestock outputs in recent years particularly for export. Among the livestock sector, cattle and buffalo play a major role in India's economy. In India, cattle and buffaloes are reared primarily for milk production and secondarily for meat production. These livestock is generally maintained by rural family with small land holdings, and landless labourers for milk, draught and meat production. Utilization of cattle and buffaloes for draught purpose has gone down due to progressive development of technologies. In India, the rearing of livestock in organized way for commercial purpose has been increased in recent years. Sheep and goats are maintained both under confined (stall feeding) and extensive grazing system. Sheep are reared for meat and wool production, while goats are reared mainly for meat production. In India, 40 per cent poultry production occurs from commercial sectors and 60 per cent from small farmers as backyard system. In India, pig production traditionally occurs through scavenging but organized pig farming has become more common in recent years.

Indian Meat Industry

India ranks 5th and contributes about 3 per cent of total meat production in the world. In India, buffaloes contribute about 31 per cent of total meat production. The major constraints in achieving the goal of quality meat production are poor infrastructure of slaughterhouses, lack of proper training of slaughterhouse personnel and thereby unhygienic meat production. Indian meat industry associated with type of abattoirs and retail meat markets, licensing and registration of abattoirs/ meat processing plants, animal health certification, by-product processing conditions and availability of different meat products, initiatives for export of red meat and meat export scenario are discussed as follows:

Types of Abattoirs

In India, there are two main types of abattoirs.

1. **Municipal Abattoirs:** For domestic market, there are more than 4000 registered with the Local Bodies and more than 25000 unregistered premises where animals are slaughtered.
2. **Export Oriented Abattoirs:** In India, 27 registered Export Oriented Abattoirs and 67 meat processing plants have been established as per the guidelines of the Agricultural and Processed Food Products Export Development Authority (APEDA). These abattoirs follow the Sanitary and Phytosanitary (SPS) measures which are mandatory requirements of HACCP and ISO certification.

Retail Meat Markets

Sheep and goat meat are marketed in the retail meat shops in small towns and air-conditioned meat shops in the metro towns. The meat of large animals is sold in the retail meat shops. In metro towns, the license is issued to these meat shops to maintain the hygienic conditions.

Licensing and Registration of Abattoirs/Meat Processing Plants

The Government of India has laid down standards for export of meat which include standards for abattoirs and meat processing plants for various meat products. Abattoirs and meat processing plants are registered by the Agricultural and Processed Food Products Export Development Authority (APEDA), Ministry of Commerce and Industry, Government of India. Inspection of abattoirs and meat processing plants is carried out by expert committee. An internal ministerial committee has been constituted comprising representatives from Department of Animal Husbandry, Ministry of Agriculture, Department of Animal Husbandry and Dairying of the State Government, Export Inspection Council, Veterinary Council of India and reputed Veterinary Scientists. Abattoirs and meat processing plants are registered after detailed inspection by the said committee. The inspection committee has main focus on the hygienic and sanitary conditions of abattoirs and meat processing plants, personnel hygiene, infrastructure, laboratory facilities, ante-mortem and post-mortem inspections and maintenance of records. The registration of abattoirs and meat processing plants are renewed annually after a detailed inspection by the committee.

Animal Health Certification

As per the export and import policy of Government of India, each export consignment is subject to compulsory microbiological and other tests and a comprehensive pre-shipment certificate is issued by the Government laboratory. Each export consignment is accompanied by this health certificate. This certificate also states that the meat has been prepared from healthy and disease-free livestock, which are free from infectious and contagious diseases, including foot and mouth disease and other diseases. The Health Certificate also states that the livestock has been subjected to ante-mortem inspection followed by post-mortem inspection of slaughtered livestock, and that meat is fit for human consumption. In addition to providing services of highly qualified and experienced veterinarians and microbiologists, the government agencies also carried out the constant monitoring and inspection from procurement of livestock from registered markets to the final export of the meat consignment.

The Government of India has nominated agencies such as all State Directorates of Animal Husbandry, Directorate of Marketing and Inspection (DMI), Export Inspection Agencies (EIAs) and Deonar abattoir, Mumbai (for chilled sheep and goat Meat only) for assurance of health certificates. Meat consignments conforming the notified standards and specifications are allowed for export. Each meat consignment is accompanied by a test certificate issued by any of these agencies. The certificate is based on microbiological examination of the consignment. Ante-mortem and post-mortem inspections of the animals are carried out by qualified government veterinarians to ensure that meat derived from healthy animals is exported.

Availability of Meat and Meat Products

India has a good potential for meat production because of large livestock population. In India, the main meat producing animal species are cattle, buffalo, sheep, goat, pig and poultry. India exports the meat mainly to Vietnam, Malaysia, Thailand, UAE, Saudi Arabia, Egypt and Australia. Uttar Pradesh state has emerged as the major exporter of buffalo meat followed by Punjab and Maharashtra.

There has been a great rise in the production of livestock products and this is expected to continue in the future. The majority of meats consumed in India are beef, mutton, chevon, pork, poultry and fish. In Indian context, culture, traditions, customs, and taboos influence meat consumption to a great extent. The rise in demand for meat products has been raised due to urbanization. Meat consumption, in particular, is determined by the religions where pork is forbidden to Muslims and beef is prohibited to Hindus. The preference and consumption of chicken meat is greatly accepted by consumers in India as compared to the other meat consumption. The increase of chicken meat consumption is due to the versatility of the meat, relatively low cost in comparison to other meat, and the acceptance of the chicken meat to all religions.

In India, the meat products are usually prepared from beef, carabeef, chevon, mutton, pork, poultry and rabbit. The processed and semi-cooked products include corn beef, meat loaf, sausages, curries, bacon, ham, chicken-ham, cutlet-mix and salami. Sausages are one of the most popular meat products in India, which is prepared and available in different varieties such as cocktail sausage, pork sausage, chicken pepperoni sausage, masala sausage and the world famous kransky. India also exports a meat product "rissole" (labeled as continental sausage) to Australia.

Initiatives for Export of Red Meat

The demand for Indian buffalo meat is growing in export meat market due to its cost competitiveness, perceived organic nature and low fat content. Indian meat exporters have been able to meet international demand of meat by utilizing modern integrated abattoirs and meat processing facilities and thereby working to improve the quality of Indian product. The major steps have been taken for quality assurance includes licensing and registration of abattoirs/meat processing plants, and animal health certification.

Meat Export Scenario

International trade of livestock and livestock products in India is mainly in live animals, meat and meat products, dairy products and eggs. All the Export Oriented Units (EOU) are registered with the Agricultural and Processed Food Products Export Development Authority (APEDA), Government of India, after due inspection. India is the largest buffalo meat (Carabeef) exporting country in the world. The quantity of buffalo meat export and number of countries to which it is exported are continuously increasing. At present, India exports the meat to about

64 countries. Indian meat exporters are strictly following the guidelines mentioned in the OIE Terrestrial Animal Health Code, in Volume II (8.5.25). There is no record of any disease outbreaks in livestock of importing countries as a consequence of the import of the meat from India. India exports approximately 50 per cent of total buffalo meat production and about 500000 MT of deboned buffalo meat annually.

Slaughterhouse By-product Processing

India ranks first in the world in livestock holding and has the potential to utilize slaughterhouse by-products to partly meet the growing requirement of animal feeds. The total availability of offal/bones in the country generated from large slaughterhouses is estimated to be more than 21lakh tons annually. Meat by-products are used for preparation of various products such as bone meal, blood meal, meat-cum-bone meal (MBM) *etc.* These products significantly contribute to fulfill the animal feeds. MBM is a protein and phosphorous supplements for animal feed manufacturers. Leading manufacturers of MBM in India are Standard Agro Vet (P) Ltd., Allanasons Ltd., Hind Agro Ltd. and Al Kabeer, Hyderabad.

2
The Abattoir

Abattoir is an enclosure where the food animals are slaughtered for meat production. A modern abattoir has facilities of slaughtering of the food animals, processing of meat (deboning, cutting, packaging *etc.*), storage of meat under refrigeration conditions, rendering of slaughterhouse by-products and effluent treatment plant for safe disposal of solid and liquid wastes. It is important to think at early stage to landscaping the site to limit the impact of the establishment on the surrounding environment. The site plan must show the complete premises, the outline of the curtilage of the site and the location in relation to roads, railways, waterways and adjoining properties and their function. Since specialized knowledge is required in the design and construction of a meat plant. It is important to employ the competent and experienced architects, engineers and veterinarians along with reputable contractors. The premises must be fenced to control access of people and animals and provided with separate gates for clean and dirty functions; a facility where livestock transport vehicles must be cleaned and sanitized after unloading a crush connecting the lairages to the restraining area must be constructed of cleanable, non-absorbent and durable material.The ideal layout of an abattoir is one which encourages the operatives to follow the standard hygienic practices.

Establishment of a modern abattoir is important for quality meat production by conducting various efficient inspection procedures like ante-mortem inspection, post-mortem inspection and other procedures. There are many advantages to improve the conditions for livestock destined for slaughter. These are helpful in improvement of productivity, animal welfare and personnel safety. Increased production through humane treatment of slaughter animals can be achieved by reducing stress in the animals, minimizing the carcase damage and waste due to bruising and other injuries, preventing damage to hide and skin and decreasing the mortality in the animals.

Abattoir should be constructed in such a way that it has all essential components and well-equipped. There should also be facilities of isolation pen in which suspect animals can be segregated and kept for detailed inspection of individual animals. A modern abattoir can be established with the help of competent architects, engineers and other experts having experience in abattoir construction. The objectives of scientific design and construction of an abattoir are as follows:

1. To help in improvement of the present condition of an abattoir.
2. To improve hygiene and sanitation in guiding the construction of an abattoir.
3. To implement the animal welfare regulations and prevent the cruelty to animals.
4. To promote export by efficient utilization of slaughterhouse by-products.

Technical Consideration for Establishment of a Modern Abattoir

The following points should be taken into consideration while establishing a modern abattoir:

1. Site of an Abattoir

Abattoirs are best sited on the outskirts of a town or village, at a distance from built-up areas to prevent possible inconvenience to dwelling-places due to pollution from slaughter wastes and nuisance from noise, stench or the presence of scavenging animals such as vultures, stray dogs *etc.* The premises of the abattoir should be away from environmentally polluted areas and industrial activities which produce pollutants such as soot, dust, smoke; chemical or biological emission and pollutants that pose a serious threat of contaminating the meat. The location of an abattoir should be free from floods, pests and wastes. Abattoir premises near waterlogged areas must be avoided to prevent the problems of stagnation of wastes, problems of sanitation and breeding of mosquitoes. A suitable site of an abattoir is one that can ensure good availability of stock nearby and also ensure the availability of sufficient land for possible future expansion of abattoir. In addition to these, the site of an abattoir should have the following facilities:

i. The ground should be suitable for good foundations and piling and freedom from flooding.
ii. Abattoir should be established in areas where public transport and labourers are easily available.
iii. It should be distant from residential areas and other development to avoid nuisance of noise and smell.
iv. There should be mains water and electricity supply and mains sewerage for disposal of effluents released from abattoir.

2. Area Size and Capacity of Abattoirs

The requirement of the land area size for establishing an abattoir depends on its capacity. The following area size of land is required for establishing abattoirs of different capacity:

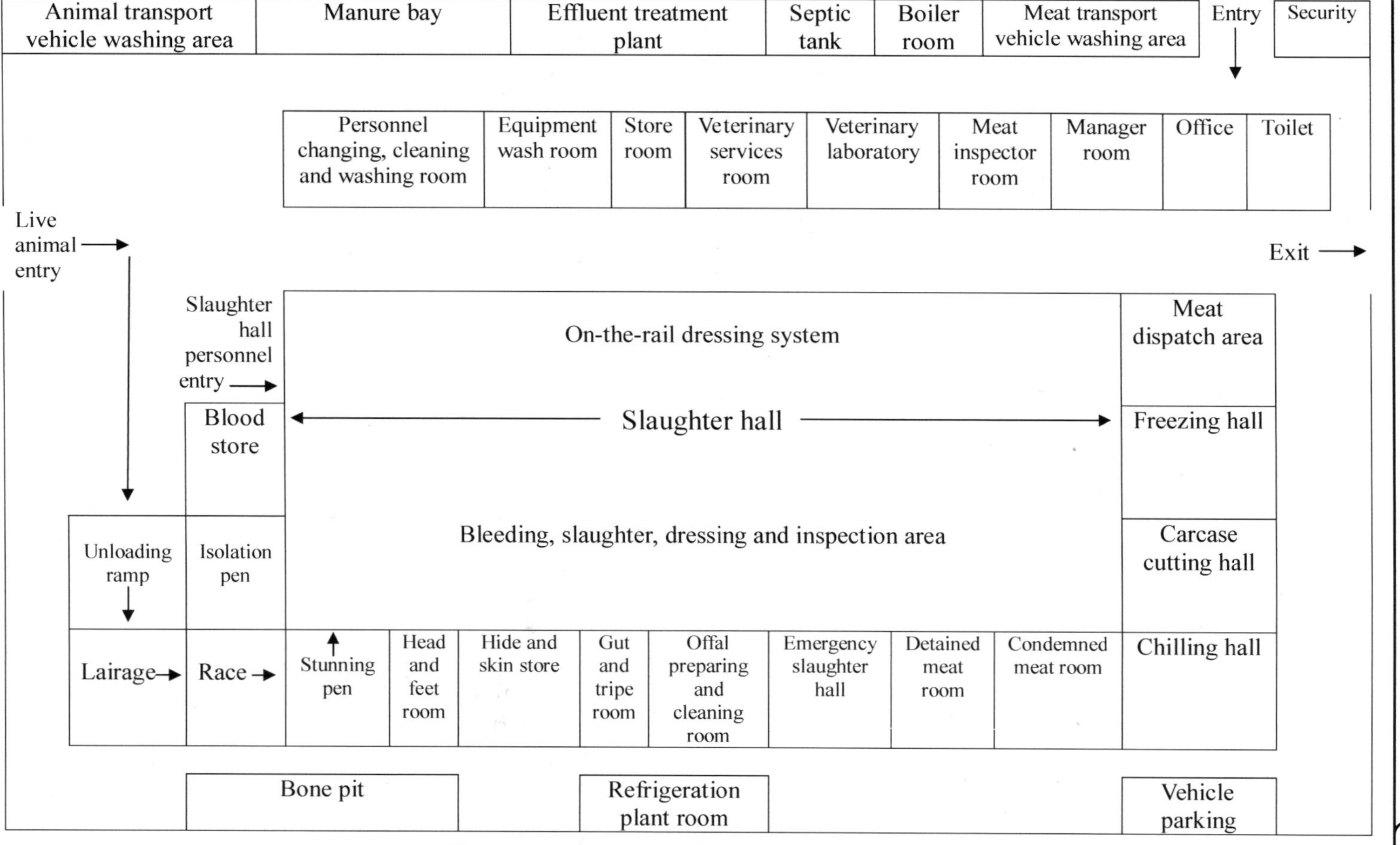

Figure 2.1. Flowdiagram of a Modern Abattoir.

i. Small abattoir (up to 30000 units/year) - 1 to 2 acres.
ii. Medium abattoir (over 50000 units/year) – 2 to 4 acres.
iii. Large abattoir (over 100000 units/year) – 4 to 6 acres.

Note: One unit/one bovine is equivalent to 2 pigs, 3 calves or 5 sheep.

3. Design and Components of an Abattoir

The layout of an abattoir and meat processing plant should permit Good Hygienic Practices (GHP) including protection against cross contamination between and during the operations. The activities in lairage, abattoir, processing, packaging and loading should be compartmentalized. The design of abattoir and equipment should be simple so that their sanitary conditions can be effectively maintained. Unnecessary features in the building and equipment should be avoided from sanitary as well as economic point of view. Sharp corners, crevices, pits *etc.* should be avoided in the building. The type of building depends on the climatic condition in which abattoir is to be constructed. In tropical countries, more open type of building may be required. Impervious materials should be used for construction of walls and floor. The floor should be non-slippery, while, the walls should be smooth. Adequate lighting facilities should be provided throughout the abattoir especially in inspection areas. The important components such as lairage, race, slaughterhall, slaughter lines, chillers, deboning and packing area, freezing, cold storage *etc.* are included in the layout plan of a modern abattoir as follows:

i. Lairage

Lairage is an enclosure in the abattoir intended for providing rest to animals arriving at abattoir for slaughter. In order to conduct efficient ante-mortem inspection of animals arrived at the slaughterhouse; a well-constructed lairage with all essential facilities should be available so that the animals can be kept clean and provide shade and shelter with adequate drinking water supply. Animal behaviour is an important consideration in lairage design and operation, particularly the reaction of animals to extreme change in environment. Sharp corners and projections should be avoided in order to prevent injury. Gates should be placed at the end, not in the middle of the pen side to facilitate the easy movement of animals especially during their exit from the pen. Lairage floor gradients should be at least 1:50. In UK, the Ministry of Agriculture, Fisheries and Food recommended the following pen sizes:

Cattle (loose): 2.3 - 2.8 m^2

Heavy pigs, calves and sheep: 0.7 m^2

Pig (bacon and small pokers): 0.6 m^2

The optimum floor space requirement for tied cattle: 3.0 m^2.

Cattle Lairage

In a modern abattoir, the lairage has several pens with wall or tubular partitions and 2.4 m wide gates at the end of the each pen. A pen size of 7.6 m x 6 m is enough to hold 20-25 cattle. Drinking water availability must be ensured to all animals in a pen. In the tie-up type of a lairage, the long water troughs in front of the cattle are more suitable and easier to clean than individual troughs. Hydrant points should be placed conveniently, so that the sufficient water supply can be reached in all parts of the lairage for cleaning and other purpose. Hayracks must be provided and the animals must be fed twice daily except on morning of the day of intended slaughter or afternoon preceding the morning of intended slaughter.

Bulls and boars should be individually penned. If they are tied, they should be able to lie down. Water must be easily available to them. Troughs should be high enough or protected to prevent the animals from falling in and drowning. If animals slip, this may cause injuries such as bruises, fractures, dislocations and/or damage of skin. Concrete floors should have patterns engraved, or covered in mesh to provide traction, and also facilitating cleaning.

ii. Isolation Pen

The abattoir should have the isolation pen where suspected animals can be kept for detailed examination. After ante-mortem inspection, diseased/abnormal animals are isolated and kept separately in isolation pen for rest/treatment.

iii. Emergency Slaughterhall

Emergency slaughterhall should be situated near to the suspect meat detention room.

iv. Slaughterhall

It should be an open hall with sufficient floor space and well-ventilated and lighted. Size and type of slaughterhall for cattle depends on the system adopted for slaughtering.

Stunning Area

It should be in front of stunning pen at least 3 m in width to opposing wall or bleeding trough and be fitted with upright bars of 5 cm in diameter and 1.2 m high, and spaced at 40 cm intervals for safety purpose. The floor must be properly drained and non-slippery.

Bleeding Area

Bleeding trough should be at least 1.5 m wide and has a proper gradient, and side wall of the same height. There should be two separate drains that are one for blood only and other for water when cleansing only.

On-the-Rail Carcase Dressing Systems

The modern abattoir has provision of line dressing system for slaughtered animals. Carcases can be conveyed by gravity or power along an overhead rail after stunning and bleeding. Dressing process can be divided into various stages, each undertaken by a separate operator as carcase reaches him. High rates of slaughter and proper dressing of carcases can be carried out with the help of a combination of several machines, tools and correlated items of equipment such as brisket saw, carcase splitting saw, aitch bone cutter, hock cutter, mechanical hide puller *etc.* An adequate number and efficient sterilizers operating at 82°C for hand tools, shackles *etc.* are provided on all slaughter/dressing floors. Various systems of line dressing are in operation, the type depending mainly on level of throughput, equipment design and animal species. There are four main types of line dressing for cattle. Slaughterhall can be designed with the following on-the-rail dressing system of carcases:

a. **Gravity Rail System:** In this system, the carcases are suspended from a spreader and single wheel trolley or runner. The carcases are gravitated to each station and manually stopped for dressing operation. In this system, various equipment such as moving-top viscera inspection table, paunch truck *etc.* may be used. Hide puller is rarely used due to small throughput. The design of system is simple; therefore, there is little chance of serious breakdown with consequent loss of production. This system is most compact and economical but slaughter rates are lower (10-40 cattle/h).
b. **Intermittent Powered System:** This system involves mechanical moving of carcases suspended on a spreader (gambrel) and trolley along a rail at intervals by means of a device which can be pre-set according to the rate of slaughter. In this system, the rate of slaughter is comparatively higher (10-75 cattle/h).
c. **Continuous Powered System:** In this system, the dressing line is kept in a continuous motion during slaughter. This system is associated with the use of more sophisticated equipment such as mechanical hide puller, moving top inspection table as with the Canpak system. This system is used for higher slaughter rate (40-120 cattle/h).
d. **Canpak System:** This is a continuous conveyorised system in which the carcases are suspended by heavy beef trolleys or runners from the overhead rail. In this system, spreader or gambrel is not used. This system is used for higher slaughter rate (50-150 or more cattle/h).

Advantages of Line Dressing System

Line dressing system has several advantages such as faster rate of slaughter, safer for operatives than traditional slaughter system, more hygienic dressing, reduces tedious labour and increases throughput and enhances the value of carcase, offal and hide.

Disadvantages of Line Dressing System

Disadvantages of line dressing system include requirement of high standard of maintenance, complete cessation of production in case of breakdown of system and sometimes difficulty in efficient meat inspection.

v. Refrigeration Accommodation

Fresh meat intended for local consumption must be chilled immediately after post-mortem inspection and kept at a constant temperature below 7°C for carcase and cuts or 3°C for offal. The EEC recommends the freezing of beef quarters below –7°C and pork below –15°C, and storage of beef below –17°C and pork below –20°C. The temperature of chilling hall must be recorded on continuous basis using charts or computerized system to control/maintain the chilling temperature. The carcase must be hung in such a way so that free movement of cold air around them can be ensured. The spacing of rail should be 0.9 m for beef, 0.7 m for pigs and 0.5m for lambs. The space between carcases on rails should be 0.3-0.4m. The internal finishes of the refrigeration accommodation should be durable, impervious, properly insulated and have good floor drainage.

vi. Detained Meat Room

Suspected carcase requiring more detailed post-mortem inspection, laboratory testing is required to be detained in separate room. Detained meat room should be located adjacent to main slaughterhall in order to close liaison over disease findings. The lighting intensity should be at least 540 lux and should be of such quality that does not distort colours.

vii. Condemned Meat Room

It should be adjacent to detained meat room with a suitable rail linkage.

viii. Gut and Tripe Room

In the gut and tripe room, the stomachs and intestines are initially separated and emptied.

ix. Cutting Rooms

Initial carcase dressing is carried out in the cutting room. During cutting process the temperature of the building should not exceed 12°C and the rooms should have sufficient refrigeration accommodation to keep the meat at an internal temperature below 7°C. The facilities of suitable equipment and adequate supply of hot, potable water are necessary for maintaining hygienic conditions.

x. Offal Room

Offal such as lung, liver, kidneys *etc.* are kept in a chilling or freezing room known as offal room. The temperature of offal room must not exceed 3°C.

xi. Edible Fat Room

This is a separate room, usually situated near the gut room where edible fat is held pending dispatch.

xii. Equipment Wash Room

This should be situated adjacent to work rooms to avoid buggies, bins and other equipment being washed in work rooms or other inappropriate places. There should be a separate way through the wash room to avoid the mixing of clean and dirty equipment.

4. Facilities in an Abattoir

i. Ante-Mortem Inspection Facilities

Ante-mortem inspection requires ample natural or artificial lighting which should be even and diffuse. Isolation pen and crush for clinical examination of suspected animals are also required.

ii. Facilities for Personnel

Sufficient number of water closets, showers and wash basins must be provided (optimally one for every 15 employees). Separate lockers should be provided to each employee. Dressing rooms must be separated from toilets and these must not open directly into the working areas.

iii. Other Facilities

Other facilities include the following:

Water

Ample and potable water supply should be available at adequate pressure (20 psi) throughout the premises. Non-potable water may be used in exceptional cases for steam production provided that pipes installed for the purpose do not permit this water to be used for other purposes. The minimum quantity of water required for slaughter of cattle and pig is 1000 liters and 450 liters, respectively.

Electricity

Industrial three-phase electricity should be supplied with an additional facility of generator.

Lighting

Natural or artificial lighting of adequate intensity must be provided throughout the meat plant. The type of lighting must not distort colours which may cause problems in inspection procedures. The recommended intensity of light for various locations such as inspection points, work rooms and other areas are 540 lux, 220 lux and 110 lux, respectively.

Ventilation

Proper ventilation in the abattoir must be ensured to prevent excessive heat, steam and condensation.

Drainage

For proper drainage, the slope of the floor should be uniform with a minimum gradient 1:50. Catch basins for grease recovery and traps and vents on drains should also be provided.

5. Manure Disposal and Washing of Livestock

Lairage waste in the form of dung must be removed periodically. Animals must be washed by foot bath spray system or other bath system before slaughter. It is also important to regularly clean and wash the lairage to maintain the sanitary condition.

6. Effluent Treatment Plant

Effluent released from a slaughterhouse contains huge amount of organic matters. Therefore, the proper treatment of effluent is essential before its disposal into any water bodies. Effluent treatment plant in a slaughterhouse should be constructed as per the capacity of slaughterhouse that is according to load of effluent due to daily slaughter of animals. Effluent treatment plant should be constructed to treat the effluent both aerobically and anaerobically to check the pollution. The effluent with 30 BOD can be utilized either for in-house agriculture purposes or let into the farmers field for growing crops. Effluent treatment plant in a slaughterhouse must be working efficiently.

7. Pest Control

The entrance of birds, rats, mice and insects such as flies and cockroaches in the slaughterhouse can cause serious problems since in addition to the dirt they may carry the pathogenic microorganisms which may lead to contamination of premises and meat. Therefore, pests should be controlled to prevent their access to slaughterhouse, production areas and storage departments. Pest can be controlled by the construction of buildings and working places where access of insects, rodents and birds is hindered, but it is almost impossible to secure buildings totally against pests. However, good design and construction may prevent the entry of pests.

It is important to prevent the birds from having access to slaughterhouse. Bird attractants may be food supplies; water, special vegetation around buildings, *etc.* and these attractants must be removed or modified. Manure bay must be sited away from meat areas to prevent the propagation of flies, insects *etc.* Pest can be controlled by ensuring cleanliness, absence of food scraps and use of specialist pest control firms. Pest can also be control by using mesh screens, electrocutors *etc.* Insecticidal spray can be used with discretion and confined to non-meat sectors.

8. Fresh Meat Dispatch Area

It must be sited away from the dirty part and access to it restricted to vehicles associated with meat and offal for human consumption.

9. Veterinary Office

Well-equipped lockable rooms are necessary for meat inspectors. Rooms should be provided with hand washing and shower facilities and lockers for clothing and meat inspection equipment.

10. Veterinary Laboratory

A well-equipped laboratory is essential for diagnosis of animal diseases, monitoring of potential pathogens and maintenance of the overall hygiene standards. There should be a microbiological laboratory with sterilization room, media preparation room, incubation room, laminar flow and washing room for microbiological examination of meat, water, air and personnel working in the plant.

11 Rendering Plant

A modern abattoir is required to have a rendering plant to treat bones, non-edible offal, blood *etc.* for their efficient utilization in the form of by-products such as meat-cum-bone meal, blood-cum-meat meal, tallow *etc.* for animal feed and soap manufacturing. Rendering plant also reduces the problem of waste disposal and pollution in the plant.

Booth System of Slaughter

In booth system, the slaughter, the bleeding and dressing operations are carried out on the floor of same booth from where dressed carcases are dispatched. There is no provision of hoists or hooks that can help in the dressing of carcases off the floor. This usually leads to heavy contamination of carcases. The system can be improved with restraining facilities for stunning of animal, shackling and hoisting facilities for bleeding and dressing.

Type of Abattoirs and their Management

In developing countries, the slaughter premises normally occur as modern abattoirs, old slaughterhouses and slaughterslabs, and makeshift premises.

1. **Modern Abattoirs:** These abattoirs have scientific and recent design, equipment and services. These abattoirs are operated on industrial lines with a wide range of services featuring cold storage, processing, by-product utilization and waste recycling activities. Some of modern abattoirs have export objectives primarily in chilled and frozen meat, although some of their manufactured products/ by-products are channeled into local sale in substitution for imports. Few modern abattoirs in developing countries slaughter directly for public consumption. These abattoirs are generally controlled by central governments with foreign technical assistance and management.
2. **The Old Slaughterhouses and Slaughterslabs:** These premises merely make facilities available for use by licensed butchers and traders for the slaughter of livestock at stipulated fees, and in accordance with public health, inspection and marketing regulations. Slaughterhouses and slaughterslabs thus operate as service establishments under the management of municipal and local authorities. The field of activities of municipal and local authorities often being limited to the larger towns and built-up areas.
3. **Makeshift Slaughter Premises:** These include all kinds of places such as converted buildings or rooms, shade of trees as well as open bare-grounds that a butcher or a community may find convenient for the operation. Makeshift slaughter premises occur generally at village and rural locations, and occasionally may occur in the suburbs or on the fringes of larger towns. Makeshift slaughter premises in the suburbs or on the fringes of larger towns are sometimes considered to have links with illegal livestock trading and the slaughter of sick and diseased animals. Makeshift slaughter premises in the suburbs or on the fringes of larger towns defy obvious norms in slaughterhouse construction, equipment services and hygiene, their existence and operation, therefore, such premises in these locations is not advised. Makeshift slaughter premises are mostly private-owned and under no formal authority or licensing. These premises and their products are not inspected, quantified and subjected to trade and health regulations.

3

The Food Animals

The animals that are used primarily for milk and meat purpose are known as food animals. Food animals include domesticated and a variety of wild animals. Recently, the efforts have been made to domesticate certain wild animals for food purpose, *e.g.*, elands and antelope are being domesticated in Russia, and elands in Africa, kangaroo in Australia and capybara in South America.

Generally cloven footed animals like ruminants are efficient converter of plant material into muscles that serve as a rich source of proteins, energy, mineral, vitamins *etc.* for humans. These animals are generally known as food animals. Meat is normally regarded as the edible parts of the food animals which consume mainly grass and other arable crops, namely, cattle, sheep, goats, pigs, horses, deer and other animals. Recent innovations have included the breeding of wild boar in England and buffalo in Germany, France and Poland.

Food Animal Species

The meat from various animal species such as cattle, buffalo, sheep, goats, horse *etc.* that consume plant foods or vegetation is preferred by people of civilized countries. However, the unconventional meat from animals such as dogs and cats is less desirable and it was consumed by people in old times but now ceased in all civilized countries. At present the great majority of animals slaughtered in almost all countries for human consumption consist of cattle, buffaloes, sheep, goats and pigs. In India, cattle, buffaloes, sheep, goats, pigs and poultries are the main food animals which are used for meat production. The importance of food animal species varies with country, geographic area, time period, culture and religious preferences as well as economic conditions. Different species of food animals are discussed as follows:

Cattle

Cattle were domesticated in Mesopotamia after settled agriculture was established about 5000 BCE, and several breeds were established by 2500 BCE. Modern domesticated cattle include the groups *Bos taurus* (European cattle) and *Bos taurus indicus* (zebu). Different breeds of cattle are used for meat purpose throughout the world. Zebu cattle and crossbred cattle of European origin are most commonly used for this purpose. In India, cattle are not primarily reared for meat production but some old aged cattle are slaughtered in few states. Some people engaged in meat trade believe that the meat of bulls is inferior due to inferior conformation, tenderness and being subjected to development of dark firm dried (DFD) meat or dark cutting meat. But there is no scientific background about this because the carcases of young bulls are heavier and leaner than the carcases of castrated male cattle (bullock) of the same age.

Buffaloes

Buffalo (*Bubalus bubalis*) belongs to family Bovidae. Buffaloes are being domesticated since more than 5000 years in the Indus valley. Buffaloes are efficient in converting roughages into valuable products. Buffalo meat is lean and contains less cholesterol. The nutritional value as well as palatability of buffalo meat is almost same as cattle meat. The dressing percentage of buffalo is about 3 per cent less than the cattle due to heavier hide and head. The important breeds of buffaloes are Murrah, Nili-Ravi, Bhadawari, Jaffrabadi, Nagpuri, Manda, Kalahandi *etc.*

Buffaloes are reared mainly in Asian countries. At present, buffaloes are reared in about 30 countries in the world but more than 50 per cent buffalo population of the world occurs in India. Buffalo rearing other major countries are China, Pakistan, Bangladesh and Nepal. In India, buffaloes are reared primarily for milk purpose; however, generally they are sold for slaughter purpose especially in old age or at the end of their productive life. In India, buffalo meat is one of the most preferred meats, especially in low income groups. India ranks first in buffalo meat production in the world. In India, buffaloes contribute about 31 per cent of total meat production. India exports buffalo meat about 95 per cent of the total meat export.

Sheep

Sheep (*Ovis aries*), originating from western Asia, were domesticated with the help of dogs prior to the establishment of settled agriculture, likely as early as the 8th millennium BCE. Several breeds of sheep were established in ancient Mesopotamia and Egypt by 3500-3000 BCE. At present, more than 200 sheep breeds exist worldwide. China, Australia, New Zealand, Turkey, Iran, India and Pakistan are the main sheep producing countries in the world. Sheep has greater potential of survival in adverse environmental conditions. Sheep can be maintained in marginal areas that are unsuitable for cultivation. Sheep are reared for multi-purposes such as meat, milk, wool, skin, manure *etc.* Sheep meat is one of the most preferred meats in many countries of the world. Sheep are important animals to nomadic pastoralists.

India ranks fifth in sheep population in the world and contributes 5 per cent of world sheep population. Sheep meat contributes 4 per cent of total meat production in India. In India, there are 42 breeds of sheep. The important Indian sheep breeds are Marwari, Malpura, Jaisalmeri, Muzaffarabadi, Nali, Chokla, Gaddi, Rampur Bushair, Ganjam, Bhakarwal, Magra, Pugal, Deccani, Bellari, Nellore, Madras Red, Mandhya, Coimbatore, Kashmir Marino *etc.*

Goats

Goat (*Capra hirius*) is an economically important domesticated animal in Asia and Africa. Goats can survive in a variety of climates ranging from extremes of tropical rain forests to dry deserts (where survival of sheep is difficult). Rearing of goat is more economical than the cattle and sheep under natural grazing/browsing. Goats are generally raised on natural vegetation, and stubble and weeds on cropped lands. Goats are reared primarily for meat purposes; however, their other valuable products are milk, fiber, skin, manure *etc.* Goat meat is naturally lean and has low saturated fat content; therefore, it is preferred by the consumers.

There are more than 100 breeds of goats in the world and half of them are found in Asia. China, India, Pakistan, Bangladesh and Nigeria are the main goat producing countries in the world. In India, the goat meat is one of the most preferred meats. India contributes 10 per cent of total goat meat production in the world. In India, important goat breeds for meat production are Black Bengal, Barbari, Beetal, Sirohi, Sangamneri, Marwari, Ganjam, Kutchi, Assam Hill, Osmanabadi *etc.* Among Indian goat breeds, Black Bengal has the best reproductive performance owing to first kidding at early age, short kidding interval and higher percentage of multiple births.

Pigs

Pig (*Suis scorfa*) is an important animal species for meat production which contributes about 40 per cent of total meat production in the world. Pig meat is consumed in many countries of the world; however, it is not consumed by certain community. Domestic pigs, which are descended from wild boars, are known to have existed about 2500 BCE in modern-day Hungary and in Troy; earlier pottery from Jericho and Egypt depicts wild pigs. Pork sausages and hams were of great commercial importance in Greco-Roman times.

Pigs are well adapted in respect of food and climate. Pigs continue to be bred intensively as they are being optimized to produce meat best suited for specific meat products. Pigs are efficient in conversion of feed energy to the body energy and proteins. They are important meat producing animals due their high efficiency of feed conversion into body weight gain, short gestation period (about 112 days) and large litter size (7-12). There are various breeds of swine, of which important breeds are Large White or Yorkshire, Hampshire, Berkshire, Pietrain, Landrace, Duroc *etc.*

In India, pig rearing in earlier time was mostly confined to poor section of people but nowadays it is reared on commercial scale for meat production for local consumption as well as export. In India, there are several indigenous breeds of swine that are reared for meat production. Exotic breeds of swine such as Yorkshire, Hampshire, and Landrace are most commonly reared for commercial meat production in India.

Pork obtained from boars has an objectionable odour and flavour when cooked. This condition is known as boar odour or boar taint. Boar taint is caused mainly by a pheromone androstenone which is an unsaturated steroid "5α-androst-16-en-3-one". This steroid is produced in the testes of the boar and accumulated in the body fat. Boar taint is also caused by amino-acid decomposition products (skatole and indole). Moreover, swine breeds especially Pietrain and Poland China are more prone to development of pale soft exudative (PSE) meat and dark firm dry (DFD) meat.

Poultry

The term poultry includes a number of domesticated avian species such as chicken, turkey, duck, goose, guinea fowl, pigeon and quail. Poultry is commonly known as chicken. The principal poultry producing countries are USA, China, Brazil, France, Mexico, UK, Japan and Italy. Important breeds of chicken are Rhode Island Red, Plymouth Rock, Australorp, Brahma, Cochin *etc.* Poultry meat production is important from economic point of view because of high efficiency of food conversion into body weight gain, and short duration to achieve body weight for slaughter/ meat production.

In India, domestication of poultry (*Gallus gallus domesticus*) started around 2000 BC. In India, poultries are reared in backyards as well as in the form of well-organized and commercial poultry farms. Backyard poultry keeping in India is popular due to need of less care and management, less amount of feed requirement; earning economy and availability of meat. Indigenous breeds are also reared on free range system in rural areas for meat production. The important indigenous poultry breeds are Aseel, Chittagong, Denki, Ghagus, Karaknath, Nicobari and Tellichery.

Duck

The growth rate is faster in ducks due its high feed conversion efficiency. In India, ducks are primarily reared for egg production; however, they are also utilized for meat purpose. Some objectionable flavour may be found in eggs and meat of ducks. Commercial duck farms are less common in India. Duck meat production is popular in United Kingdom.

Quails

Quail is an important avian species for egg and meat production. It requires less space and feed in comparison to poultry. Japanese quail (*Cotornix cotornix japonica*) is popular in India and commercially reared for egg and meat production.

Ratites

Ratites commonly used for meat production includes ostriches, emus and rheas. However, cassowary and kiwi are also reared for meat production in some countries. The Ratite family includes flightless birds with a flat, keel-less breastbone (the keel is where the flight muscles connect). They are mostly large and long-legged birds of the infraclass Palaeognathae. Ratites and humans have had a long relationship starting with the use of the egg for water containers, jewelry, or other art medium. Ratite hides are popular for leather products like shoes. Ostrich is native to Africa; emu, to Australia; and rhea, to South America, particularly the grasslands of Argentina.

Ratites meat is popular especially in developed countries. Most of their muscle is in their legs and thighs. In the wild, ratites eat seeds, herbaceous plants, insects, and small rodents. Ostriches, rheas, and emus are the ratites most commonly raised as livestock in the United States.

Ratites produce red meat that is similar to beef or venison, and the hide makes fine leather products. Although ratites are poultry, the pH of their flesh is similar to beef. Therefore, they are classified as "red" meat. The raw meat is a very dark cherry red. After cooking, the meat looks like beef and the flavour is similar but a little sweeter. The meat of ratite is characterized by low fat and cholesterol contents, darker than beef and having high final pH (due to stress). Ratite meat is also rich in copper, selenium, vitamin B and biologically active peptides such as creatine (emu), anserine (ostrich) and has low content of sodium (ostrich). Some important ratites are discussed as follows:

Ostriches

Ostrich is the world's largest bird and one of the oldest (having existed as a species for over 40 million years), is adapted to living in open, arid country. Four distinct geographic subspecies are recognized, ranging from the Arabian and Saharan deserts southward throughout Africa. The first ostrich farm was founded in 1838 in South Africa, and that country is still the dominant producer of ostriches. Male ostrich feathers were popular for hats during the 18th century, which led to hunting and sharp declines in populations. Ostrich farming grew out of this need, and humans harvested feathers, hides, eggs, and meat from the ostrich. Ostriches were present in Asia as recently as the Holocene, although the genus is thought to have originated in Africa. The African ostrich is the largest living ratite. A large member of this species can be nearly 2.8 m tall and weigh as much as 156 kg. Ostriches are reared in Australia, North and South America, Israel and European countries. Ostriches are reared for meat, hide and feathers. Ostrich farming has been found more economical than cattle and sheep farming. Ostrich meat is comparable in taste and texture to calf meat and beef. Ostrich and emu farming have been started in India for commercial meat production.

Emus

Emu farming also became popular for similar reasons and for their emu oil. Of the living species, the Australian emu is next in height, reaching up to 1.9 m tall and about 50 kg. Its life span is between 30 and 40 years, and it starts its reproduction cycle at about two to three years of age. Production practices for emus differ only slightly from those for ostriches. Emus breed and lay eggs mainly during the winter months, while ostriches and rheas breed and lay eggs in late spring, summer, and early fall. Emu meat is apparently lean, low in cholesterol, high in iron and vitamin C and has a taste similar to filet mignon.

Rheas

Rhea is a large fast-running bird. The rhea, native to South America, is smaller than the ostrich but similar in appearance, with feathered necks and heads but with three toes. The rhea family consists of two separate species that is the white rhea and the common or gray rhea (gray or light brown with white bottom). Rheas reach 1.5 m in height and weigh 22 to 36 kg. The larger American rhea grows to about 1.4 m tall and usually weighs 15 to 40 kg. Rhea feathers are popular for dusters, and eggs and meat are used for chicken and pet feed in South America.

Cassowary

Cassowary is a ratite that belongs to the genus "*Casuarius*". Cassowary is the native to the tropical forests of New Guinea (Papua New Guinea and Indonesia), nearby islands, and northeastern Australia. It is shorter than an emu, but heavier and solidly built. Cassowaries prefer thickly vegetated tropical forest. They can be very dangerous when surprised or cornered because of their razor sharp talons. In New Guinea, cassowary eggs are brought back to villages and the chicks are raised for eating as a much-prized delicacy. They reach up to 1.7 m tall and weigh as much as 59 kg. The meat of the cassowary is said to be very strong in taste. Cassowaries are still part of the diet in New Guinea, where they are either hunted or kept as domestic animals.

Kiwi

Kiwi is the smallest ratite, mostly found in New Zealand. Kiwi is chicken-sized, shy, and nocturnal. They nest in deep burrows and use a highly developed sense of smell to find small insects and grubs in the soil. Kiwis are notable for laying eggs that are very large in relation to their body size. A kiwi egg may equal 15 to 20 percent of the body mass of a female kiwi. The smallest species of kiwi is the little spotted kiwi, at 0.9 to 1.9 kg and 35 to 45 cm in height. Kiwi has thick layer of subcutaneous fat. The meat of kiwi is somewhat unpleasant in taste.

Horse

Horses are commonly eaten in France, Italy, Germany and Japan. Horses and other large mammals such as reindeer were hunted during the late Paleolithic in Western Europe. Horses have long been domesticated for draft, ride and a food source, and are still very important in many countries. In some countries including Japan and Korea, horsemeat is considered a delicacy. Horse bones have also been used as a traditional medicine for bone diseases including bone fracture and arthritis in Korea.

Among red meats, horsemeat has been known to contain high levels of unsaturated fatty acids such as α-linolenic and monounsaturated fatty acids, indicating that consumption of horsemeat may be more beneficial for health than that of beef. The horse carcase is characterized by its dark colour, with absence of large fatty adipose tissues and a peculiar yellow fat. In comparison to beef, pork or poultry, the horsemeat has low levels of fat and cholesterol (about 20 per cent less), relatively high concentrations of *n-3* fatty acids and heme iron indicating that its consumption may be beneficial for health.

Camel

The camel is a multipurpose animal. It can be used for milk, meat, tourism and agriculture. Camel meat is usually eaten as an alternative source of protein in arid countries such as Somalia, Egypt, Saudi Arabia and Kazakhstan and Kenya, camel blood is mixed with milk as a source of iron, vitamin D, salt and other minerals. Camel meat has been consumed for centuries. Ancient Greeks recorded it being offered at Persian banquets, where it was usually roasted whole. Roman Emperor Heliogabulus is said to have enjoyed camel heels. Heliogabulus aside, most people prefer the brisket, ribs and loins of the camel, while the humps are considered a particular delicacy. The hump contains "white and sickly fat" that can be used to preserve other meat (known as Khli) such as mutton, beef and camel.

The low rate of camel slaughter observed in southern Asia is linked to religious aspect, the Hindus being mainly vegetarian. Globally, the farming systems are still not very well developed or organized to ensure good camel meat supply. Camel meat has low fat content and is highly nutritious, and has potential to be used to combat hyperacidity, hypertension, pneumonia and respiratory disease.

Rabbits

Domestication of rabbits has been found since Roman times. Commercial rearing of rabbits started in sixteenth century. Domestic breeds of rabbits are descendants of European wild rabbit (*Oryctolagus cuniculus*). Rabbits have high fertility rates and fast growth rates. They are highly efficient in food conversion. Rabbits are efficient in converting roughages into high quality protein. Rabbit meat is of high quality due to low fat and high protein contents. The rabbit forms a subsidiary but important source of meat food. Rabbit meat contributes a significant role in meeting the demand of meat in Great Britain by importation of rabbit meat from Ireland and Australia.

There are about 40 recognized breeds of rabbits. New Zealand White is the most common breed for commercial meat production. Other commercial meat producing breeds of rabbits are Grey Giant, White Giant, Soviet Chinchilla and California. In UK, rabbits are commercially reared for meat production. In India, rabbits are reared in small farms mainly for wool and pelts, but recently efforts are being made for rearing of rabbits for meat production purpose.

Deer

The domestication of deer species for meat and hunting has been taking place for an estimated 2,000 years. Meats from the deer species (fallow, whitetail and red deer) as well as elk, moose, caribou (reindeer), and sometimes antelope are known as **venison**. Deer meat is of high quality due to low fat content. Game-farmed venison has been proven to have a lower fat and cholesterol content than most red meats. Elk is generally tenderer than beef. Ranched elk is a naturally tender and healthy meat with a mild, distinctive flavour, although some people refer to wild elk meat as the queen of game meats. The demand for venison has increased greatly in the last few years and continues to grow rapidly. Red deer are reared in the UK and New Zealand for commercial meat production. The New Zealand venison industry is currently the largest in the world. However, in the last few decades the deer farming for meat production has grown in, Germany, Great Britain, Ireland, the United States, Canada and South Africa.

Other Animals

Some non-conventional food animals are also used for their meat in many countries of the world. Historically, dog meat has been consumed in various part of the world such as Switzerland, Mexico, Hawaii and Japan. Dog meat is frequently consumed in Vietnam, China and South Korea. Dogs are occasionally consumed in the Arctic regions. Cat meat is consumed in Southern China, Peru and Northern Italy. Guinea pigs are raised for their flesh in the Andes. Dolphins and whales are hunted for their flesh in Norway, Greenland, Iceland, Alaska, Siberia, Canada, the Faroe Islands, Saint Vincent, the Grenadines, Indonesia and Japan. Polar bear meat is consumed by Eskimos in northern America. Kangaroo meat is consumed in Australia. Rabbit meat is consumed in UK. Rhinoceros and hippopotamus meat is consumed in Africa. Deer meat is consumed in UK and New Zealand.

4
Meat Hygiene Practice

The role of meat hygiene team in the slaughterhouse is very important in independent verification of the food safety management system. Accidental puncture of the stomach and intestine may cause spillage of contents which serve as a source of contamination. Care must also be taken to prevent the contact of carcase and offal with the floor or structures of the slaughterhall. The slaughterhouse and meat processing personnel must be trained regarding their personal hygiene and safety measures. Insectocutor can be implemented in the slaughterhouses and meat processing plants to prevent the contamination of meat. The wearing of rubber and chain mail gloves is important in maintaining the hygiene and safety of personnel. In developing countries; some traditional methods of handling, processing and marketing of meat cause its contamination and lead to a considerable loss of product as well as risk of food-borne diseases.

Sources of Contamination of Meat

1. **Animal:** Diseased animals serve as potential source of microorganisms. The meat from diseased animals may act as source of contamination for meat from normal and healthy animals. Therefore, animals should be healthy and disease-free for production of hygienic quality of meat.
2. **Hide, Hair and Fleece:** Animal itself, particularly in the winter season is one of the main sources of contamination in the slaughterhouse. Soil, manure *etc.* adhering to the hide, hair and fleece of the animal are main sources of contamination of carcase. Therefore, excessively dirty animals should be detained and separately slaughtered to minimize the contamination of carcase.

3. **Gastrointestinal Contents:** Careless or accidental puncture of the stomach and intestine may lead to spillage of gastrointestinal contents and thereby increase in the microbial load in the carcase.
4. **Sticking Point:** During the act of sticking the microorganisms can enter the jugular vein or anterior vena cava and reach in the bloodstream and thereby in the body tissues.
5. **Surfaces of Slaughterhall and Other Structures:** The design of line dressing system in a slaughterhall should be in such a way that the animal body, carcase, visceral organs and other parts should not touch with the surfaces of slaughterhall floor and walls, supporting structures, tables *etc.*
6. **Equipment/Instruments:** Improperly cleaned and inadequately sanitized slaughtering/processing equipment/instruments serve as potential source of contamination of meat. Equipment like knives, knife sharpener, carcase splitting saw, brisket cutter, hock cutter *etc.* come into direct contact with the carcase and other parts. Therefore, equipment/instrument must be regularly cleaned and sterilized.
7. **Personnel:** Butchers and meat handlers may be an important source of contamination. The cleanliness and personal hygiene in butchers and meat handlers cannot be obtained by regulations alone. Therefore, their health education and training about cleanliness and maintaining the personal hygiene is important for quality meat production. Butchers and meat handlers should be regularly examined for their health. They should be examined for some important diseases like typhoid, paratyphoid, tuberculosis and other infectious diseases. They should be provided suitable clothing which should be regularly cleaned, washed and disinfected. There should be provision of ample and potable water supply and soap for cleaning of hands and bathing. Moreover, clean towel should be provided to them for wiping and drying the hands. Control of movement of personnel is also important in controlling the contamination of meat. The movement of personnel should not be allowed from dirty area to clean area to avoid contamination of clean premises and meat.
8. **Water:** The contaminated water supply in the slaughterhouse may lead to spread of diseases in the animals and slaughterhouse personnel. Moreover, the water is used in the slaughterhouses in various processes like cleaning and washing of lairage, cleaning of animals before their slaughter, cleaning of slaughtered animals and blood, cleaning and washing of carcases, cleaning and washing of slaughterhouse equipment/instruments *etc.*
9. **Environment:** Slaughterhouse environment plays a very important role in maintaining the sanitary and hygienic conditions of slaughterhouse. Soil and dirt adhering to the animal body surfaces serve as important source of contamination of meat. The atmosphere of the slaughterhouse should be free from contamination. Sterilizers in the slaughterhall are usually maintained above 82 °C to sterilize and remove the fats adhering on the surface of the equipment like knife, knife sharpener, other equipment

etc. There should be proper ventilation in the slaughterhouse buildings, especially in the slaughterhall to avoid accumulation of excess heat, moisture, steam *etc.* Slaughterhouse wastes must be properly disposed and the effluents must be properly treated before their disposal into the water bodies.

10. **Rodents:** Rodents may be reservoirs of many pathogenic organisms. They can spread organisms in the slaughterhouse premises and can contaminate equipment, instruments, slaughter and processing areas.
11. **Flies and Insects:** Flies and insects can contaminate the meat. They must be prevented from entry in the slaughterhouse buildings by applying appropriate screens on doors, windows, ventilators *etc.* Insectocutor can also be installed to kill them.
12. **Chemical Contamination:** The carcase/meat may become contaminated when comes in contact with surfaces and equipment which have not been properly rinsed during their sanitization.

Meat Hygiene in Relation to Public Health

Meat hygiene practice begins not only from the stage of meat production but it begins from rearing of animals under hygienic conditions. Meat may be contaminated from different sources such as animal itself, butcher, meat handler, slaughtering equipment and instruments, wall and floor of the slaughterhall, water supply in the slaughterhouse, atmosphere of the slaughtering premises *etc.* Meat being rich in protein, fat, mineral, vitamins as well as moisture; furnishes a favourable medium for the growth of many microorganisms. Microbial contamination of meat may produce either spoilage of meat or health hazard when consumed. Meat and meat products is important vehicle of food-borne illness outbreaks worldwide. Therefore, the strict hygienic measures must be adopted at all the stages of production, processing, storage, distribution and marketing of meat to ensure its high quality in order to protect the public health as well as to compete the international meat market for its export.

Elements of Meat Hygiene

Elements of meat hygiene includes ante-mortem inspection, post-mortem inspection, final inspection, re-inspection, prevention of adulteration, prevention of misrepresentation (prevention of false labeling), sanitation (environmental sanitation, equipment sanitation, personal cleanliness *etc.*), condemnation and destruction of waste materials. Labeling includes site, size, type, material of the label, labels such as fit or unfit for consumption.

1. **Ante-Mortem Inspection:** Ante-mortem inspection has an important role in meat hygiene. The hygienic quality of meat can be maintained by conducting efficient ante-mortem inspection. There are many diseases and conditions which can be easily detected during ante-mortem inspection. Efficient ante-mortem inspection reduces the load on post-mortem inspection of carcase, offal and other parts of the slaughtered animals.

2. **Post-Mortem Inspection:** Post-mortem inspection plays an important role in detecting and eliminating abnormalities, reducing the contamination and ensuring the wholesomeness of meat for human consumption.
3. **Final Inspection:** Carcase and viscera which are found diseased and abnormal on routine post-mortem inspection are subjected to final inspection with a view to more thorough inspection. Final inspection is carried out to determine whether the carcase and visceral organs are fit for human consumption or not. In doubtful cases, the carcase is held pending till laboratory findings.
4. **Re-Inspection:** Re-inspection is carried out to ensure that the product is being handled in a clean environment; no harmful ingredients are being added to the product and have not been mislabeled.
5. **Condemnation:** As a general rule the acute and generalized conditions of carcase and visceral organs require total condemnation from public health point of view. Chronic and localized conditions require only local condemnation. The condemned carcase/meat, visceral organs and other parts should be properly disposed in order to prevent the contamination of environment and spread of disease in animals and humans.
6. **Prevention of Adulteration:** Adulteration is a fraudulent practice which involves substitution or mixing of flesh of cheaper variety which is objectionable for the reasons of health, religion and economy. Food adulteration is becoming a widespread problem in recent years. The incorrectly declared composition of products on labels and the failure to declare the presence of a specific ingredient are common problems. Despite the binding nature of national and international food regulations, it is impossible to prevent unfair business practices from harming the consumer. Beef for buffalo, horse (UK) and kangaroo (Australia) meat; mutton for goat meat; dog/cat for calf meat; and rabbit meat for chicken are few examples of substitution of meat. Mixing of meat involves pooling of meat from different animal species or good quality meat with poor quality/infected meat.
7. **Prevention of Falsification:** It is a fraud practice in selling of meat. Falsification of meat includes the selling of meat of different animals or mixing the inferior or undesirable meat with superior or costly meat, *e.g.*, female animals instead of male animals. This practice is carried out for economic reasons or to increase the selling of meat.
8. **Prevention of Misrepresentation:** Misrepresentation of meat includes the selling of meat by replacing/adding a particular part of the carcase/animal body part (head, tail *etc.*)/organ (testes).
9. **Supervision of Distribution of Meat:** Supervision of meat distribution from slaughterhouse to the retail meat shops is essential in order to prevent contamination of meat and its spoilage.
10. **Environmental Sanitation and Personal Hygiene:** Environmental sanitation is very important for hygienic meat production. Unclean

slaughterhouse environment contain large number of microorganisms. Therefore, proper cleaning and disinfection of slaughterhouse and its surroundings are essential for hygienic meat production. Condemned meat, effluent and other slaughterhouse wastes must be properly disposed. Effluent released from slaughterhouse must be subjected to effluent treatment plant for its essential treatment before its disposal into the water bodies. Adequate ventilation must be provided in the slaughterhouses to prevent accumulation of excessive heat, steam and condensation. Windows and ventilators should be fitted with screens to prevent the entry of insects, flies *etc.*

General Principles of Meat Hygiene

The OIE (2009) has given the 12 General Principles of Meat Hygiene as follows:

1. Meat must be safe and suitable for human consumption and all interested parties including government, industry and consumers have a role in achieving this outcome.
2. The competent authority should have the legal power to set and enforce regulatory meat hygiene requirements and have final responsibility for verifying that these requirements are met.
3. It should be the responsibility of the establishment operator to produce meat that is safe and suitable in accordance with regulatory meat hygiene requirements.
4. Meat hygiene requirements should control hazards to the greatest extent practicable throughout the whole food chain. Information from primary production should be taken into account so as to tailor meat hygiene requirements to the spectrum and prevalence of hazards in the animal population from which the meat is sourced.
5. The establishment operator should apply Hazard Analysis and Critical Control Point (HACCP) principles. The competent authority should define the role of those personnel involved in meat hygiene activities where appropriate, including the role of the veterinary inspector.
6. The competent authority should verify that the establishment operator has adequate systems in place to trace and withdraw meat from the food chain.
7. Competent authorities should recognize the equivalence of alternative hygiene measures where appropriate, and promulgate meat hygiene measures that achieve required outcomes in terms of food safety and suitability and facilitate fair practices in the food trade.
8. Principles of meat hygiene applying to primary production include hygiene of slaughter animals, killed wild game, feed and feed ingredients, the primary production environment, transport of slaughter animals and transport of killed wild game.

9. Principles of meat hygiene applying to animals presented for slaughter include condition of lairage, ante-mortem inspection (design of ante-mortem inspection systems, implementation of ante-mortem inspection and ante-mortem judgement categories) and information of animals presented for slaughter.
10. Principles of meat hygiene applying to inspection of killed wild game presented for dressing and inspection.
11. Principles of meat hygiene applying to establishment, facilities and equipment include design and construction of lairage, slaughter areas and areas where bodies of animals are dressed or meat may otherwise be present, construction and design of equipment where bodies of animals are dressed or meat may be present, water supply, temperature control, facilities for personal hygiene and transport control.
12. Principles of meat hygiene applying to transportation, product information and consumer awareness and training in meat hygiene.

Objectives of Meat Hygiene

1. To ensure the supply of clean and wholesome meat to the consumers.
2. To protect the public health from meat-borne diseases.
3. To ensure the efficient meat inspection (ante-mortem inspection, post-mortem inspection *etc.*) for hygienic meat production.
4. To ensure the humane method of slaughter of animals.
5. To promote the meat trade in the local meat markets by ensuring hygienic meat.
6. To promote the export of meat by ensuring hygienic meat production.
7. To prevent the adulteration of meat.
8. To protect the interest of honest butchers and meat retailers.

Method of Reducing Contamination

It is important to prevent the contamination of meat and poultry whenever possible. Contamination may occur inadvertently or cross-contamination from the live animal, processing procedures and equipment, employees, the environment *etc.* Contamination can be minimized or prevented by the following appropriate measures:

1. **Dealing with the Dirty Animal:** It is generally challenging to deal with dirty animals for hygienic dressing after their slaughter. There is always risk of contamination in this situation. The dressing of the dirty and dry animals (like cattle and sheep) is easier than the dirty and wet animals. The slaughter of the dirty and wet animals are delayed and keep them dry by resting them in strawed yards or in the case of sheep and lamb, on expanded-metal floors.
2. **Protecting the Meat from Meat Plant Workers:** Meat plant workers engaged in handling, packaging and transportation, must wear the clean

and protective clothing including head cap, and footwear in order to prevent or minimize the level of contamination in meat. There should be facility of hand washing unit with germicidal soap and disposable paper towels for operatives at each slaughterfloor. The use of rubber and chain-mail gloves by the meat plant worker also serve as source of contamination. Rubber gloves exposed to hide with faeces are contaminated which is almost impossible to render these free from microorganisms by cleaning and washing. Chain-mail gloves also become contaminated which can be cleaned effectively only after removal from the hands. Cut-resistant polyester yarn gloves can be used instead of chain-mail gloves and these can be washed and used many times. Diseased person may serve as a potential source of contamination; therefore, they must be prohibited from working on it and handling it. At the time of recruitment of persons for working in the meat plant, they must be subjected to medical checkups and certified by the medical experts.

3. **Basic Hygienic Techniques:** It is important to keep the equipment and instruments in clean and sanitized conditions in order to reduce the contamination of meat. The most common method of sterilizing/ decontaminating equipment and instruments is by keeping these in a cabinet containing hot water maintained at 82°C for at least 2 minutes. The knife, knife sharpener and other equipment should be completely emerged in hot water. The equipment is preferably subjected to flow-through of water to avoid adherence of filth released from the equipment. For sterilization of carcase splitting saw, its blade is placed in a cabinet and sprayed with stream of water. Knife scabbards are used for safety reasons. These are more liable to contamination, therefore their sterilization is essential. Knife sharpeners should be made up of stainless steel and kept free from contamination. When knife sharpeners are not in use, these should be kept clean and stored in hygienic conditions.

4. **Layout and Flowlines:** The layout of the slaughterfloor and the flowlines for entry and exit of carcase/meat, operatives, waste bins and other equipment is important and must be properly designed at planning stage. There should be physical separation between clean and dirty parts of the slaughterhouse. There should be a proper provision so that the operatives can reach to their work-station without contaminating themselves or meat by walking through or under carcases on the lines or passing through sensitive areas like green offal rooms. Moreover, the waste bins and the hide conveyor should never cross the slaughterline because of the risk of contamination. The layout of the individual work-station is also important and it should be located at appropriate site so that it can be conveniently used.

5. **Dressing Techniques:** Dressing is the removal of hide/fleece/hair from slaughtered animals. The carcase and offal contamination by faces, urine, dirt, debris and hair can be prevented or minimized by good and hygienic dressing techniques. The animal should be clean and dry before stunning.

It should be hoisted on the rail and bled before removal of the hide. When an incision is made on the skin, the knife receives contamination from the dirty exterior of the skin, and knife gradually moves towards cleaner interior part of skin. Therefore, effort should be made to make the cut small and minimum as possible. All incisions from the initial cut should be made using a "spear cut technique" in which the blade of the knife is revered so that the back of the knife is against the carcase and the cut is made from the inside of the skin towards outside. This effort ensures moving of the knife from clean inside towards dirty outside. On the rail dressing system, the freeing of hide should be started from the rear quarters. The used knife should be cleaned and sterilized in the sterilizer, and the hands should be washed properly. An initial short incision is made on the ventral aspect in the midline between the hind legs of the suspended animal. The initial incision is extended downwards the umbilicus, and upwards along the free hind limb towards the hock, using the spear cut technique. Great care must be taken to avoid in-rolling at the edge of the hide to prevent contamination.

6. **Prevention of Contamination from the Gastrointestinal Tract (GIT):** Gastrointestinal content is one of the most important sources of contamination for meat. The contamination of carcase/meat from gastrointestinal content can be prevented by careful and technical dressing of the slaughtered animal. Great care must be taken to avoid puncture of the gastrointestinal tract and spillage of its contents. The contamination of carcase/meat from gastrointestinal tract can also be prevented by sealing of the rectum (bunging) and sealing of the oesophagus (rodding). The rectum of the slaughtered animal can be sealed by using plastic bag and elastic band. An incision is made around the rectum to make it free from the surrounding tissues. The rectum is grasped with one hand and the plastic bag is then unfurled over the rectum with other hand and secured in place with the elastic band. The oesophagus in cattle can be sealed by an elastrator ring applied at the oesophageal-ruminal junction. The rubber ring can be applied on oesophagus by using a stainless instrument for placing it at oesophageal-ruminal junction.

Post-Slaughter Decontamination

The decontamination of meat and poultry carcases can help to reduce human food-borne infections. However, process hygiene to prevent contamination should never be neglected. Decontamination substances are applied to foods of animal origin intended for human consumption to remove microorganisms such as *Salmonella* and *Campylobacter* from their surface. Decontamination substances can be chemical substances, but also biological agents such as bacteriophages.

The methods of post-slaughter decontamination of poultry and red meat carcases are quite different. Post-slaughter decontamination of poultry carcases can be carried out either by adding antimicrobials in chill water tank or by spraying after chilling of carcases. Post-slaughter decontamination of red meat carcases can be carried out by de-hairing through chemicals, removal of physical contaminants using knife through trimming, spot cleaning through hot water, spray rinsing of carcases with water or with chemical solutions and pressurized steam. Some important methods of post-slaughter decontamination of beef and poultry carcases are discussed as follows:

1. **Spray Washing with Water**: It was a common practice in recent past to use a large volume of hot or cold water to clean and wash the carcase for removal of any visible contamination. Nowadays the hot water maintained above 74°C is applied for 10 seconds for decontamination of carcase. Hot water also gives carcases a better bloom than cold water sprays. Carcases can also be decontaminated by washing with water maintained at temperatures 85-90°C containing 15 ppm chlorine. This can be preferably applied through a fan-shaped nozzle at a pressure of 100 psi, at a rate of 18 litres per minute for cattle and 9 litres per minute for sheep.
2. **Steam Pasteurization/Sterilization:** The surface of the carcases can be decontaminated and sterilized by using steam vacuum sterilization process which uses water maintained at 88°C, vacuum and steam at 45 psi. The carcases can also be decontaminated by steam sterilization system which involves removal of carcase surface water followed by passing of split carcases through sealed chamber where they are exposed to low-pressure steam above 85°C for 8 seconds. This is followed by cooling of carcases with chilled water spray.
3. **Trimming:** Surface contamination of the carcases can be reduced or minimized by trimming followed by washing with water. Frequent sterilization of knives and other equipment is essential to make the trimming process more effective in view of reducing the contamination.
4. **Chemical Treatment:** Various types of chemicals such as chlorine, organic acids (acetic acid, lactic acid, citric acid, fumaric acid *etc.*), hydrogen peroxide, antimicrobials like nisin and bactericin, and phosphates like trisodium phosphate can be used to reduce the bacterial load on meat after slaughter. The efficiency of a chemical to reduce the number of pathogenic microorganisms from meat surface depends on the particular chemical used, the concentration of chemical, the contact time, the type of tissue, the time period lapsed between contamination and chemical treatment, the organism, temperature of chemical solution and method of delivery.

   Post-slaughter decontamination of carcases can be carried out by spraying chlorine solution to reduce the level of contamination in meat. Chlorine was one of the first chemical treatments to be used for decontamination of carcase in the beef industry, and good reductions in microbial count have been achieved using water containing 200-500 ppm chlorine. However, such high levels of chlorine are not permitted in the food industry, and lower concentrations are not effective. Chlorine (hypochlorite, ClO_2) water is the most common and widely used antimicrobial in poultry processing. The incorporation of 18-25 ppm of chlorine into chill water can reduce *Salmonella* significantly. The efficacy of chlorine increases with increase in the concentration but higher concentration may results in discolouration, off-odor and off-flavor in carcases. Higher concentration of chlorine solution (350 ppm or above) has been found effective in reducing the contamination. But such higher concentrations of chlorine are not recommended due to its carcinogenic effect. Another disadvantage of using chlorine is its inactivation by organic matter and combines with amino nitrogen to form the less effective chloramine.

Organic acids have been found more effective on adipose tissues than lean meat. These acids are successful in reducing the number of spoilage bacteria present on the meat surface but these acids are less commonly used for removing pathogenic bacteria. Organic acid solution mainly acetic acid, lactic acid, or citric acid at 1.5-2.5 per cent is used for decontamination of beef carcases. The spray of organic acid such as lactic acid and acetic acid at 1-2 per cent are also effective in reducing bacterial load in poultry, but higher concentration of these acids tends to bleach the carcases.

Spraying or adding in chill water containing chemicals such as 8-12 per cent trisodium phosphate, 0.5 per cent cetylpyridinium chloride, 0.03 ppm ozonated water for chill water, 5 per cent hydrogen peroxide for spray and 0.5-1.5 per cent hydrogen peroxide for chill water can be used for decontamination of poultry carcases. Bacteriocins and activated lactoferrin are also permitted for decontamination of poultry carcases.

Trisodium phosphate has been found effective in reducing the total viable counts on poultry carcases by 50 per cent and the incidence of *Salmonella* contamination of carcase from 9.5 per cent to 0 per cent. The post-processing chemical treatment of carcases to reduce the level of contamination in them is not suitable because these practices sluggish the process of implementation of strict sanitary measures. Another problem existing with chemical treatment of carcases is that the chemical residue-free meat cannot be obtained.

5. **Washing of Edible Offal:** Edible offal such as liver, heart, kidneys *etc.* need washing with continuous flow of clean and wholesome water.
6. **Irradiation:** Irradiation is very effective technology to control food-borne pathogens and can be applied in meat based products to increase safety and shelflife by improving quality and maintaining nutrient content during storage. Irradiation is also very effective technique to control the microbial growth and to enhance the microbial safety of meat. Irradiation involves preservation of food without using chemicals. X-rays, gamma rays and electron beams can be used to reduce pathogenic bacteria of raw and processed meat. A dose of up to 4.5 kGy is approved for use on refrigerated meat products whereas a dose of up to 7.0 kGy is approved for frozen meat products.

   The number of bacteria such as *Campylobacter jejuni*, *Escherichia coli* O157:H7, *Staphylococcus aureus*, *Salmonella* spp., *Listeria monocytogenes* and *Aeromonas hydrophila* can be significantly decreased or eliminated by low-dose (less than 3.0 kGy) irradiation. However, high dose of radiations can cause detrimental effects on meat quality by producing off-flavour such as rotten egg, fishy, barbecued, burnt, and acetic acid.

The bactericidal action of ionizing irradiation is linked to bacterial DNA damage by free radicals produced during the irradiation process. Irradiation can be applied in combination with other chemicals and technologies that yield better decontamination of meat. Combinations of irradiation and organic acid have been found more effective than the alone use of irradiation for retarding total microbial counts and coliforms in pork during storage.

Monitoring of Hygiene

Monitoring of cleaning and disinfection effectiveness is partly a matter of trained visual assessment and partly of microbiology. A plant that is not visually clean represents a risk of microbial contamination. The control of these risks is best achieved by a HACCP (Hazard Analysis Critical Control Point) approach.

The critical control points (CCPs) in the cleaning and disinfection programme should be identified and measurement and protocol set up. Some of the CCPs can be readily measured, *e.g.*, the solution concentration of a detergent or disinfectant, while others involves checking the procedure itself. After each CCP stage in the cleaning procedure, the operator should check the effectiveness and thoroughness. This check should be backed up by the hygiene supervisor, who, if not satisfied, should ask for a repeat of the stage.

Quality assurance and/or production personnel should also check the plant regularly, looking in particular for old soil deposits not removed at the last clean and buildup of fungi, corrosion or scale deposits. The supervisor and his or her team should allow be periodically audited to ensure that the cleaning and disinfection procedures are being followed, chemical solution strengths should be regularly checked by titration, fast kit or conductivity meter.

Food safety is related to all steps in the food chain. It starts from livestock producer who should bear the responsibility to provide clean and healthy animals which are free from undesirable residues followed by thorough manufacturing processes to retail outlets, hotels and restaurants and the consumer's kitchens. Food poisoning organisms can be introduced into the food chain at any stage, but the most important is judged to be at primary production where intensive system of husbandry and with the breeding of stock solely for productivity purposes without regard for disease resistance constitutes the main offenders.

Assessment of Operational Hygiene

The standard of operational hygiene was traditionally assessed by meat hygiene inspection team by utilizing a programme of regular checks and combined with a random visual and microbiological examination of the end products. In most meat factories the standard of operational hygiene was assessed by regular inspection of the hygiene of structure and equipment, usually utilizing check lists. The checks carried out, and the standards of hygiene structure and procedures required were laid down in codes of practice such as those developed to encourage Good Manufacturing Practices (GMP), regulations, directives and by-laws enforced by local, national and international agencies. These quality control programmes can only identify problems after they had occurred and make them correct. Therefore, to render the meat fully safe for human consumption, the implementation of HACCP in meat industry is essential.

Hygiene Assessment System (HAS)

Hygiene assessment system is a tool used in abattoirs to conduct audits regarding adherence to legislation and good practices, by focusing on the slaughtering process as a whole, abattoir maintenance, pest control, personnel matters as well as the hygiene management system. A standard HAS document is used on a regular basis to determine conformance and deviations. Each aspect in the abattoir is rated from critical to good. Based on the rating, suggestions are made in order for corrective measures to be taken. The hygiene management system (HMS) is a part of the HAS audit. The abattoir needs to implement the HMS as per legislative requirements and is audited to ensure that all documentation and checklists are in place and completed as per the relevant requirements.

Risk assessment principles are also applied as a tool for assessing and monitoring the hygienic standards. Most of the developed HAS includes observation of each steps in the process and awarding the points according to the degree of compliance with the required standard. The score gives a measure of the hygienic compliance. There is a correlation between operational HAS system and total viable bacterial counts. HAS scores are useful in predicting the ability of an abattoir to produce carcases of sound microbiological status. The HAS score gives a snapshot of the hygienic standard of the process and provides a useful tool to improve Good Manufacturing Practice (GMP).

Status of Hygienic Meat Production in Developing and Developed Countries

1. Slaughter Facilities

In developing countries, there are improper slaughtering techniques and lack of appropriate slaughtering facilities that cause unnecessary losses in meat as well as in valuable slaughterhouse by-products. While these techniques and facilities are available in the developed countries.

2. Processing

The facilities for an uninterrupted cold chain for meat are insufficiently available in many developing countries, while these facilities are available in developed countries.

3. Training

In developing countries, the personnel are usually untrained or inadequately trained in the meat industry. This is one of the main constraints limiting the improvement of the hygienic and technical quality of meat. While, the personnel in the meat industries in developed countries are adequately trained.

Meat Hygiene Practice in India

There are more than 4000 authorized slaughterhouses registered with Local Bodies, several unauthorized private slaughterhouses and 27 Export Oriented Abattoirs in India. Export Oriented Abattoirs follow the Sanitary and Phytosanitary (SPS) measures which are mandatory requirements of HACCP and ISO certification. These regulations and measures are not properly followed in other slaughterhouses.

In India, the food animals are not raised especially for meat purpose. Facilities for transportation of animals from purchase place to slaughterhouses are inadequate and uncomfortable. Most of the slaughterhouses do not have proper infrastructure. The slaughterhouses generally do not have proper ventilation and adequate light. The hygienic conditions of slaughterhouses especially municipal slaughterhouses are poor. Slaughterhouse personnel generally do not follow the proper hygiene due to lack of awareness and training on hygiene. The clean and unclean operations are carried out in the same hall. Refrigeration facilities for transportation of meat are inadequate. There is lack of hygienic water supply in most of the slaughterhouses. Most of the slaughterhouses do not have facilities of waste treatment and disposal.

Maintaining the Hygiene in Retail Meat Shops

The quality of meat in retail shops can be maintained by keeping it covered to prevent contamination from flies, insects, rodents and environment. Meat can be safely stored under refrigerated conditions. Cleaning and disinfection of floor and walls should be carried out regularly. The table top, knife and other equipment should be thoroughly cleaned and sanitized to prevent the contamination of meat. There should be provision of screens on the door, windows, ventilators *etc.* to prevent the entry of flies, insects and other pests.

5

Hygiene and Management of the Food Animals from Farm to Slaughter

Meat hygiene safety management systems should take into consideration of entire process in production and processing involved in producing meat or meat products as food from farm animals. The monitoring of all aspects of husbandry practices on the farm should be the first step in a longitudinally integrated meat hygiene management system. A high level of stockmanship is essential, especially when animals are first housed. The animals should be in a clean condition while leaving the farm and be transported under clean conditions. The inspection veterinarian on duty at the slaughterhouse decides which animals have to be slaughtered in the sanitary slaughter department or in the common slaughterhall after the slaughter of the clean animals. Among the food animals, the fat pigs are the most commonly affected during transport as their heat-eliminating powers are limited and they soon succumb to overexertion.

Both the quality and quantity of meat are very much affected by pre-slaughter conditions. In developing countries, the meat animals are transported from the farm to the slaughterhouse on foot and by road or rail. The quality of meat is affected due to travelling of animals on foot or insufficiently rested during transport before slaughter. Often holding pens are overcrowded which causes unnecessary stress to the animals. Carcase and its storage depend greatly on the care taken prior to slaughter. The animal body temperature rises due to nervousness, tiredness, and excitation which cause imperfect bleeding. Muscular fatigue reduces the glycogen content in the blood, which after slaughter changes into lactic acid, thus causing favourable conditions for spoilage and the growth of food-borne microorganisms.

Fatigue and excitement in animals also cause penetration of bacteria from the intestinal tract to the meat.

Rearing of Food Animals

In many developing countries, livestock keeping is a multifunctional activity. Beyond their direct role in generating food and income, livestock are a valuable asset, serving as a store of wealth, collateral for credit and an essential safety net during times of crisis. The livestock sector has been undergoing change at an unprecedented pace over the past few decades. Booming demand in the world's most rapidly growing economies for food derived from animals has led to large increases in livestock production. This surging demand has been mostly met by commercial livestock production and associated food chains. The hygiene and management system at the animal farms contribute significantly in the quality meat production. Various food animals, *viz.*, cattle, buffalo, sheep, goats, pigs *etc.* are reared at farms for commercial meat production.

Animal Welfare on the Farm

Animal welfare is important for quality meat production as well as to prevent cruelty to animals. The following points should be taken into account in order to effective implementation of animal welfare on the farm:

1. The animal should not suffer from hunger, thirst and malnutrition.
2. There should be provision of appropriate comfort and shelter for animals.
3. There should be provision of prevention, rapid diagnosis or treatment of injured, diseased or infested animals.
4. The animal should be free from fear and distress.

Transport of the Food Animals

The food animals need to be moved for various purposes such as marketing, slaughter, re-stocking, from drought areas to better grazing and change of ownership. Typically, methods used to move animals are on hoof, by road motor vehicle, by rail, on ship and by air. Generally, the majority of livestock in developing countries are moved by trekking on the hoof and by road or rail. Historically, livestock has been moved on foot, but with increasing urbanization of the population and commercialization of animal production, livestock transport by road and rail vehicles has surpassed trekking. Transport of livestock is the most stressful and injurious stage in the chain of operations between farm and slaughterhouse and contributes significantly to poor animal welfare and loss of production.

The most appropriate methods of moving cattle are on hoof, by road motor vehicle or by rail wagon. Moving cattle on the hoof (trekking) is suitable only where road and rail infrastructure does not exist, or when distances from farm to destination are short. This method is slow and fraught with risks to the welfare and value of the animals. Rail transport is useful for short-haul journeys where loading ramps are available at railheads and communication is direct to destination. Road motor transport is by far the most versatile, the method of first choice and the most users friendly. The most satisfactory method of transporting cattle is by road motor vehicle. Moving by rail wagon requires more careful management.

Upon reaching a predetermined age or weight, livestock are usually transported in mass to the slaughterhouses. Depending on its length and circumstances, this may exert stress and injuries on the animals and some may die during transport. Unnecessary stress in transport may adversely affect the quality of the meat. In particular, the muscles of stressed animals are low in water and glycogen, and their pH fails to attain acidic values, all of which results in poor meat quality. Moreover, campaigning by animal welfare groups, laws and industry practices in several countries tend to become more restrictive with respect to the duration and other circumstances of livestock transports.

Loading and unloading of animals are often the most stressful part of transportation. Long distance transportation especially in inconvenient conditions causes stress in the animals. Therefore, proper care, management and transportation of food animals are important in order to prevent economic losses that may incur due to injury or death of animals and deterioration of the quality of meat. These are also important for implementation of animal welfare regulations and prevention of cruelty to the animals. Ramps and platforms are necessary for loading and unloading livestock from transport vehicles or walking them to slaughter facilities. Ramps should have cross-slating or steps (10 cm high x 30 cm deep) to facilitate walking and prevent slipping. The ramp should be sloped at an angle of 20degrees or less. Poor offloading facilities result in injuries from mishandling of animals.

Design and Construction of Transport Vehicle

Injuries in the animals occur during transport are mostly associated with the design and construction of the vehicle. It is important that a vehicle used to transport livestock be designed for the purpose for which it is to be used. To minimise unnecessary injuries and stress, the floor of the vehicle must be solid and impervious. This is important especially in the case of double and multiple decked vehicles. If there are gaps in the floor, animals on the lower decks are soiled by excreta from the upper deck. In the case of single decked trucks, excreta pollute the road. The sides of vehicles should be high enough to prevent animals, particularly pigs, from jumping out and injuring themselves. Insides should also be padded at hip level with old tyres to reduce bruising of cattle and ostriches. Narrow entry doors can lead to considerable bruising of hips. A roof is not necessary on a transport vehicle for bovines and small ruminants provided the animals are not exposed for hours in the hot sun. Vehicles for pigs should have roofs unless the pigs are to be transported in the early morning or late evening.

Transport vehicle for slaughter animals should have adequate ventilation, non-slippery floor with proper drainage, provision of protection from rain and heat from the sun, particularly for pigs. The side of the vehicle surfaces should be smooth and free from sharp edges and projections. The transport vehicle should not be totally enclosed. Improper ventilation in the transport vehicle may lead to suffocation and stress in the animals. Pigs are particularly susceptible to excessive heat, poor air circulation, high humidity and respiratory stress. Therefore, the transport vehicles should never be totally enclosed. Proper ventilation in the transport vehicle is also important in removing ammonia gas emitting from the urine. The free flow of air at floor level is important to facilitate removal of ammonia from the urine.

The floor of the transport vehicle should be non-slippery to reduce the risk of animals falling. A grid of cross slating made from wood or metal is suitable. The grid can be removable, so the vehicle can be used for other purposes. The floor of the transport vehicle should be at the level with off-loading platforms to prevent falling and injuries to the animals. There should be sufficient floor space in the transport vehicle for comfortable journey. Overloading of the vehicle may cause injuries or even death of animals.

Table 5.1. Approximate Floor Space in the Vehicle or Wagon Required for Transporting Different Class of Animals

Sl.No.	Class of Animal	Floor Area/Animal (m^2)
1	Cattle	0.62-2.0
2	Calf	0.28-0.56
3	Pig	0.3-0.8
4	Sheep/goat	0.15-0.4
5	Horse	1.0-1.75
6	Ostrich	0.8

Table 5.2. Approximate Floor Space in the Vehicle or Wagon Required for Transporting Different Class of Cattle

Sl. No.	Class of Cattle	Body Weight of Cattle (kg)	Mean Dimension of Cattle (Width x length) (m)	Space Required for Unhorned Cattle (m^2)	Space Required for Horned Cattle (m^2)
1	Calves	Up to 50	0.27 x 1	0.28	0.28
		50-100	0.46 x 1.32	0.56	0.56
2	Cattle/ Buffaloes (Young/ Adult)	100-200	0.46 x 1.33	0.62	0.73
		200-300	0.56 x 1.52	0.86	0.96
		300-400	0.64 x 1.65	1.06	1.20
		> 400	-	>1.27 to 1.73	>1.59 to 2

Table 5.3. Space Allowances for Road and Rail Transport of Horses

Sl.No.	Class of Horse	Required Floor Space in the Vehicle or Wagon/Animal (Dimension)
1	Adult horses	1.75 m^2 (0.7 x 2.5 m)
2	Young horses (6-24 months)	1.2 m^2 (0.6 x 2 m)
3	Ponies (under 144 cm at the whither)	1 m^2 (0.6 x 1.8 m)
4	Foal (0-6 months)	1.4 m^2 (1 x 1.4 m)

During long journeys the foals and young horses should be able to lie down. These figures may vary by a maximum of 10 percent for adult horses and ponies and by a maximum of 20 percent for young horses and foals, depending not only on the horses' weight and size but also on their physical condition, the meteorological conditions and the likely journey time.

Table 5.4. Approximate Floor Space in the Vehicle or Wagon Required for Transporting Different Class of Sheep/Goat

Sl. No.	*Class of Sheep/ Goat*	*Body Weight of Sheep/Goat (kg)*	*Mean Dimension of Woolen Sheep/ Goat (Width x Length) (m)*	*Space Required for Woolen Sheep/ Goat (m^2)*	*Space Required for Shorn Sheep/ Goat (m^2)*
1	Lambs	Up to 15	0.24 x 0.6	0.15	0.15
		16-20	0.25 x 0.7	0.17	0.16
2	Sheep/ Goat (Adult)	21-25	0.26 x 0.74	0.18	0.16
		26-30	-	0.22	0.20
		31-40	-	0.25	0.24
		41-55	0.35 x 0.91	0.30	0.28
		> 55	-	> 0.40	> 0.40

Table 5.5 Maximum Distances for Trekking

Sl.No.	*Animal Species*	*One Day Journey*	*More than One Day Journey*	
			First Day	*Subsequent Days*
1	Cattle	30 km	24 km	22 km
2	Sheep/goat	24 km	24 km	16 km

Mixing of Animals

The mixing of animals in the transport vehicle may lead to bruising and biting marks in the animals which cause deterioration in the quality of skin. Bruising may occur during loading, transport and unloading of animals. Pigs from different farms should be loaded on the same vehicle, behave comparatively quietly once the vehicle is in motion but start fighting when vehicle stops. The incidence of bite marks in the pigs can be reduced by 25 per cent by reducing the stopping periods by 50 per cent. Mixing of pigs from socially unfamiliar groups in the pen, two days prior to transport reduces the injuries from fighting.

Pre-Loading Precautions

Pre-loading precautions should be taken in order to prevent or reduce the risk of injury and stress in the animals. The following pre-loading precautions can be taken:

1. Pre-mixing of animals of the same species before their transportation reduces the risks of fighting among them because they become more familiar.

2. Cattle should be mixed in a pen 24 hours before loading. Victimised or aggressive animals can be weeded out during this period.
3. Fighting amongst pigs that are strangers is common, resulting in injuries, skin damage, wounds and stress. Pigs from different pens can be mixed together before loading when smearing pigs with litter or excreta from the same pen so that they smell similar.
4. Animals should be fed and watered before transportation. However, feeding of pigs before transportation should be avoided because gas may be produced due to fermentation of feed which causes pressure on the heart in the thoracic cavity, leading to heart failure and death.
5. Mixing of horned and hornless animals in the same vehicle should be avoided as this causes bruise and injury.
6. Mixing of animals of different species should be avoided in order to avoid fighting and injuries.
7. Transportation of diseased, injured and emaciated animals should be avoided in order to prevent the deterioration of conditions.
8. Vehicles should be fitted with a portable ramp to facilitate emergency unloading.

Pre-Loading Care of Animals

Most animals can be fed and watered before their transportation. This has a settling effect. Cattle, sheep and goats should be provided with food and water up to the time that the journey commences. Pigs should have access to frequent drinks of water during long journeys, particularly in hot and humid conditions.

Pigs are difficult animals to transport, and the only satisfactory method is by road, although rail can be used under careful circumstances. Fighting occurs mostly when a vehicle loaded with pig stops, or amongst horned and polled cattle. Pigs should not be transported with other animal species unless separated by a partition. Bulls should not be carried together with other stock unless separated by a strong partition. Where possible, journeys should be short and direct, without any stoppages. If the vehicle stops, pigs will tend to fight. Prior to the start of any journey, pigs should be placed in a well-ventilated area for at least three hours and loading should not be earlier than one hour before the proposed departure time. Unloading facilities at the abattoir play an important role in the prevention of injuries and the reduction of stress factors. It is at this point especially that animals require calm and quite handling.

Other pre-loading care of animals includes various aspects. Transport vehicle should not be overloaded. Sick, week and tired animals should not be loaded in the vehicle. Horned and hornless animals should not be loaded in the same vehicle. Mixing of strange animals with existing stock should be avoided. A combination of high humidity and high environment temperatures is especially deadly to pigs. Heat can rapidly build up to lethal levels in a stationary vehicle. Wetting pigs with water helps to keep them cool. Pigs are susceptible to high environment temperatures and humidity. Exposure to sun affects pigs seriously which may cause sun burn.

Factors Associated with Transport Operations

A number of factors must be taken into account during the journey in order to prevent injury, suffering and death of animals.

1. **Mode of Transport:** Animals such as cattle, sheep and goats can be successfully moved on foot but certain risks are involved. The journey should be planned, according to the distance to be travelled, opportunities for grazing, watering and overnight rest. Animals should be walked during the cooler times of the day and, if moving some distance to a railhead, they should arrive with sufficient time to be rested and watered before loading. The maximum trekking distances by these animals depends on weather, body condition, age *etc.* The maximum trekking distance per day for cattle and sheep/goat are 30 km and 24 km, respectively. If the animals are trekked for more than one day, the maximum distance for trekking of cattle and sheep/goat is 24 km on the first day, and on subsequent days 22 km and 16 km, respectively.
2. **Time of Transport:** High environment temperatures increase the risk of heat stress and mortality during transportation. The optimum time of transportation of animals particularly pigs in vehicles is in the cooler mornings and evenings or even at night. A high environment temperatures and high humidity is dangerous especially for pigs.
3. **Duration of Journey:** Preferably the journey should be short and direct, without any stoppages. Cattle and sheep/goats should not travel for more than 36 hours. If the journey is longer than this period, they should be unloaded after 24h for feeding and watering.
4. **Careful and Smooth Driving:** Driving of the vehicle should be performed very carefully and smoothly to minimize the stress in the animals and avoid injuries and accidents. Vehicles should be driven smoothly, without jerks or sudden stops to avoid falling of the animals in the vehicle. The vehicle should be driven slowly and gently especially at turns. It is also important to monitor the downer animals so that the vehicle can be stopped and the animal lifted.
5. **Wind Chill Factor:** Wet animals exposed to cold wind during transport in cold weather causes a wind chill factor, where the body temperature is considerably reduced, resulting in severe stress or death of animals.

Animal Transport Legislation in India

Transport of Animals Rules, 1978, Chapter IV-Transport of Cattle include Rules 47 to 56 for the transport by rail of cows, bulls, bullocks, buffaloes, yaks and calves. In these rules, various aspects have been specified, which include animal health certification, veterinary first-aid equipment facility accompanying transport vehicle, consignment details, the average space provided per cattle in railway wagon or vehicle, suitable rope and platforms, feeding and watering prior to loading of animals, precautions for transport of pregnant cattle, adequate ventilation, number of cattle to be loaded in the transport vehicle and design of the vehicle and required facilities. The Central Government amended the Transport of Animals Rules 1978. These Rules may be called Transport of Animals (Amendment) Rules, 2009. The BIS has provided detailed guidelines for transportation of animals by rail and road. The BIS guidelines include (IS: 4157, Part II- 1968) for cattle and buffaloes, (IS: 4157, Part III, 1983) for sheep and goats, and (IS: 5238, Part II, 1982) for poultry.

Indian Standard for Transport of Livestock: Code of Practice, First Revision (IS 14904: 2007)

This Indian Standard (First Revision) was adopted by the Bureau of Indian Standards. This Code is aimed to serve as a guide to commercial transporters, producers, hobbyists and industries that utilize animals in assessing their facilities, equipment and practices with regard to livestock transportation and may provide guidance for improvement in deficient areas. This Code has been revised to take into account changes in practices of animal management and knowledge of animal welfare. Earlier there were three standards on transport of livestock: IS 4157 (Part 1): 1981 'Transport of Equines (horses, ponies, mules, and donkeys) by rail, road, air and sea', IS 4157 (Part 2): 1968 'Transport of cattle by rail and road', and IS 4157 (Part 3): 1983 'Transport of sheep and goats by rail and road'. These standards were amalgamated into one standard which makes it more comprehensive and user-friendly and also to update the standard during formulation of this Indian Standard. This Code, in general, is intended for use as a guide and educational tool in promoting sound animal transportation and welfare practices. This standard prescribes the conditions for the transport of livestock (cattle, sheep, goat and equines) by rail and road, of live domesticated animals (cattle, buffalo, deer, Camelids, sheep, goats, pigs, equines and others) by sea, and of various species of animals by air. This standard prescribes the transport of livestock by land (road and rail) as follows:

1. Responsibilities

The welfare of animals during transport by road and rail is the joint responsibility of all people involved such as animal keeper, animal handler, business agents, transporters, owner, driver, official veterinarian *etc.* Animal keepers and business agents or buying/selling agents are responsible for careful selection of livestock for their fitness for the journey. Animal handlers shall be responsible for the humane handling and care of the animals, especially during loading and unloading. Transporters shall be responsible for planning the journey to ensure the care and welfare of the animals during transport until discharged at the place of destination, selection of appropriate vehicles or containers for the species being transported and their maintenance, and preparation of any required documentation.

Drivers and animal handlers shall be responsible for loading animals which are fit to travel and the animals care and welfare during transport and frequent inspection of animals during the journey. Owners and managers of assembly centres at the start or end of the journey and at resting points used during the journey shall be responsible for providing suitable premises for loading, unloading, securely holding animals and protecting them from extremes of weather by providing appropriate shade or shelter, appropriate facilities for feeding and watering of animals, emergencies including the treatment of sick, injured or diseased animals by a veterinarian or timely euthanasia or slaughter by a veterinarian or suitably trained individual and facilities for cleaning out transport vehicles.

The official veterinarian shall be responsible for supervision of the unloading and arrival of animals at the assembly centre and inspection of all animals as soon as possible after arrival, monitoring of the welfare of all animals within the assembly centre area, providing shade, shelter, food and water to all animals, supervision of the loading of animals from the assembly centre, checking of the assembly centre facilities, supervision of the unloading and arrival of animals at the slaughterhouse and inspection of all animals, and appropriate dealing of any ill, sick or injured animals. When animals are transported by rail, the Railway Board is considered to be the transporter which is responsible for ensuring the firefighting equipment and fire precautions.

2. Skills

Animal handlers and others such as farmers, transporters, managers, owners and drivers responsible for animals during journeys should be competent according to their functionalities. They must meet standards set out by the competent authority for their jobs and undergo any required training developed by the competent authority.

3. Documentation

Each consignment of animals shall be accompanied by appropriate documentation or bear a label (in bold letters clearly visible from distance) showing the name, address and telephone number (when available) of the consignor, consignee and driver; the date and time of loading of first animal, last watering and feeding and next feeding and watering; and likely date and time of unloading of last animal. Animals shall be accompanied by valid veterinary certification issued by an official veterinarian. This should include identification of the animal, health status including test, treatment and vaccination status, statement of fitness to travel and details of any at risk animals. Animals shall not be loaded until the required documentation is complete.

4 Pre-Travel Rest Period/Preparation of Animals

Adequate rest is essential to all animal species after collection or assembly and before loading. This is beneficial and should be added to the pre-transport timetable. The animals may be under excessive stress in a large gathering area, therefore, they should be provided with 24 h of rest and access to food and water. Food should be withdrawn for the last 2 - 4 h before transport. Behaviour modifying compounds such as tranquilizers should not be used prior to or during transport except when a problem exists in an individual animal.

5. Fitness to Travel

Owners of animals or their agents have a responsibility to select only fit and healthy animals for transport. The points should be taken into account when selecting animals for transport include body condition, clinical evidence of disease or parasitism, injuries or other physical defects, stage of pregnancy, and age. Unfit animal shall not be transported. New born (where the navel has not healed), diseased, emaciated, lame, fatigued, or having given birth during the preceding 72 h or likely to give birth during transport shall not be transported.

An animal with any leg broken shall not be transported except under the supervision of a veterinary surgeon to a place of further treatment or slaughter. Where an animal is not fit to travel and is experiencing pain, suffering or distress immediate veterinary assistance must be obtained for the animal or it must be humanely destroyed without delay. Occasionally, where an animal that would not normally be selected for transport may be transported in an emergency for further treatment or slaughter in order to relieve pain, suffering and distress.

When animals become ill or are injured during transport, they should be separated from others and receive appropriate treatment as soon as possible. Veterinary first-aid equipment, with an attendant trained in rendering first-aid, shall accompany all batches of animals. When euthanasia or slaughter is necessary at any point during transportation or at a rest stop or assembly centre it should be carried out by a veterinarian or suitably trained individual.

6 Means of Transport

Means of transport shall be designed, constructed, maintained and operated so as to avoid injury and suffering and to ensure the safety of the animals. Means of transport shall be appropriate for the species, size and weight of the animals to be transported, maintained in good mechanical and structural condition and shall not be used to transport any other goods or merchandise during the transport of animals. Any external doors and internal gates shall be large enough to permit animals to pass through easily without bruising or injury. Special attention should be paid to the avoidance of injury to animals through the use of secure, smooth fittings free from sharp protrusions.

Floors of animal areas should be non-slippery. In order to minimize the risk of spread of disease, means of transport shall be designed to permit thorough cleansing and disinfection. Provision should be made for the drainage and/or absorption of urine. Means of transport should be designed so that the faeces or urine from animals on upper levels does not soil animals on lower levels, nor their feed or water. Means of transport shall provide protection from inclement weather, extreme temperatures and adverse changes in climatic conditions. Sufficient space shall be provided inside the animal compartments and at each level to ensure that there is adequate ventilation and enough room overhead to enable animals to travel in a natural position without risk of injury. The Railway Board must ensure that wagons are suitable for the species of animals being transported.

7. Loading and Unloading

Loading is one of the most stressful parts of the transport process for the animals.

Loading and unloading should be carried out or supervised by experienced animal handlers. The facilities for loading and unloading (collecting areas, races and ramps) should be designed and constructed to take into account the needs and abilities of the animals with regard to dimensions, slopes, floor surfaces and the absence of holes, gaps, sharp edges or projections which might injure animals. Ramps should be positioned to present the minimum possible incline to the animals.

The width of any ramp should be suitable enough for animals to safely use and the top shall be level with the floor of the vehicle. The ramps should have side protection and non-slip floors. Ramps shall be aligned so that there are no gaps where an animal might become trapped and injured. Where animals are to be transported in hot or humid conditions loading should take place in the coolest part of the day to reduce the risk of heat stress.

Animals shall be unloaded as soon as possible after arrival at the place of destination. At the destination, there should be appropriate facilities and equipment such as a sturdy drag mat for the humane unloading of animals that are not able to walk due to fatigue, injury or sickness. These animals shall be unloaded in a manner which causes the least amount of pain, distress or suffering. After unloading these animals shall be penned separately and given appropriate care by the animal handler or other responsible person.

8. Handling

Animals become alert and inquisitive in a new situation or location and behave very differently; therefore, they should be handled by competent people. Animals should not be rushed when being moved, loaded or unloaded. Painful procedures (including whipping, tail twisting, use of nose twitches, pressure on eyes, ears or external genitalia) or the use of sticks with sharp ends, lengths of metal pipe or leather belts shall not be used. Animals shall not be lifted or dragged by their tail, head, horns, ears, limbs, wool or hair. Any goad, stick or other handling aid shall only be used where animals have refused to move forward when there is room for them to do so. The use of electric prods is discouraged for all species; should only be used on the hindquarters of adult cattle and shall not be applied to the head, genitalia or any other sensitive part of the body.

Permitted handling aids such as flags, bags or sacks should be used in a way that is sufficient to encourage and direct forward movement but without physical contact with the animal. Rump straps (easy- loaders) can be used for animals that are reluctant to load.

Injurious objects or irritant substances shall not be applied to any part of the animal as a means of getting the animal to move. All animals must be inspected by the official veterinarian at the point of unloading. Sick or injured animals that can move unaided off the vehicle must be moved to the designated isolation/treatment area without delay at the assembly centre or destination point.

9. Separation

Wherever possible animals reared together should be maintained in that social group for transport or compatible groups established at least one week prior to the journey. Animals with a strong social bond should be transported together. The animals shall be transported separately in conditions such as animals of different species, animals of significantly different size or age (with the exception that dam and offspring should be transported together), adult breeding stallions, sexually mature males and females, sick or injured animals, animals which are hostile to each other, horned and polled animals, tied and untied animals. Separation of animal groups on the vehicle shall be achieved by the use of partitions or by tying compatible groups together.

10. Leading, Tying and Tethering

Animals must be led, tied and tethered during handling and transport in a way that does not cause them unnecessary pain, injury or distress. Animals shall not be hobbled or tied by the legs within an assembly centre or for transport. Sheep and goat must not be tied but instead must be transported in a secure vehicle with adequate ventilation and from which they cannot fall out or escape.

11. During Transport

Animals shall be inspected frequently during the journey to ensure their safety and welfare. Means of transport shall be provided with sand, saw dust, straw bedding or other suitable material to absorb faeces and urine, provide insulation for young animals, protect animals from injury from hard flooring and prevent slipping of animals. The layer of saw dust or sand shall not be less than 3 cm thick and the thickness of straw bedding not less than 10 cm. Means of transport shall be padded with suitable material where there is risk of animals being rubbed, bruised or injured by coming into contact with the sides of the vehicle or internal partitions.

12. Space Allowance

All animals shall have sufficient space to stand normally. Animals must not be caused pain or injury as a result of being stocked too tightly, in particular through impact or rubbing against the vehicle side.

13. Temperature and Ventilation

Transport of animals should be avoided in extreme weather conditions. Transportation during the night may reduce the risk of heat stress. Adequate airflow throughout the vehicle or container must be provided to keep the animals comfortable. In hot weather, the duration of journey stops should be minimized and vehicles parked under shade with maximal ventilation. In extreme, high summer temperatures, water shall be sprinkled over the wagons by the railway authorities to bring down temperature. Ice slabs in specially made containers may be placed inside the wagon, if recommended by veterinarian.

14. Water and Feeding Intervals, Journey Times and Resting Periods

In planning the need for, and provision of food and water, account must be taken of the total transport time from initial gathering and assembly prior to the start of the journey until unloading at the final destination, including all transit rest periods. Unweaned calves, lambs, kids and foals which are still on a milk diet should, after 9 h of transport be unloaded, rested, fed and watered for a period of at least 15 h. Adult cattle, sheep and goats should after 12 h of transport be unloaded, rested, fed and watered for a period of at least 12 h. Adult horses may be transported for a maximum period of 24 h. During the journey they must be given liquid and if necessary fed every 8 h. Clean drinking water should be provided for all animals while being held prior to loading. Water should be provided in troughs so that all ages and classes of animals can drink. It is acceptable to withdraw feed 2-4 h before transport. Any animals in milk, without suckling young at foot, should be milked at least every 12 h.

15. Driving

Drivers and animal handlers should check immediately before departure to ensure that the animals are properly loaded. Animals should be checked again early in the journey and at regular intervals thereafter. To reduce the risk of injury to animals', road and rail vehicles carrying animals must be driven steadily, avoiding rapid acceleration and braking as far as possible.

16. Assembly Centres

All places where animals are temporarily assembled for sale, show, feed, water and rest, prior to or during any journey, must have adequate facilities for the care and safe loading, unloading, handling, holding, feeding and watering of animals. Assembly centres should provide appropriate shelter for animals from heat, wind and cold.

17. Control of Disease

Animal transport is often a significant factor in the spread of infectious disease, transporters must ensure the means of transport is adequately cleansed and disinfected at appropriate points in the journey. All means of transport shall be visibly clean before loading any animals. They should be properly disinfected and sprayed with a solution of a reliable disinfectant before loading. Once animals have been unloaded at a place of rest or transfer or the place of destination the means of transport should be thoroughly cleansed and if possible disinfected before animals are reloaded or any other consignment of animals is carried.

International Animal Transport Legislation

Animal transport legislations for humane transportation of animals through road, rail, air or sea have been laid by most of the countries. The International Transport Association (IATA) has designed a manual having sets of standards which are mandatory and binding on all IATA members- the IATA Live Animals Regulations. The OIE has also described the transport of animals by land in its Terrestrial Animal Health Code. This is applicable to live domesticated animals such as cattle, buffaloes, camels, sheep, goats, pigs, poultry and equines. This is also largely applicable to some other animals such as deer, other Camelids and ratites.

Effects of Transport

Poor transportation causes serious deleterious effects on the animal welfare and can lead to significant loss of quality and production. Effects of transport and movement include stress, bruise, fighting, injuries, trampling, heart failure, heat stroke, sun burn, bloat, poisoning, suffocation, predation, dehydration, exhaustion, *etc.*

Conditions Induced Due to Transport of Food Animals

Long duration of transportation of animals, poor design of transport vehicle, improper transportation operation, inadequate rest, improper feeding and watering of animals due to continuous journey that is without halt may lead to heavy stress, suffocation, dehydration, starvation, injuries and even death of animals. Such prevailing conditions during transport are against the animal welfare rules and also lead to poor meat quality and loss of production due to injuries and death of animals. Conditions induced due to transport of food animals are discussed as follows:

1. **Transit or Shipping Fever:** It is a catarrhal and often fatal disease caused by *Pasteurella haemolytica* and characterized by an acute, toxaemic bronchopneumonia with high fever. It mainly affects stored cattle in a poor condition that have become fatigued due to a long journey by road, rail or ship without providing sufficient rest to them. Extreme weather conditions act as predisposing factors for development of transit or shipping fever.
2. **Transit Tetany:** Transit tetany may develop due to long and stressful journey. This condition is observed mainly in cows particularly in advance pregnancy. Warm weather acts as a predisposing factor for development of transit tetany.
3. **Transport Mortality:** Transport mortality in livestock is low (less than 0.5 per cent) if they are transported under standard conditions. Transport mortality rate may be very high due to several predisposing factors like long distance transportation without providing sufficient rest to the animals during transport, transportation stress, extreme weather conditions, overloading of animals in transport vehicle, inadequate supply of food and water to animals, poor cleansing and disinfection of transport vehicle and poor design of the vehicle.
4. **Travel Oedema:** Travel oedema may be analogous to PSE meat. This condition is sometimes encountered in New Zealand in lambs slaughtered after an arduous journey in which the carcase particularly the shoulders is unduly watery. This problem can be alleviated by providing proper rest to the animals before its slaughter and immediate chilling of carcase especially in summer months.
5. **Traumatic Injury:** Traumatic injury may occur in the form of bruise, open wounds and fracture during loading, transport and unloading of animals. This usually occurs due to poor design of vehicle, overloading of animals in the vehicle, high speed of the vehicle especially at the turns, frequent and strongly use of brakes, long duration of journey *etc.*

   Bruising is the escape of blood from damaged blood vessels into the surrounding muscle tissue. This is caused by a physical blow by a stick or stone, animal horn, metal projection or animal fall and can happen anytime during handling, transport, penning or stunning. Bruises can vary in size from mild and superficial, to large and severe involving whole limbs, carcase portions or even whole carcases. These conditions considerably reduce the carcase quality because the injured parts or in extreme cases the whole carcase requires condemnation. Moreover, the secondary bacterial infection occurs in bruise and other injury which causes abscess formation at the injured site and may lead to septicaemia which requires condemnation of the entire carcase. Bruising is a common cause of meat wastage and can be significantly reduced by correct techniques of handling, transport and slaughter of animals. Bruising in meat requires condemnation of affected portion which is not suitable for consumption due to the following reasons:

i. It is not acceptable to the consumers.

ii. It is not suitable for processing or manufacture of meat products.

iii. It is more prone to spoilage due to blood content in the meat which favours the growth of microorganisms.

iv. It has short shelflife.

6. **Loss of Body Weight:** Loss of body weight in transported animals occurs due to dehydration, inadequate feed and water to the animals, long distance of journey, extreme hot weather *etc.* The average body weight loss during transport for a period of 24 hours is 2.2-5.4, 3.6, 4.0 and 30-40 kg in pig, sheep, calf and heavy bullock (610 kg), respectively. The rate of body weight loss is higher during the first day of travel and it becomes lower in subsequent days of travel.

7. **Dark Firm and Dry (DFD) Meat/Dark Cutting Meat:** Both the DFD and dark cutting meat are the same abnormal condition of meat. The term DFD is used for pork, while, dark cutting meat is used for beef. This condition develops due to stress or exhaustion of animals. In this condition the glycogen content is used up so that less amount of lactic acid is formed after slaughter resulting in an abnormal muscle condition in which it remains dark purplish-red on exposure to air instead of a bright red colour. The condition is frequently observed in young bulls due to enhanced temperament and behaviour. DFD meat is observed in physically exhausted pigs. In deer, stag carcases are more often affected than those from hinds. The condition is rarer in lambs. DFD meat is characterized by unpleasant purplish-red colour, direr and darker than the normal meat, lower shelflife, less desirable colour and flavour but fit for human consumption. Exposed surface of meat appears bright red due to oxygenated myoglobin. Development of DFD meat is due to partly conversion of myoglobin into oxymyoglobin. In post-mortem glycolysis, the muscle glycogen is converted into lactic acid and carbon dioxide. The amount of lactic acid determines the ultimate pH of the meat. DFD meat has a shorter shelflife due to the abnormally high pH-value of the meat (6.4-6.8).

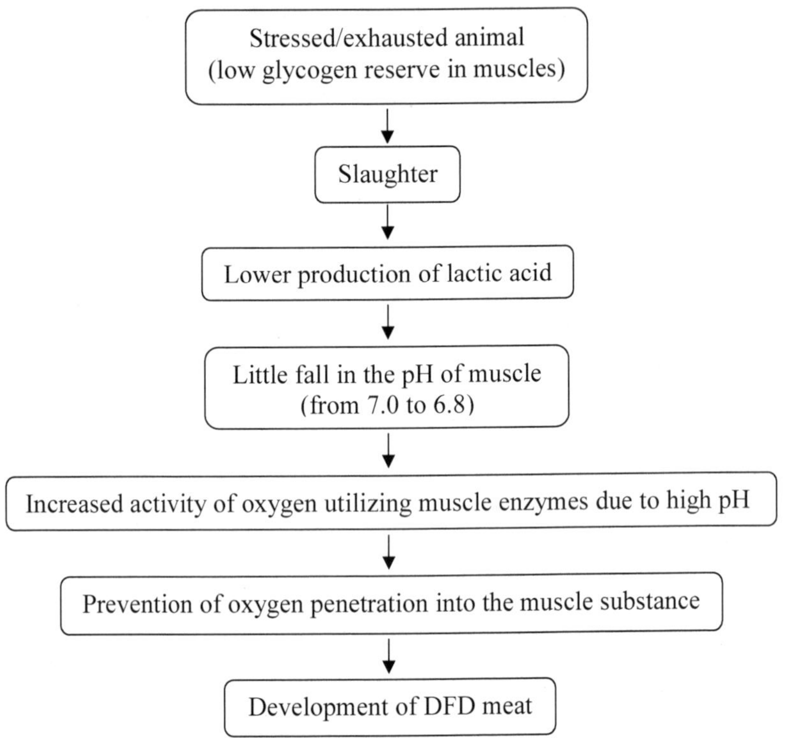

Figure 5.1. Mechanism of Development of DFD Meat.

8. **Pale Soft Exudative (PSE) Meat/Watery Pork:** Development of PSE meat in pigs is caused by severe, short-term stress just prior to slaughter, *e.g.*, during off-loading, handling, holding in pens and stunning. Here the animal is subjected to severe anxiety and fright caused by manhandling, fighting in the pens and bad stunning techniques. All these may result in biochemical processes in the muscle in particular in rapid breakdown of muscle glycogen and the meat becoming very pale with pronounced acidity (pH values of 5.4- 5.6 immediately after slaughter) and poor flavour. However, in healthy and adequately rested pigs, the lactic acid is gradually produced over a period of 12 hours which causes fall in the muscle pH from 7.0 to 5.5. But these changes rapidly occur (within an hour) in carcases of exhausted animals.

   High environmental temperature, rough ante-mortem handling, fighting among animals, physiological differences between breeds of swine and inefficient slaughter techniques are predisposing factors for development of PSE meat. Certain breeds of swine such as Danish Landrace, Pietrain and Poland China are more prone to rapid post-mortem glycolysis and development of watery pork. PSE meat is generally observed in leg, loin and fillet. PSE meat is a serious problem if pork is to be sold fresh because of amount of drip which causes weight loss up to 10 per cent. PSE meat become tougher when cooked and has higher cooking losses due to the reduced water-binding capacity. Curing of PSE meat also takes up more brine than the normal pork.

PSE and DFD meat are perfectly safe to eat but limited in their processing capacity. PSE meat has less desirable colour and flavour than the normal meat. This type of meat is difficult to use or cannot be used at all by butchers or meat processors and is wasted in extreme cases. The problem of PSE meat can be reduced by gentle handling of animals, adequate rest to exhausted and fatigued animals prior to slaughter, providing sufficient water and feed to pigs in the lairage, feeding of sugar in liquid form to pigs in the lairage and use of overhead fine sprays of water in lairage (for cooling purpose).

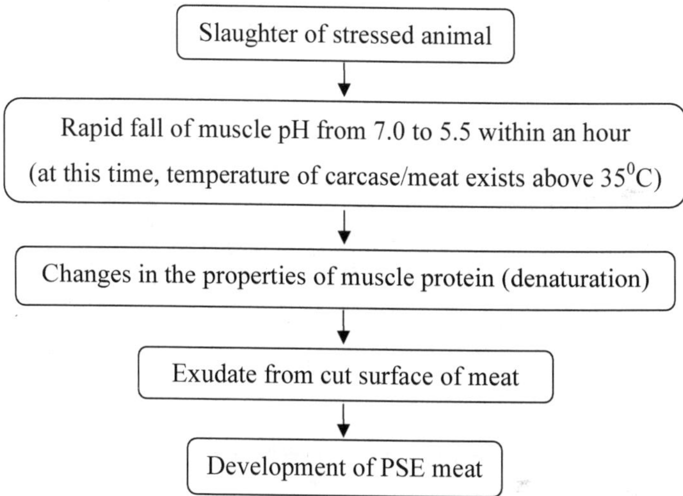

Figure 5.2. Mechanism of Development of PSE Meat.

9. **Porcine Stress Syndrome (PSS):** It is also known as herztod, fatal syncope or shock heart failure syndrome. This condition is observed in exhausted and fatty pigs. This is characterized by fever, skin blanching, erythema and dyspnea followed by sudden death. Post-mortem finding shows diffuse and severe degeneration of skeletal muscles with excess pericardial fluid. Fatal syncope is common in Pietrain and Landrace breeds of swine.
10. **Salmonellosis:** It is observed mainly in young animals especially calves and lambs. Transportation stress, cold weather, intensive rearing, lack of food and water, poor cleansing and disinfection of transport vehicle and poor design of vehicle are predisposing factors for development of salmonellosis.

Supervision of Unloading and Handling of Animals in the Abattoir

The unloading and general handling of animals in the abattoir should be supervised by trained and responsible personnel in order to prevent injuries and stress in the animals. The floor of the vehicle must be fitted with raised ridges in the form of grids or cleats to prevent slipping and injury. Gaps or holes in the floor and/or between the floor and side panels are often the cause of injury, especially bone fractures. Sharp points or corners must be eliminated in the construction of the vehicle.

Pre-Slaughter Care of Animals Resting in Lairage

The meat obtained from exhausted, stressed, fatigued and frightened animals adversely affect its quality. Therefore, efforts should be made to keep them free from these conditions to maintain the quality of meat.

1. **Resting:** Livestock, who have travelled long distances on foot or in transport frequently are insufficiently rested before slaughter, negatively affect the quality of the meat. Overcrowding of animals in holding pens causes unnecessary stress to the animals. The quality and condition of the carcase and its storage depend greatly on the care taken prior to slaughter. Nervous, tired, and excited animals may have a raised body temperature, causing imperfect bleeding. Muscular fatigue reduces glycogen content in the blood, which after slaughter changes into lactic acid. Meat obtained from inadequately rested animals has poor keeping quality due to insufficient development of lactic acid in the muscles. Such meat is prone to microbial putrefaction. Fatigue and excitement also cause penetration of bacteria from the intestinal tract to the meat.

   The quality of meat obtained from inadequately rested animal is poor. Therefore, it is necessary to provide adequate rest to fatigued, stressed and excited animals before their slaughter. Resting period depends on many factors such as species, age, sex, class and condition of animal; length of journey and method of transportation. The optimum resting period for cattle in the lairage before slaughter is 24 hours, but it should not be less than 12 hours. A rest period of 2-3 hours is recommended for swine before their slaughter.

2. **Watering:** Adequate drinking water should be provided to animals arrived at slaughterhouse after transportation. This lowers the microbial load in the intestine and facilitates the removal of hide or pelt during dressing of the slaughtered animal. Ample water supply to the animals during their detention in the lairage also facilitates efficient electrical stunning.

3. **Feeding and Fasting:** It is essential to provide regular feed to the animals kept in the lairage. This is important from humane point of view as well as to prevent economic losses incur due to starvation and body weight losses. Prolonged fasting of animals in the lairage causes significant body weight losses and lower carcase yield. However, it is also important to withhold the feed from animals for a certain time period before their slaughter. Withholding of feed or fasting of animals before slaughter minimizes the emigration of bacteria from intestinal tract during digestion.

It is recommended to withhold the feed from cattle for a period of 6 hours before their slaughter.

4. **Handling and Keeping Stress-Free:** Improper and rough handling of animals may lead to loss of weight and bruising. Animal should not be beaten or subjected to rough handling when they are being moved from the lairage pens to the slaughterhall. Use of stick for handling or sifting of animals from one place to other place in the abattoir should be avoided. Instead of prods, other droving aids can be used such as flat straps, rolled-up plastic or newspaper, sticks with flags on or panels for pigs. Hesitant animals can often be enticed into pens or vehicles by first leading in a tame animal and the others will follow.
5. **Maintaining Hygiene:** Holding animals in vehicles or lairages without adequate litter and/or drainage frequently results in faecal soiling of the skin. Cattle entering slaughterhouses often are very dirty, their legs covered with manure. In these cases, the knife used for bleeding and de-hiding will have to cut through manure and faecal residues, resulting in a great possibility for meat contamination. Therefore, maintaining hygiene such as proper disposal of animal faeces, urine, discharges, leftover feed and other wastes from the lairage and its proper disinfection are important for quality meat production.

Pre-Slaughter Feeding of Sugars

Pre-slaughter feeding of sugars to the animals is important to minimize the stress, fear, excitement, fatigue or excessive exertion on the animals. Sugars are fed to the animals to ensure adequate pre-slaughter levels of glycogen in the muscle which is required for sufficient lactic acid production in the meat after slaughter of animal. It has been observed that the pre-slaughter feeding of sugar solutions to pigs resulted in increase in the carcase weight by 2.8 per cent and liver weight by 27 per cent and decline in muscle pH by 0.2-0.3 units.

Principles of Animal Handling

The first principle of animal handling is to avoid getting the animal excited. It is important to follow the principles of animal handling in order to prevent the development of poor quality meat. After rough handling the animal become normal that is calm down with its normal heart rate in about half an hour. Therefore, animal handlers should move with slow, deliberate movements and refrain from yelling. Sometimes electric prodders (prods) can be used as little as possible or only on stubborn animals. This is more humane and causes less damage to give an animal a mild electric shock than to hit it with a stick or twist its tail. Battery-operated prods with a maximum 32V is suitable for this purpose. It should not be used on sensitive body parts such as eyes, muzzle, anus, vulva *etc.*

Handling of Animals in Crowd Pens and Races

The animal caretaker should be properly trained to handle the animals humanely for slaughter. While moving the animals on race, the use of electric rods should be avoided. The animals can be moved with plastic sticks. Overcrowding in pen is one of the most common animal handling mistakes. Animal handlers must be careful to ensure that the animals should not be forced to move towards the gate with overcrowding. Animals should walk up the race without being forcibly pushed. If they are pushed up too tightly with a crowding gate, handling becomes more difficult. Animals generally move from a darker to a lighter place. Extra, indirect lighting may help in moving animals in pens.

Animals may hesitate and refuse to move when they see things in the race that scare them, such as sparkling reflections, dangling chains, moving people or equipment, shadows or water dripping. Cattle, sheep and ostriches have very sensitive hearing, particularly to high frequency sound. Sounds that do not bother people, such as intermittent high-pitched noise, may damage animals' ears. Stress and the risk of injury in the animals can be minimized by reducing the noise from equipment, instruments, machines, and personnel.

Pre-Slaughter Handling of Animals and Meat Quality

Careless handling of animals may lead to stress, injuries and even death of animals. Moreover, rough handling of animals also causes deterioration in the quality of meat produced from them. Pre-slaughter stress in the animals causes very significant effect on the quality of meat. Handling of the animals before slaughter is a major concern to the livestock industry, especially the pork industry. Stress in the animals before slaughter can lead to adverse effects on the meat produced from these animals, including both PSE and DFD. Pre-slaughter stress can be reduced by preventing the mixing of different groups of animals, by keeping livestock cool with adequate ventilation, and by avoiding overcrowding. Before slaughter, the animals should be allowed access to plenty of water but held off-feed for 6 hours to assure complete bleeding, ease of evisceration and reduce the microbial loads in the meat produced from them.

6
Comparative Anatomy of Tissues and Organs of the Food Animals

The anatomical structures of the body vary in different species of the food animals. For a veterinarian or meat inspector, the comparative anatomy of different species of the food animals is important for identification of the carcase of a particular species of animal. The anatomy of animal body and its organs is also important in differentiating the normal structure of animal body and its organs, and the pathological conditions. Therefore, the comparative anatomy of the food animals is important in sound meat inspection. There are four main types of tissues *viz.*, epithelial tissues, muscular tissues, connective tissues and nervous tissues in animal bodies. The important characteristics of different systems in the body of different species of food animals are discussed as follows:

Skeletal System

Skeletal system acts as frame of the body which support and protect the soft tissues of the body. This system also provides a system of levers for locomotion and body movement. Bones, especially the long bones act as haemopoetic organ. The long bones in very young animals are relatively very long, slender, and smooth with their prominences less pronounced. In the very old animals, the bone organic matter is decreased. The average percentage of bone in the beef, lamb, bobby calves, veal calves, pork and poultry carcases are 15 per cent, 17-35 per cent, 50 per cent, 25 per cent, 12-20 per cent and 8-17 per cent, respectively. The bones in the carcase are valuable means of identification of different species of food animals, *e.g.*, when substitution is suspected.

The examination of the carcase bones becomes more important in determining the age, especially when the teeth are unavailable in the bovines. This is based on the degree of ossification of certain parts of skeletal system, the most valuable of which are the cartilaginous extensions of the spines of the first five dorsal vertebrae. In cows, these changes take place more rapidly and the cartilage is ossified after 3 years. In bovines, the ossification of the cartilaginous extensions of the spines of the first five dorsal vertebrae varies according to the age. At the age of 1 year, the extension is entirely cartilaginous, soft, pearly white and sharply delineated from the bone which is soft and red. At the age of 2 years, there are small red islets of bone appear in the cartilage. At the age of 3 years, the cartilage becomes greyish, and red areas are more numerous. At the age of 4-5 years, the area of ossification within the cartilage extends until the proportion of bone is greater than that of cartilage. At the age of 6 years, the cartilage is ossified into compact bony tissue, though the line of junction between cartilage and bone can still be defined. The skeletal system of domestic fowl differs from other animals. In the domestic fowl the bones are very light in weight and contain air sacs.

Digestive System

Tongue

In the ox, the filiform papillae of the tongue are horny and directed backwards. The posterior part of the dorsum is prominent and anteriorly by a transverse depression. In the ox, the black pigmentation of the tongue is frequently observed which is a normal condition. In the sheep and goat, the tongue is similar to that of cattle but the centre of the tip is slightly grooved and the papillae are not horny. The tongue of the sheep is narrower, more marked dorsal eminence, smoother surface and more rounded tip than that of the tongue of the calf. In black-skinned sheep, the black pigmentation on the surface of the tongue is common. In the pig, the tongue is long and narrow and there is no dorsal ridge. In the horse, the tongue is long and flat with spatulate end. There is no dorsal ridge on the tongue. There is only one circumvallate papilla on either side of the tongue. The epiglottis is pointed. Pigmentation on the tongue is not found.

Oesophagus

In the ox, the oesophagus is comparatively short and wide. It is about 1 m long and 5 cm wide.

Stomach

In the ox, the stomach (paunch) consists of four compartments, namely, the rumen, the reticulum, the omasum and the abomasum. The rumen occupies about 75 per cent of the abdominal cavity. The rumen is bounded on the left side by the abdominal wall, on the anterior extremity by reticulum and part of the omasum and on the right side by the remainder of the omasum, abomasum and intestine. The reticulum is placed transversely between the anterior extremity of the rumen and the posterior surface of the diaphragm to which it is adherent, cause a depression on the posterior aspect of left lobe of the liver. The omasum and abomasum are attached to the posterior aspect of the liver. The average capacity of the stomach is 150 litres.

In the sheep and goat, the structure of stomach is similar to that of ox. The average capacity of the stomach is 18 litres. In the pig, the stomach is simple and semilunar-shaped with small diverticulum at the cardiac end. The average capacity of the stomach is 6.5 litres. In the horse, the stomach is simple and average capacity is 12 litres. In the domestic fowl, the alimentary tract consists of oesophagus, fusiform crop, proventriculus or glandular stomach, thick muscular gizzard, small intestine (duodenum, jejunum and caecum), large intestine (two caeca and one rectum) and a cloaca, which is a common end of the digestive and urogenital tracts.

Liver

The liver of all food animals are reddish-brown in colour, except in horse, which is purplish. It is the largest gland of the body, which lies mainly to the right of the midlines in all animals. In the ox, the liver is poorly divided into three lobes, *viz.*, thin left, a thicker right and a caudate lobe. The left and right lobes are divided by a slight notch, the umbilical fissure. The weight of ox liver is about 5.4 kg. In the sheep, the shape of liver is similar to that of ox but the caudate lobe is more pointed, and its edges are more clearly defined. These characteristics are useful in distinguishing between sheep and calf liver. In the calf, the caudate lobe of the liver is more rounded and has a blunter extremely which frequently extends beyond the lower edge of liver. The weight of sheep liver is about 453-680 g. In the pig, the liver has five lobes, *viz.*, two smaller inner, two smaller outer and a caudate lobe. The oesophageal notch in the liver is prominent. The weight of liver varies from 0.9 kg in pork pigs to 2 kg in sows. In the horse, the liver has three distinct lobes and a thumb piece which terminates in a point. The gall bladder is absent in horse. The weight of horse liver is about 4.5 kg. In the fowl, the liver consists of right and left lobes. The gall bladder is fusiform-shaped.

Pancreas

In the ox, the pancreas is reddish brown, loosely lobulated and roughly the oak leaf-shaped. It is attached to the back of the liver and is deeply notched to accommodate the portal vein. The average weight of the pancreas in the cattle, sheep, pig and horse are 226-340 g, 85-142 g, 28-56 g and 340 g, respectively. In the fowl, the pancreas has dorsal, ventral and splenic lobes. The pancreas of fowl is reddish in colour.

Respiratory System

Organs like lungs, heart and associated large blood vessels occur in the thoracic cavity. The thoracic cavity is separated from the abdominal cavity by a strong muscular-membranous diaphragm which is convex anteriorly and concave posteriorly. The respiratory system comprises the nose, nasal cavity, part of pharynx, larynx, trachea and lungs.

Lungs

In the ox, the cartilaginous trachea meets at an angle and forms a distinct ridge along the dorsal aspect. The left lung has three lobes, namely, apical, cardiac and diaphragmatic. The right lung has four to five lobes. The lobulation in the lungs is well marked by the large amount of interlobular tissues and is particularly evident in old cows. The pair of the ox lungs weighs about 2.2-3.0 kg. In the sheep, the divisions of the lobes in the lungs are similar to those of ox, but the consistency of the lungs is denser, leathery, and duller in colour and less distinct lobulation. The pair of the sheep lungs weighs about 340-907 g.

In the pigs, the number of lobes in the lungs varies. The left lung has two to three lobes and right lung has three to four lobes, because the apical and cardiac lobes can be subdivided. The lung tissue is very spongy and compressible. The surface lobulation of the lungs is well marked. The lungs of the pig show greatest variation among all the food animals. The colour of pig lungs vary from red to a light pink due to the amount of blood left in the lungs after bleeding. The pig lungs weigh about 340-453 g. In the horse, the lobar divisions in the lungs are very indistinct, only two left lobes and three right lobes can be clearly distinguished. The horse lungs can be differentiated from ox lungs by absence of surface lobulation and absence of accessory bronchus. The horse lungs weigh about 2.2-3.0 kg.

In the fowl, the respiratory system is distinguished by the presence of an anterior or cranial larynx which opens into the floor of the pharynx and a posterior larynx or syrinx, located at the terminal part of trachea and partly formed by the bronchi. There are 10 pairs of air sacs and one clavicular sac associated with respiratory system of fowl. These sacs form a communication between a bronchus and the interior of some of the pneumatic bones.

Pluck

In the pig, sheep and calf, the internal organs comprising larynx, trachea, lungs, heart and liver constitute the pluck. In the pig pluck, the oesophagus remains attached and related to the trachea. The trachea of the pig is short and consists of 32 rings. In the sheep, the oesophagus is removed with stomachs in the dressing of the carcase. The trachea of sheep is long and consists of 50 rings.

Circulatory System

The circulatory system consists of heart, arteries, veins and capillaries.

Heart

The heart lies in the mediastinal region of the thorax between the lungs. In the ox, the heart shows three ventricular furrows on its surface. The ox heart weighs about 1.8-2.2 kg. In the sheep, there are three ventricular furrows. The sheep heart weighs about 85-113 g. In the pig, there are two ventricular furrows on the surface of the heart, although a rudimentary posterior furrow may be present. The pig heart weighs about 170-198 g. In the horse, the heart has two ventricular furrows. The horse heart weighs about 2.7 kg. In the domestic fowl, the heart possesses a very pointed apex.

Portal Circulation

The portal circulation is important in the study of the spread of certain bacterial and parasitic infections throughout the body. The portal vein is formed by two main branches, *viz.*, the gastric and mesenteric veins which drain the stomach, intestine and pancreas. The portal vein conveys the venous blood from these organs to the liver. The liver is drained by the hepatic vein, which enters the posterior vena cava, wherein the blood is conveyed to the heart.

Lymphatic System

The lymphatic system constitutes the following:

Lymph

Lymph is the medium which transfers the oxygen and nutrients from blood to the body tissues, and remove the waste products. Lymph is similar to that of blood plasma but is thinner, more watery, and poorer in protein, which cannot pass through the walls of the capillaries.

Lymphatics

The lymph (after performing its function of feeding the tissues) is forced by the animal's muscular movements into the fine-walled lymphatics, which rise as blind-ended vessels in the tissues. Lymphatics are similar to veins, but have thinner walls and more valves. Lymphatics become beaded appearance when distended with lymph. All lymph vessels discharge their contents into lymph nodes. The lymph passes through at least one lymph node before it returns into the blood circulatory system. The lymph flows from the centre of an organ towards its surface.

Lymphatic vessels which convey the lymph to a lymph node(s) are known as afferent lymphatics, and the area drained by the particular lymph node is known as its drainage area. The lymph passes through lymph node, where some impurities are removed and further conveyed by lymphatic vessels known as efferent lymphatics. The lymph from efferent lymphatics is further discharge into larger lymph collecting vessels, which all flow towards the heart.

Spleen

In the ox, the spleen is related to the left dorsal side of the rumen and also to the diaphragm. In the young bovine, it is reddish-brown, elongated and slightly convex with rounded edges. In the sheep, the spleen is usually attached to the pluck. It is oyster-shaped, soft and elastic. The spleen of the sheep weighs about 56-83 g. In the pig, the spleen is connected to the greater curvature of the stomach by gastrosplenic omentum. The spleen of pig is elongated, tongue-shaped and triangular in cross-section. The spleen of the pig weighs about 113-125 g. In the equine, the spleen is flat, sickle-shaped and bluish in colour. The spleen of the equine weighs about 453-907 g. In the fowl, the spleen is spherical and reddish-brown in colour.

Haemal Lymph Nodes

Haemal lymph nodes are deep red or almost black in colour, oval-shaped and up to the size of a pea. The afferent and efferent lymphatics are absent in the haemal lymph nodes. The haemal lymph nodes are supplied by arteries which break up in the gland substance and discharge their blood into tissue spaces, in this respect the haemal lymph nodes resembles to the spleen, therefore, they are known as accessory spleen. Haemal lymph nodes are numerous in the ox and sheep but are not found in horse and pig. In the cattle, the haemal lymph nodes are found along the course of the aorta and in the subcutaneous fat, while in the sheep, they are found mostly beneath the peritoneum in the sublumber region. The red lymph nodes of the head and neck of the pig are frequently mistaken for haemal lymph nodes. In the poultry, mural lymph nodes are present on the lymphatics vessels.

Lymph Nodes

Lymph node consists of a reticular framework of elastic and smooth muscle fibres enclosing lymphatic tissues which contain lymphocytes. The colour of lymph nodes varies considerably, which may be white, greyish-blue or almost black. The mesenteric lymph nodes of ox are invariably black. In the pigs, the lymph nodes are lobulated and almost white in colour, except head and neck, which are reddish in colour. The consistency of the lymph nodes varies in different parts of the body. The lymph nodes of abdomen are generally softer than those of the thorax. On exposure to irritants, the lymph nodes may become enlarged, congested, and possibly breakdown of tissues. These pathological conditions are important in determining the disease status in the animals. The lymph nodes are absent in most poultry, although lymphatics are present.

Lymph Nodes of the Ox

The lymph nodes of various parts of the body is discussed as follows:

Lymph Nodes of the Head and Neck

The lymph nodes of the head and neck include parotid, retropharyngeal and middle cervical. Parotid lymph nodes are found one on each side on the edge of the masseter muscle. Retropharyngeal lymph nodes are divided into two groups, *viz.*, internal retropharyngeal and lateral pharyngeal lymph nodes. The internal retropharyngeal lymph nodes are two to four in numbers and situated between the hyoid bones. The lateral pharyngeal lymph nodes are situated beneath each wing of the atlas. The middle cervical lymph nodes are situated in the middle of the neck on each side of the trachea, and often absent in cattle. The middle cervical lymph nodes may vary in number (1-7), size and position.

Lymph Nodes of the Chest and Forequarter

The lymph nodes of the chest and forequarter includes prepectoral (lower cervicals), costocervical, prescapular, intercostal (dorso-costal), subdorsal, suprasternal (sternocostal), bronchial, anterior mediastinal, posterior mediastinal, axillary (bronchial) and xiphoid (ventral mediastinal). The prepectoral lymph nodes are found two to four in numbers on each side, along the anterior border of the first rib. The costocervical lymph nodes are found on the inner side or just anterior to the first rib and close to its junction with the first dorsal vertebra. The prescapular lymph nodes are elongated and lie about 10 cm in front of the point of the shoulder. The intercostal lymph nodes are situated in the intercostal spaces at the junction of the ribs with their vertebrae. The subdorsal lymph nodes are superficial which lie in the fat between the aorta and the dorsal vertebrae. The suprasternal lymph nodes lie between the costal cartilage and are covered by muscles.

The bronchial lymph nodes are found mainly two in numbers, *viz.*, left and right. The left bronchial lymph node is found close to the left bronchus, while the right bronchial lymph node is related to the right bronchus. The anterior mediastinal lymph nodes are found in many numbers and lie in the mediastinal space anterior to the heart. The posterior mediastinal lymph nodes are found in 8-12 in numbers and situated in the fat along the dorsal wall of the oesophagus. The axillary lymph nodes are situated in the muscle external to and about midway along the second rib. The xiphoid lymph node is found in the loose fat at the junction of the sternum and diaphragm at the level of the sixth rib. This lymph node is absent in 50 per cent of cases.

Lymph Nodes of the Abdomen and Hindquarter

The lymph nodes of the abdomen and hindquarter include lumbar, portal (hepatic), renal, mesenteric, splenic, gastric, iliacs, superficial inguinal, supra mammary, deep inguinal, ischiatic, sacral, precrural (sub-iliac) and popliteal. The lumbar lymph nodes are situated in the fat covering the lumbar muscles and are related to aorta and posterior vena cava. Some of these lymph nodes are superficial and others are embedded in the loin fat; haemal lymph nodes are common in this region. The portal lymph nodes form a group around the portal vein, hepatic artery and bile duct, and are covered by pancreas. Another group of lymph node draining the pancreas lies between the edge of the pancreas and the caudate lobe of the liver. The renal lymph node is found in the fat at the entrance to the kidney.

The mesenteric lymph nodes are found in large numbers, which lie between the peritoneal folds of the mesentery and receives lymph from the intestine. The splenic lymph nodes are present in the horse and pig, while, absent in the ox and sheep. Lymph drained from the spleen passes to the gastric chain of lymph nodes. Gastric lymph nodes are numerous in numbers and form a chain along the left and right longitudinal grooves of the rumen. The iliac lymph nodes are situated near the terminal branches of the aorta and are embedded in fat. The internal iliac lymph nodes lie at the junction of the sacrum and the last lumbar vertebrae. The external iliac lymph nodes are found in one or two in numbers and situated laterally to the internal iliac. The superficial inguinal lymph nodes are found in the males and lie in the fat about the neck of the scrotum and behind the spermatic cord.

The supramammary lymph nodes are found in the females which lie above and behind the udder. They are usually two present on each side, one large and one small. The deep inguinal lymph nodes are found in the inguinal canal. It is generally absent and its function carried out by internal iliac lymph node. Ischiatic lymph node lies on the outer aspect of the sacroschiatic ligament. Sacral lymph nodes are not constantly present. If present they are difficult to distinguish from medial iliac lymph nodes. Precrural or sub-iliac lymph node is also known as prefemoral lymph node. It is embedded in fat which may be exposed by an incision at the edge of the tensor facia lata. Popliteal lymph node is deeply seated in the muscles which can be exposed by a deep incision along the superficial seam or division of which connects ischium and os calcis.

Lymph Nodes of the Pig

The lymph nodes of various parts of the body is discussed as follows:

Head and Neck

The lymph nodes of head and neck region includes submaxillary (mandibular), anterior (upper cervical), parotid, pre-scapular (superficial cervical) *etc.* Submaxillary lymph nodes lie anterior to the submaxillary salivary gland near to the angle of jaw. It is covered by the lower part of parotid salivary gland. These lymph nodes are found one pair on either side in which one is large while other is small. Anterior lymph nodes are also known as accessory submaxillary or submandibular. These lie a short distance behind and above the preceding nodes, being separated from them by the submaxillary salivary gland. Parotid lymph nodes are found in several numbers on each side. These are red in colour and situated just posterior to the masseter muscle of the lower jaw and partly covered by the parotid salivary gland. The neck of the pig is short; therefore, prescapular lymph nodes lie close to the parotid salivary gland, being partly covered by its posterior border. These lymph nodes form an oblique chain which is directed downwards and backwards to the shoulder joint.

Other Lymph Nodes

Precrural lymph node is situated anterior to the stifle joint. Popliteal lymph nodes are superficial but absent in 50 per cent pigs. Gastric lymph nodes are situated on the lesser curvature of the stomach. Bronchial lymph nodes are situated one on the bifurcation of the trachea and another on the apical bronchus of the right lung. Portal lymph nodes are present about the portal vein.

Urogenital System

Urogenital system comprised of kidneys, ureters, bladder and urethra. The kidneys are found in pair. In the ox, the kidneys are reddish-brown and composed of about 15-25 lobes which are fused at their deeper portions. The left kidney lies to the left of the vertebral column. The left kidney is freely movable. The right kidney is elliptical. The weight of each kidney is about 283-340 g. The kidneys of sheep and goat are dark brown, bean-shaped and unlobulated and possess single renal papilla. The left kidney of sheep and goat are movable. Each kidney weighs about 56-85 g. The kidneys of pig are reddish brown, smooth, bean-shaped, thinner and flatter than the other food animals. Each kidney weighs about 85-170 g. In the pig, the bladder is long with a narrow neck. In the horse, the left kidney is bean-shaped and the right kidney is triangular or heart-shaped. Each kidney weighs about 680 g. In the domestic fowl, the kidneys lie down on either side of the vertebral column from the level of the sixth rib to the iliac fossa. Each kidney consists of four dark red lobes.

Reproductive System

In the cow, the uterus consists of a small body, about 2.5 cm long and two cornnua or horns, about 38 cm long. The uterus of cow has about 100 oval-shaped cotyledons with prominences on the mucous membrane of the body and horns. In the sheep, the uterine horns are relatively longer and the cotyledons are circular, pigmented and much smaller than in the cow. In the pig, the uterine horns are very long and arranged in coils. The cotyledons are not found on the mucous membrane of uterus of the pig. In the cow, the left and right side of the udder are separated anatomically by a tendinous septum. In the sheep and goat, the udder is composed of two halves, each with one small teat. In the pig, there are 10-16 mammary glands, arranged in two parallel rows.

Nervous System

Nervous system consists of two main parts, *viz.*, central nervous system (brain and spinal cord) and peripheral nervous system (12 pairs of cranial nerves, approximately 40 pairs of spinal nerves and the autonomic nervous system). The brain is situated in the cranial cavity. It consists of cerebrum, cerebellum and medulla oblongata. Cerebrum is divided into two hemispheres by a longitudinal fissure.

Endocrine System

Endocrine system comprised of thyroid, parathyroids, thymus, adrenals and gonads like testes and ovaries.

Thyroid

In the ox, the thyroid gland consists of two lateral lobes, lying on either side of the trachea, immediately posterior to the larynx. The thyroid lobes are flat, dark red and about 7.5 cm long. In the sheep, the thyroid gland is dark red and about 5 cm long. In the pig, thyroid lobes are large, dark red and about 5 cm long.

Parathyroids

The position of the gland is relative to the thyroid gland. In the pig, one pair gland is found in front of the thyroid gland.

Thymus

The thymus is pinkish-white and distinctly lobulated. It consists of two portions, *viz.*, thoracic portion which is rich in fat, which is hand palm shaped in the ox, and the second portion is poor in fat. In the calf, the thymus is at the largest at age of 5-6 weeks, but gradually atrophies.

Adrenal Glands

Adrenal glands are reddish-brown in colour, and found one on each kidney. In the ox, adrenal glands are related to the kidneys and lie anterior to them. In the sheep, both the adrenal glands are bean-shaped. In the pig, the adrenal glands are long and narrow, each lying on the inner aspect of the kidney.

Pituitary Gland

Pituitary gland is situated on the lower aspect of the cerebrum and connected with it by a short stalk. It is situated in the pituitary fossa of the sphenoid bone. It consists of two lobes, *viz.*, anterior lobe and posterior lobe.

Testes

In the bull, the testes are oval-shaped, elongated and about 12.5 cm long and weigh about 283-340 g. The epididymis is narrow and closely related to testicle along the posterior border. In the ram, testicles are long and pear-shaped and more rounded than in the ox.

Ovaries

The ovary comprises an outer cortex containing the follicles and an inner medulla containing the main blood vessels which enter the hilus. In mammals, the ovary is paired and is situated in the dorsal part of the abdomen, caudal to the kidneys. In birds, only the left ovary normally develops. In gilt and sows, the ovaries are rounded with irregularly lobulated surface.

Muscular System

Muscular system is composed of skeletal, cardiac and smooth muscles. Skeletal muscles along with other tissues like associated connective tissues and intermuscular fat forms the flesh or meat. Flesh or meat represents 25-45 per cent of the live weight of the animal. Muscles are made up of spindle-shaped, multi-nucleated muscle cells or fibres. Each muscle cell is encased in a thin membrane known as sarcolemma. Group of muscles joined together to form muscle bundles. Muscle bundles are sheathed in connective tissues (perimysium) and fats. Muscle is completely covered by connective tissue known as epimysium.

Determination of Age of the Food Animals

Determination of age of animal is important in keeping the records of disease in the animals during inspection. The age of the food animals can be accurately determined by dentition and horns and in carcase by bones. The age of the animal can be determined by the presence of permanent incisors.

7

Composition, Quality and Nutritional Importance of Meat

Meat provides an excellent source of protein in human nutrition. The protein requirement in humans makes it necessary to conserve and utilize meat supplies to the fullest possible extent. Meat can make a valuable contribution to the diets in developing countries. The importance of meat in the diet is as a concentrated source of protein which is not only of high biological value but its amino acid composition complements that of cereal and other vegetable proteins. It is also a good source of iron and zinc and vitamin B complex, and liver is a very rich source of vitamin A. The composition of lean meat is given in the Table 7.1:

Table 7.1. Composition of Lean Meat

Constituent	*Per cent Net Weight*
Water	75.00
Protein	19.00
Lipids (fats and oils)	2.50
Carbohydrates (glycogen)	1.20
Non-protein nitrogen	1.65
Minerals	0.65
Vitamins	Minute quantities

The animal carcase consists of muscle, connective tissue, fat and bone and about 75 per cent water. The composition of meat varies according to the proportion of lean to fat depending on species, breed, size, age *etc.* The muscle (lean meat) is relatively constant in composition in a given species. The greatest variable in the carcase is the amount of fat, which may vary from 2 per cent in some free-living animals to 15-40 per cent in intensively reared domesticated animals. The lean meat of various animal species has similar values for macronutrients, inorganic constituents and vitamins. Pork exceptionally has very high levels of thiamine.

Nutrient Composition of Lean Meat

Lean red meats are an excellent source of protein, vitamin B6, vitamin B12, niacin, iron, zinc and phosphorus. Lean red meats are also a source of long-chain omega- 3 polyunsaturated fats, vitamin D, riboflavin, pantothenic acid and selenium. In general lean red meat has a low fat content, is moderate in cholesterol and rich in protein and many essential vitamins and minerals. The nutritional composition of lean red meats varies according to breed, feeding regimen, season and meat cuts.

Proteins

Protein is the most important constituent of the meat which is made up of myofibrils and connective tissue proteins. Meat proteins are of high quality in comparison to vegetable proteins. Meat proteins are easily digested. The protein of typical mammalian muscle contains about 19 per cent protein (11.5 per cent structural protein like actin and myosin, 5.5 per cent soluble sarcoplasmic protein, 2 per cent connective tissue protein like collagen and elastin). Myoglobin is present in relatively large quantities in cardiac muscle because of heavy oxygen demand. The highest amount of myoglobin in mammals is found in the whale which permits prolonged submersion under water. Human requirements for protein are estimated to be 55 g per day for adult man and 45 g for woman. There is a higher requirement in various disease states and stress conditions.

Meat protein provides all essential amino acids like lysine, leucine, isoleucine, methionine, threonine, phenylalanine, valine, and tryptophan. Creatine and its phosphorylated derivative creatine phosphate play an important role in muscle energy metabolism. Red meat contains approximately 350mg/100g. It is the principal dietary source for humans. Meat is a rich source of taurine amino acid which is considered a conditionally essential amino acid during lactation, during times of immune challenge, and may offer protection against oxidative stress.

Non-Protein Nitrogen

Non-protein nitrogen is mainly represented by free amino acids, creatine, nucleotides, inosinemonophosphate and carnosine. The free amino acids and nucleotides of the meat give a characteristic flavour on cooking.

Lipids

The lipids are mostly composed of triglycerides (fats and oils). The level of cholesterol is low in carcase meat but much higher in liver. There is a wide variation in the amount of total separable fat between the different beef and lamb cuts, ranging from 37 per cent in loin lamb chops to only 1 per cent in veal steak.

Saturated fatty acids comprise about 40 per cent of total fatty acids in the lean component and 48 per cent in the fat component of red meat. In beef and veal, approximately half of the saturated fatty acid in both the lean and fat component of red meat is palmitic acid and about a third is stearic acid. In lamb and mutton, the proportions of these two fatty acids are almost same. There is little variation between cuts in the proportion of fatty acids. Polyunsaturated fatty acids (PUFA) range from 11 per cent to 29 per cent of total fatty acids. Pasture fed beef is better source of omega-3 fats than grain feed beef. Beef and lamb also have more omega-3 fatty acids than either chicken or pork, although fish is still a significantly better source than any of the red meats.

Trans fatty acids are found in ruminant fat as a result of biohydrogenation by rumen bacteria. Trans fatty acids in raw meat varies from 22mg in veal to 123 mg in lamb per 100g. Trans fatty acids in meat are generally less than 3 per cent of the total fatty acid content. The levels of trans fatty acids in meat are higher in lamb and mutton than in beef and veal. Choline is a precursor of a number of compounds including neurotransmitters and membrane phospholipids. The best dietary sources of choline are milk, eggs, and liver, but meat is also a significant source and beef contains 78mg per 100g meat.

Lipids Storage Sites in the Body

Lipids (fats) are found in the body at three main sites as follows:

1. **Storage Deposits like under the Skin and Around the Organs:** The largest amount by far is in the storage deposits under the skin and around the organs. These storage sites contribute approximately 40-50 per cent of the total weight in fatty meat or fatty bacon.
2. **Between the Bundles of Muscle Fibres (Intermuscular) Fat:** Small streaks of fat are visible between the bundles of muscle fibres. This is also known as "marbling". Bundles of muscle fibres are the lean part of the meat. This site contributes approximately to 4-8 per cent of the weight of lean meat.
3. **Within the Muscle Structure:** This is known as intramuscular or structural fats which are found within the muscle structure. Intramuscular fat deposition contributes about 1-3 per cent of the wet weight of muscle and 5-7 per cent of the weight of the liver.

Carbohydrates

Carbohydrates in meat are mainly present in the form of glycogen. Approximately 2.5 per cent carbohydrate (lactic acid, glucose and derivatives) is present immediately after rigor mortis.

Minerals

Meat is an excellent source of minerals such as iron, copper, zinc and manganese. Other important minerals in the meat are calcium, phosphorus, sulphur, sodium, potassium, magnesium, cobalt *etc.* Beef and lamb meat are among the richest sources of the minerals like iron and zinc. The absorption of zinc from a diet high in animal protein is greater than from plant foods, and the requirements for zinc may be as much as 50 per cent higher for vegetarians. Red meats are also good sources of selenium. Lean meat is low in sodium with a potassium/sodium ratio of greater than five. The copper content in mutton is relatively higher than the beef.

A wide variety of mineral salts are present in the meat and offal. The contents of iron, copper and zinc vary considerably in different animal species. Liver is the richest source of these minerals in comparison to muscle tissue. High levels of minerals in the feed do not necessarily increase the level of that mineral in the meat due to the reason of a complex inter-relation between minerals. For example, the molybdenum content of sheep meat increases with dietary molybdenum only when dietary sulphate levels are low. Dietary molybdenum inhibits the accumulation of copper which is partly off-set by increased manganese. Copper content in liver decreases and molybdenum increases with increasing amounts of molybdenum. Other inter-relations between minerals include calcium and zinc. Pasture deficient in minerals; especially phosphorus and cobalt lead to reduced amounts in the muscle.

The amount of iron absorbed from the diet depends on many factors including its chemical form, the simultaneous presence of other food ingredients that can enhance or inhibit absorption, and various physiological factors of the individual. About half of the iron in meat is present as haeme iron (in haemoglobin). Zinc is present in all tissues of the body and is a component of more than fifty enzymes. Meat is the richest source of zinc in the diet.

Vitamins

Meat is an important source of all the B-complex vitamins including thiamine, riboflavin, niacin, biotin, vitamins B6 and B12, pantothenic acid and folacin. Pantothenic acid and folacin are especially abundant in liver which, together with certain other organs is rich in vitamin A and supplies appreciable amounts of vitamins D, E and K. Mutton is very rich source of thiamine, vitamin B6, vitamin B12, phosphorus, iron and copper.

In comparison to other animal foods, red meat is an excellent source of bioavailable vitamin B12, providing over two thirds of the daily requirement in a 100g serve. Pork is good source of thiamine. Liver is an excellent source of vitamin A and folate, but the levels in lean meat are low. The concentration of these vitamins is higher in older animals so the levels in beef are generally higher than those in veal, and mutton has more than lamb. Levels of vitamin D in meat are low.

The content of most vitamins in the body is largely independent of diet with the exception of vitamin which is stored in the liver in amounts depending on intake, with small amounts present in the kidney. Under free range conditions of grazing there is a very high intake of carotene (pro-vitamin A) which is mostly converted into retinol that is vitamin A. Pig meat is very rich in thiamine which is nine times more than the other animals. Liver is the richest of animal tissues in all the vitamins.

Reducing Bodies

Reducing bodies have an enzyme action. Their action can be seen in carcase affected with icterus. The yellow colouration in the icteric carcase (occurs due to the presence of bilirubin) disappears after its detention for 24 hours. This occurs due the action of reducing bodies.

Endogenous Antioxidants

Carnosine and anserine are antioxidative histidyl dipeptides. These are the most abundant antioxidants in meat. Carnosine is present approximately 365mg/100g in beef and 400mg/100g in lamb. Coenzyme Q10 (ubiquinone) also has antioxidant properties. It is found approximately 2mg/100g in both beef and mutton. Glutathione has antioxidative and immunomodulatory properties and also enhances iron absorption. Glutathione is found approximately 12-26mg/100g in beef.

Other Constituents

Other constituents of the meat include pigments such as myoglobin and hemoglobin. The redness of the meat is due to the presence of myoglobin. The content of myoglobin is higher in male animals, old animals and active muscles such as muscles of heart and diaphragm.

Nutrient Composition of Organ Meats

The nutrient content of liver, kidney, heart, brains and tripe from beef and lamb are important source of several nutrients as follows:

1. All organ meats (except tripe) are extremely rich in vitamin B12.
2. Liver is a rich source of protein, iron, zinc, vitamin A, riboflavin, folate, and niacin.
3. Heart is a good source of iron and zinc, but less than the liver and kidney.
4. Kidney is rich in protein, iron, thiamine, riboflavin and folate.
5. Brains and tripe are not particularly good sources of vitamins or minerals
6. All organs meats have high level of cholesterol, especially brain, and low level of sodium.

Factors Influencing Composition of Meat

1. **Diet:** There is limited effect of diet on the nutrient composition of lean meat. With the exception of vitamin A stored in the liver, diet has little effect on vitamin content. However, the addition of thiamine to the diet of pigs can double or treble the amount of thiamine in various muscles.
2. **Age at Slaughter:** The proportions of total nitrogen, fat and iron increase as the animals approach maturity and become slow after that. At the same time the ratio of polyunsaturated to saturated fatty acids falls. As the age of animals advanced the collagen becomes less soluble and less digestible. Older animals and animals reared on poor pasture have a high proportion of water-soluble extractives in the muscle; therefore, they have long been used for the preparation of meat extract.

Effect of Cooking on Protein Quality

The quality of proteins is affected by cooking due to damage to amino acids. When food is cooked below 100°C the proteins are coagulated, which does not affect the nutritional quality of food. When food is heated to temperatures around 100°C in the presence of moisture and reducing sugars (present naturally or added to the food) a chemical reaction takes place between part of one essential amino acid, lysine and a sugar to form a bond that cannot be broken during digestion, and so part of the lysine is rendered unavailable. At a higher temperature, the lysine in the food protein can react with other chemical groupings within the protein itself and more becomes unavailable. Sulphur containing amino acids like cystine and methionine are rendered partly unavailable due to heat treatment of foods.

Toxic Compounds Formed During Processing and Cooking

1. **Oxidation Products of Meat Fat:** The oxidation of fats takes place, especially at frying temperatures, can give rise to compounds that decompose to aldehydes, esters, alcohols and short chain carboxylic acids with undesirable flavours. Unsaturated lipids present in meats are more readily oxidised due to catalysis by haeme and non-haeme iron. Oxidation products formed in meat may be carcinogenic. Cholesterol can also be oxidised and the oxidation product has been suggested as a possible factor in coronary heart disease.
2. **Carcinogens:** The pyrolysis products like polycyclic hydrocarbons are formed due to overheating and charring of organic material. These polycyclic hydrocarbons are believed to be carcinogenic. Polycyclic hydrocarbons like 3, 4-benzpyrene is formed on the surface of barbecued and broiled (grilled) and smoked meat products.
3. **Nitrosamines:** Nitrites, used in curing salts can react with amines commonly present in food, to form nitrosamines, which has been shown to be carcinogenic in all species of animals examined but it is not clear, despite years of intensive research, whether the amounts present in cured meats affect human beings. As a precaution, legally enforced in some countries, there is a tendency to reduce the amount of nitrite used in the curing mixture and to add vitamin C which inhibits the formation of nitrosamines. Erythorbic acid and tocopherol are also effective in reducing nitrosamine formation.

Comparison of the Composition of Red Meat and Vegetarian Protein

Lean beef and lamb are better protein sources than all the options except cheese, and are mostly lower in sodium. The meats contain higher amount of zinc and niacin than all the alternatives. The meats also have higher content of omega-3 fats than the vegetable sources. Meats are better source of vitamin B6 (except for the walnuts); richer in selenium (except in eggs); and are the best source of vitamin B12, which is absent entirely from the vegetable products.

Post-Mortem Changes (Conversion of Muscle into Meat)

The post-mortem changes take place when muscle is converted into meat. These changes have a marked effect on the quality of the meat. After slaughter of animal, the glycogen in the muscle is converted into lactic acid causing decline in pH from an initial value 6.8 - 7.3 to about 5.4 - 5.8 at rigor mortis. Stress in the animals immediately prior to slaughter, rough handling of animals and fighting among them cause release of muscle glycogen into the blood stream. After slaughter of animal the muscle glycogen is rapidly broken down into lactic acid while the carcase is still warm. This high level of acidity in muscles causes a partial breakdown of muscle structure which results in pale, soft and exudative (PSE) meat. This condition most commonly occurs in pigs. The PSE meat loses some of its water-binding capacity which is very important in meat processing.

Rigor Mortis

Rigor mortis is a condition that develops after death or slaughter of the animal which is characterized by the rigidity of muscles and stiffness of joints. It is the first and most considerable change in the muscle that occurs after slaughter of animals. Rigor mortis commences at the head and neck and extending backwards to involve the limbs and body. In a physiologically normal animal, the rigor mortis usually commences after 12 hours of its slaughter and attained maximum rigidity after 20-24 hours of slaughter, and then gradually declines. The onset of rigor mortis may be very rapid and may pass off rapidly in carcases of hunted animals due to decline in the level of muscle glycogen. The longer the time of appearance of rigor mortis the longer is its duration. The too rapid development of rigor mortis is indicative of early putrefactive change in the carcase. The maintenance of rigor mortis for longer period is desirable. Rigor mortis is characterized by the following:

1. Hardening and contraction of all voluntary muscles.
2. Stiffness of the joints.
3. Loss of transparency of the surface of the muscles which become dull.
4. Slight rise in temperature of the beef carcase to 1.5°C or more above normal.

Significance of Rigor Mortis

The occurrence of an adequate degree of rigor mortis and low ultimate pH of the meat due to lactic acid production in it are desirable characteristics. The concentration of lactic acid is ten times that of living muscles. In the meat, the low pH is inhibitory to microbial growth. Lactic acid also play important role in the conversion of the collagen of the connective tissues into gelatin and meat becomes tender on cooking. Lactic acid is also important in maintaining the rigidity of the carcase and hence easy making of desired pieces of meat from the carcase. In freshly slaughtered cattle the pH of the flesh is about 7.0 which declines to 5.5 or 5.8 and then gradually rises due the formation of alkaline substances from breakdown of protein.

Factors Influencing Rigor Mortis

There are various intrinsic and extrinsic factors that can influence the onset, existence and passing of rigor mortis.

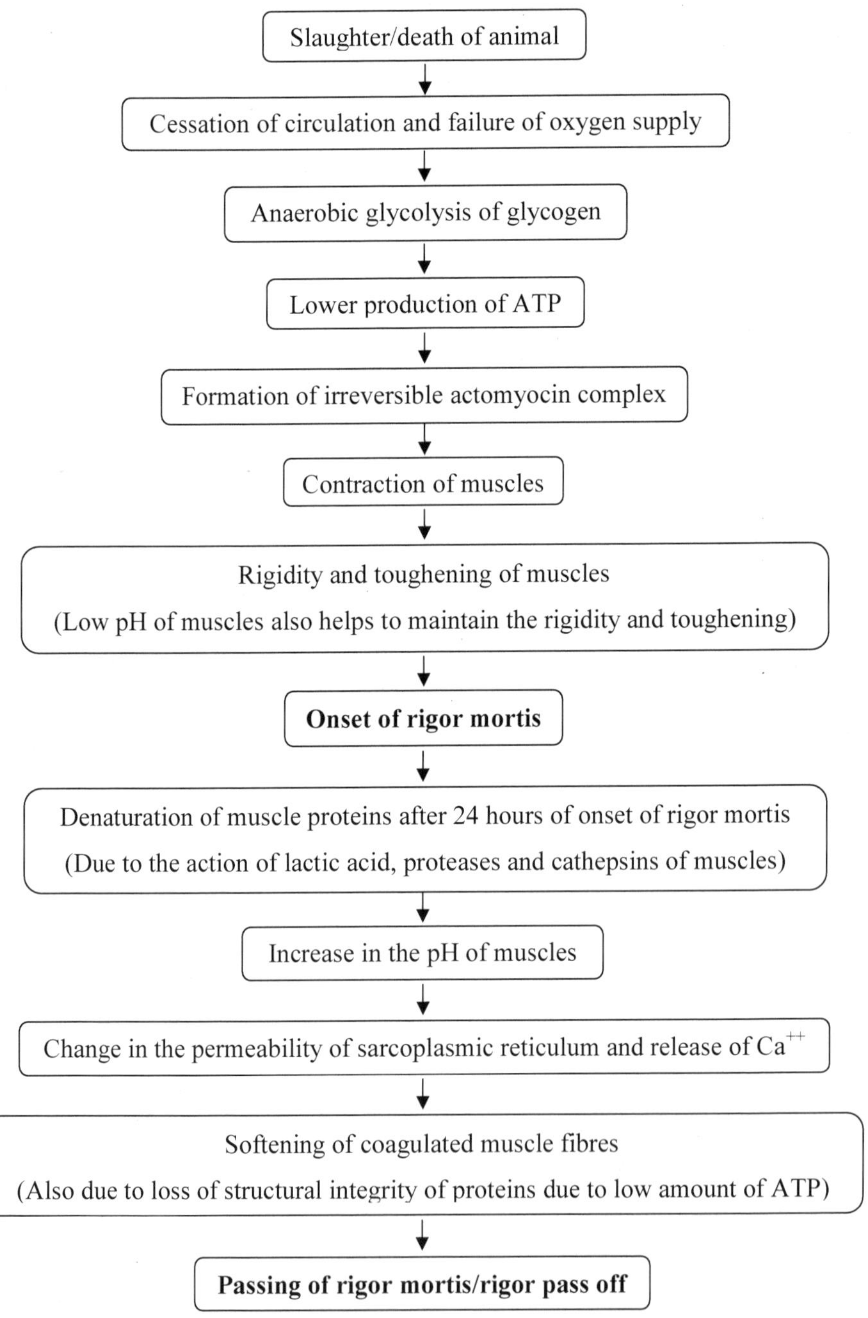

Figure 7.1 Mechanism of Onset and Passing of Rigor Mortis.

Intrinsic Factors

These factors are associated with factors internally existing in the body of the animals. The following intrinsic factors can influence the rigor mortis:

1. **Species of the Animal:** The rate of post-mortem glycolysis varies from species to species of the animal. The onset of rigor mortis occurs faster in more active animal species like cattle and horse, while slower in less active animal species like pig. The condition of rigor mortis exists for a short period and passes off early in the active animal species.
2. **Type of Muscle:** Rigor mortis affects first the most active and well-nourished muscles. Active muscles like muscles of heart, chest, diaphragm *etc.* are affected earlier that is onset of rigor mortis occurs earlier than the less active muscles. The heart muscles are affected very early, usually within an hour of slaughter.
3. **Amount of Glycogen:** The amount of glycogen is directly proportional to the production of lactic acid in the muscles and thereby development of acidity in the meat and also the onset of rigor mortis. The onset of rigor mortis occurs early in starved animals, while late in well-fed animals due to large quantity of glycogen stores. The rigor mortis is maintained for a longer period when the glycogen content is high in the muscle.
4. **Initial level of ATP and Creatinine Phosphate:** Onset of rigor mortis is dependent on ATP. In healthy and non-exhausted animals, ATP levels are sufficiently high generated during aerobic breakdown of glycogen. Creatinine phosphate and ADP combine and also form the ATP. These initial high levels cause little delay in the onset of rigor mortis. In case of anthrax, the causative agent *Bacillus anthracis* releases a toxin called oedema factor or calmodulin dependent adenylate cyclase. This toxin inhibits the stiffening of muscles by constantly stimulating the release of cAMP, so that the cAMP level in body is maintained for a prolonged period which leads to a relaxed condition in the carcase.

Extrinsic Factors

Some factors that may be present in the environment and some other factors or conditions can affect the onset, duration of existence and passing off rigor mortis. The following extrinsic factors can influence the rigor mortis:

1. **Atmospheric Temperature:** High atmospheric temperature accelerates the onset of rigor mortis whereas low temperature retards it. Therefore, the onset and passing of rigor mortis is faster in summer and slower in winter. Rigor is completed in cattle after 12-24 hours. Rigor is resolved in 10-13 days at 0°C, 4-5 days at 10°C, 30-40 hours at 20°C and 10-11 hours at 30°C. The process is twice as fast in pork as in beef or lamb. It is faster in young animals and slower in red muscles.
2. **Disease:** Rigor mortis may be absent or scarcely present in the animals suffering from fever. In animals suffering from fever the meat may become permanently alkaline until decomposition commences. Animal suffering

from hypothermia has lower level of glycogen in the body and hence the onset of rigor mortis is faster. The onset of rigor mortis is faster in the animal suffering from tetanus.

3. **Conditions of the Animal:** Animals struggling, fighting, excessive handling, long distance travel, exhaustion, fatigue, fear and other stressful conditions also deplete the glycogen level in their body. Therefore, the onset of rigor mortis is faster in such slaughtered animals. In fatigued animals the muscle may remain alkaline for 2 to 3 days before becoming acidic.
4. **Administration of Drugs:** Drugs such as sodium salicylate, strychnine, alcohol and ether tend to encourage the early onset of rigor mortis.

Nutritional Importance of Meat

From the nutritional point of view, meat's importance is derived from its high quality protein, containing all essential amino acids and it's highly bio-available minerals and vitamins. Meat is rich in Vitamin B12 and iron which are not readily available in vegetarian diets. Meat, and other animal foods such as milk, can make a valuable contribution to the diets in developing countries. Meat in the diet has a concentrated source of protein which is not only of high biological value but its amino acid composition complements that of cereal and other vegetable proteins. It is also a good source of iron and zinc and several B vitamins, and liver is a very rich source of vitamin A. Meat is an important part of human diet because of the following reasons:

1. It is the most concentrated and easily assimilable food.
2. It is stimulating to metabolism due to its high protein content which assists the body in the production of heat and energy.
3. It acts as stimulant to gastric secretion and is readily digested.
4. It consists of almost all essential amino acids.

Role of Meat in the Human Diet

Meat is held in high esteem in most communities. It has prestige value and it is often regarded as the central food round which meals are planned. Various types of meat are sometimes made the basis of festive and celebratory occasions, and from the popular as well as the scientific point of view, it is regarded as a food of high nutritive value. In the developing countries the meat is in short supply and it can be taken as a measure of the nutritional quality of the diet as a whole. Where a typical diet is heavily dependent on one type of cereal or root crop, meat, even in small amounts, complements the staple food. It provides a relatively rich source of well absorbed iron and also improves the absorption of iron from other foods. Meat is a concentrated source of B vitamins, including vitamin B12 which is absent from plant foods. Consequently there is pressure to increase the availability of meat products.

In industrialized countries, there have been slow but continuous changes over the years in the relative amounts of different types of meat consumed (beef, pork, lamb, poultry *etc.*) depending partly on price and influenced by fashion, advertising, *etc.* In more recent years the health aspects have become a factor. The concerns about public health in industrialized countries where coronary heart disease and other diseases of affluence are common have led to recommendations to the public to modify their diet, popularised as Dietary Guidelines. These particularly recommend a reduction in fat consumption, especially saturated fatty acids. This has led in some sections of their populations to a relative increase in the consumption of poultry and fish at the expense of red meat.

Residues of Drugs and Pesticides in Meat

Residues of drugs, pesticides and agricultural chemicals can be found in small amounts in meat and meat products. Pesticides may be applied specifically to the animals to control insects or intestinal parasites but may also be present in meat as a result of exposure of the animals to chemicals used on buildings, grazing areas and crops. In pigs, some tranquillizers are used immediately prior to slaughter. There is no time for their removal through the normal metabolic processes. They can persist in the human body so that repeated intakes could possibly result in accumulation of the drugs. Administration of drugs to animals and birds is always a potential risk to human health and so there is a need to control the use of these drugs and to measure the extent of any residues left in the food intended for human beings.

8

Infectious Diseases

Infectious diseases are responsible for most losses in the food animals throughout the world; therefore, these diseases are of great economic importance. Therefore, their accurate diagnosis and prevention and control are important in reducing the risks and economic losses.

BACTERIAL DISEASES

Cattle

Actinobacillosis

This is also known as "wooden tongue". Actinobacillosis is a chronic disease of cattle caused by *Actinobacillus lignieresi*. The disease is characterized by inflammation of the tongue and less frequently lymph nodes of the head and of even the viscera and carcase. Other symptoms are mouth erosions, salivation and chewing and loss of appetite.

Post-mortem findings show enlarged tongue showing tough fibrous consistency "wooden tongue"; a cluster of small yellowish nodules and erosions of tongue mucosa; granulomatous lesions in the lymph nodes; marked thickening of the lower part of oesophagus and stomach wall; and raised plaques and erosions in the mucosa of rumen and reticulum. Typical actinobacillosis lesions in the lymph nodes and organs consist of greenish-yellow thick creamy pus with "sulphur granules". These are bacterial colonies surrounded by club like structures

Carcase of an animal affected with active progressive inflammatory lesions of actinobacillosis in lymph nodes and lung parenchyma is condemned. If the disease is slight and confined to lymph nodes, the head and tongue and whole carcase are approved after the condemnation of lymph nodes. If the tongue is diseased and no lymph nodes are involved the head and carcase are approved. The tongue is condemned. Differential diagnosis should be made with neoplasms, tuberculosis, abscesses in the lymph nodes, foreign body, salivary cysts, fungal granulomas, and parasites.

Actinomycosis

This is also known as "Lumpy Jaw". Actinomycosis is a chronic granulomatous disease of cattle and pigs and rarely in sheep and horses. It is caused by *Actinomyces bovis* which is an obligatory parasite in the mucous membrane of the mouth and pharynx. The disease is transmitted through injury with a sharp object or hard feed pieces to the oral mucosa. The disease is characterized by painful swelling of the maxilla and mandible (lumpy jaw) and rarely in feet, suppurative tracts in the granulation tissue breaking towards oral cavity or skin, ulceration of cheeks and gums and wart like granulations outward on head, difficult breathing and salivation, diarrhoea and bloat.

Post-mortem findings show lesions in the mandible (Lumpy jaw) or maxilla; granulomatous lesions in lower part of oesophagus or anterior part of the reticulum; local peritonitis; and mild abomasitis and enteritis. Judgement is same as actinobacillosis. Differential diagnosis should be made with tooth infection, impacted food, bone injury, neoplasms and osteomyelitis.

Anthrax

Anthrax is a bacterial disease caused by *Bacillus anthracis*, a Gram-positive, rod-shaped, capsulated, aerobic and spore-forming bacterium. The disease is worldwide in distribution. The disease is enzootic in most countries of Africa and Asia, a number of European countries and countries/areas of the American continent and certain areas of Australia. The disease is transmitted through inhalation, ingestion and through a wound in the skin. In animals, the disease is found in peracute, acute or subacute form. In ruminants, the disease occurs in peracute form which is manifested with septicaemia, sudden death and tarry blood from the body openings of the dead animal. In pigs and horses, the disease is usually localized and chronic and is often characterized by swelling around the throat and head. In pigs, the disease is manifested with oedematous swelling of the throat and neck, swallowing and breathing difficulties, and death due to choking or toxaemia. Septicaemia is not observed in pigs.

Post-mortem findings show dark-tarry blood discharge from body orifices; absence of rigor mortis; haemorrhage of the mucous and serous membranes, lymph nodes and subcutaneous tissue; enlarged spleen; severe haemorrhagic enteritis; and bloating and rapid decomposition of carcase. The carcase and its parts are condemned and disposed of by burning or burial. If disposed by burial, the carcase should be buried at least 6 feet below the ground. The site should be surrounded by a foot thick layer of quicklime. Differential diagnosis should be made with peracute blackquarter and septicaemic form of other diseases. In splenic enlargement as seen in babesiosis, anaplasmosis and leucosis, spleen consistency is firm. In anthrax, the spleen is soft and upon incision the pulp exudes like thick blackish-red blood.

Brucellosis

This is also known as contagious abortion or Bang's disease. Brucellosis is a contagious disease caused by *Brucella* spp. and is characterized by abortion in late pregnancy and a high rate of infertility. *B. melitensis* affects goats, *B. ovis* sheep and *B. suis* swine. *B. abortus* infection may occur in horses. In animals, the disease is transmitted through contaminated feed, pasture, water, milk, by an aborted foetus, foetal membranes and uterine fluid and discharges. The disease may also be transmitted by dogs, rats, flies, boots, vehicles, the milking machine and other equipment used in the barn. The *Brucella* organism may be occasionally shed in urine. In cattle, the disease is characterized by abortion in pregnant cows in the last 3 - 4 months of pregnancy, oedematous placenta and foetus, orchitis, epididymitis, hygromas on the knees, stifles, hock and angle of the haunch, and between the nuchal ligament and the primary thoracic spines. In sheep, the disease is characterized by fever, increased respiration and depression, oedema and swelling of scrotum, infertility in rams and abortion in ewes.

Cattle and horse carcases affected with brucellosis are approved (after removal of affected parts), as *Brucella* spp. remain viable for only a short period in the muscles after slaughter. In acute abortive form (after the miscarriage), cattle carcases are condemned. Pig, sheep, goat and buffalo carcases require total condemnation. Heat treatment may be recommended in some areas for these species due to economic reasons. Affected part of the carcase, udder, genital organs and corresponding lymph nodes must be condemned. Reactor animals should be carefully handled during slaughter and dressing procedures. Gloves and goggles should be worn when known reactors are being slaughtered and hygroma lesions should be sprayed liberally with 1 per cent lactic acid at meat inspection. Differential diagnosis should be made with causes of abortion in cattle, vibriosis, leptospirosis, trichomoniasis, mycoplasma infections, mycosis, infectious bovine rhinotracheitis, nutritional and physiological causes.

Tuberculosis

Tuberculosis is a chronic contagious disease of many animal species and poultry, caused by *Mycobacterium* spp. The disease is characterized by development of tubercles in the organs of most species. Bovine tuberculosis is caused by *Mycobacterium bovis*. Tuberculosis is a significant zoonotic disease. An infected animal is the main source of infection. The organisms are excreted in the exhaled air and in all secretions and excretions. Inhalation is the main mode of transmission. Infected milk is an important source of infection for calves. The disease is characterized by low grade fever, chronic intermittent hacking cough and associated pneumonia, difficult breathing, loss of appetite, weakness, emaciation, and swelling of superficial body lymph nodes.

Post-mortem findings show tuberculous granuloma in the lymph nodes of the head, lungs, intestine and carcase. These have usually a well-defined capsule enclosing a caseous mass with a calcified centre. They are usually yellow in colour in cattle, white in buffaloes and greyish white in other animals. Active lesions may have a reddened periphery and caseous mass in the centre of a lymph node. Inactive lesions may be calcified and encapsulated. Nodules are noticed on the pleura and peritoneum. The lesions are noticed in the lungs, liver, spleen, and kidney. Firmer and enlarged udder particularly rear quarters are noticed. Lesions in the meninges, bone marrow and joints are also noticed.

Carcase of an animal affected with tuberculosis requires additional post-mortem examination of the lymph nodes, joints, bones and meninges. Carcases are condemned where an eradication scheme has terminated or in cases of residual infection or re-infection; in final stages of eradication - natural prevalence low; and during early stages in high prevalence areas. Carcase of a reactor animal without lesions may be approved for limited distribution. If the economic situation permits, this carcase should be condemned. Heat treatment of meat is suggested during early and final stages of an eradication programme. Differential diagnosis should be made with lung and lymph node abscess, pleurisy, pericarditis, chronic contagious pleuropneumonia, actinobacillosis, mystic and parasitic lesions, tumours, caseous lymphadenitis, Johne's disease, adrenal gland tumour and lymphomatosis.

Johne's Disease

This is also known as bovine paratuberculosis. Johne's disease is a chronic, infectious bacterial disease of adult wild and domestic ruminants such as cattle, sheep and goats. The disease is caused by *Mycobacterium paratuberculosis*. The disease is transmitted through ingestion of feed and water contaminated with faeces of infected animals. The agent persists in soil, pasture, manure and stagnant water for long period. Carrier animals, so called "faecal shedders", are the most important source of infection. Calves may become infected from a nursing infected dam. Incubation period ranges from 6 months to 15 years (commonly 2-3 years). The disease is characterized by gradual weight loss and emaciation, rough hair coat and dry skin, non-responsive chronic diarrhoea with watery fluid faeces, the thickening and corrugation of the wall of the intestine, submandibular oedema ("bottle jaw") and debility and death.

Post-mortem findings show thickened and corrugated intestinal mucosa and enlarged caecal lymph nodes. Carcase of an animal affected with Johne's disease is approved when generalized systemic signs of disease are not present. A poor, thin and slightly moist carcase should be held in the chiller and assessed after 24 or 48 hours. If the dryness and setting of the carcase improves during this time it can be released. The carcase with associated oedema and emaciation is condemned. Differential diagnosis should be made with other causes of diarrhoea and weight loss, malnutrition, chronic salmonellosis, ostertagiasis, winter dysentery, bovine viral diarrhoea, coccidiosis, toxic inflammation of the intestine caused by arsenic, plants and mycotoxicosis and neoplasm.

Leptospirosis

Leptospirosis is an important disease of domestic and wild animals and humans. In cattle, the disease is manifested by interstitial nephritis, anaemia and mastitis and abortion in most species. Leptospirosis is caused by *Leptospira* spp. In animals, the disease is transmitted through ingestion of feed and water contaminated with urine of infected animal or by direct contact of broken skin or mucous membranes with mud, vegetation or aborted foetuses of infected or carrier animals. Recovered animals and animals with unapparent (subclinical) leptospirosis frequently excrete billions of leptospirae in their urine for several months or years.

Acute and subacute forms of the disease are characterized by transient fever, loss of appetite and mastitis. Milk may be yellow, clotted and frequently blood stained. Severely affected animals may show jaundice, anaemia, pneumonia, abortion and retention of the placenta. In young calves, severe illness may be associated with yellowish discolouration of mucous membranes and reddish-brown urine before death. The chronic form of the disease has mild clinical signs and only abortion may be observed. Animal may show incoordination, salivation and muscular rigidity due to meningitis.

Post-mortem findings show anaemia and jaundice; subserosal and submucosal haemorrhage; ulcers and haemorrhages in the abomasal mucosa; rarely pulmonary oedema or emphysema; interstitial nephritis; and septicaemia. Carcase of an animal affected with acute leptospirosis is condemned. A chronic and localized condition may warrant an approval of the carcase. Differential diagnosis should be made with acute and subacute forms to be differentiated from babesiosis, anaplasmosis, rape and kale poisoning, bacillary haemoglobinuria, post parturient haemoglobinuria and acute haemolytic anaemia in calves. The presence of blood in the milk is a characteristic clinical sign which will differentiate leptospirosis from other infectious diseases.

Salmonellosis

Salmonellosis occurs in all animals and humans. In animals, the disease occurs in peracute septicaemic form, acute enteritis or chronic enteritis. The young, old, stressed and debilitated animals are at greater risk. In cattle, the disease is caused by *S.* Typhimurium, *S.* Dublin, *S.* Newport and *S.* Muenster. The disease is transmitted through ingestion of feed and water contaminated with faeces of infected or carrier animals. Active carrier animals shed *Salmonella* organisms intermittently. In human, the disease is transmitted through contaminated water, raw milk and meat. Pigs and poultry are more significant sources of infection in humans.

Peracute septicaemic form of the disease is most frequently noticed in colostrum deficient calves up to four months of age and the disease is characterized by fever, diarrhoea, dehydration, depression, and death within 48 hours. Approximately four weeks after the onset of diarrhoea, there may be meningoencephalitis, polyarthritis, and necrosis of extremities of limbs and tail and tip of ears. Acute enteritis form of the disease commonly occurs in adult cattle in late pregnancy and early postpartum, which is characterized by fever, watery and foul smelling diarrhoea, dehydration, depression, emaciation and death. Chronic enteritis is preceded by acute enteric form of the disease. This form of the disease is characterized by fluctuating fever, diarrhoea, dehydration and emaciation.

In septicaemic form, the post-mortem findings show submucosal and subserosal haemorrhage. In Acute enteritis form, the post-mortem findings show haemorrhagic enteritis, necrosis of ileum and large intestine, enlarged, oedematous and haemorrhagic lymph nodes, fatty change and enlargement of liver, subserous and epicardial haemorrhage and cholecyctitis. In chronic enteritis form, the post-mortem findings show necrosis of caecum and colon, swollen mesenteric lymph nodes, enlarged spleen and chronic pneumonia.

Carcase of animal affected with salmonellosis is condemned. Differential diagnosis should be made with acute diarrhoea in calves due to *E. coli, Clostridium perfringens* type C, rotavirus, cryptosporidiosis; coccidiosis; and dietetic gastroenteritis. Acute diarrhoea in adult cattle should be differentiated from bovine viral diarrhoea, winter dysentery, coccidiosis, gastrointestinal parasitism, bracken fern poisoning, arsenic and lead poisoning. Chronic diarrhoea of adult cattle should be differentiated from Johne's disease, gastrointestinal parasitism and copper deficiency.

Calf Diphtheria

Calf diphtheria is an acute infection of calves (less than 3 months of age). The disease is caused by *Fusobacterium necrophorum (Sphaerophorus necrophorum)*. The causative agent is an inhabitant of digestive tract of cattle and also found in the environment. The disease is transmitted through unhygienic conditions of feeding troughs and dirty milk pails. The disease can also be transmitted through abrasions in the oral mucosa. The disease is characterized by fever, coughing, anorexia, depression, difficult breathing, problems in chewing and swallowing, swelling in pharyngeal region, and ulcerations on the tongue, palate, and inside of cheeks.

Post-mortem findings show inflammation and ulceration with large masses of yellow-grey material in the mouth, tongue, pharynx and larynx. Pneumonia is commonly noticed. Carcase of an animal affected with local lesions is approved; while generalized diphtheric lesions associated with pneumonia or toxaemia require total condemnation of the carcase. The emaciated carcases are also condemned. Differential diagnosis should be made with vesicular diseases, abscesses and neoplasms.

Black Quarter

This is also known as black leg. Black quarter is an acute infectious disease of cattle and sheep. The disease is caused by *Clostridium chauvoei*. The disease is transmitted through grazing of pastures contaminated with organisms of blackleg. The organism is also found in the digestive tract of healthy animals. In sheep, the disease is transmitted through wounds at shearing, docking and castration and during lambing in ewes. The disease is characterized by high fever, severe myositis, discoloured, dry or cracked skin, crepitating swellings often on the hips and shoulder, lameness, stiff gait and reluctance to move, and high mortality. In sheep, gaseous crepitation cannot be felt before death.

Post-mortem findings show dead cattle laying on one side with affected hind leg stuck out. Other post-mortem findings show dark red to black colour of muscles in the loin, back or leg regions; bloating of carcase and blood stained frothy exudates from the nostrils and anus; yellowish, gelatinous subcutaneous tissue and associated gas bubbles; and blood stained fluid in body cavities. The carcase and parts from animal affected with black quarter is totally condemned.

Botulism

Botulism is a toxaemic, paralytic disease of animals, birds and humans, caused by ingestion of toxins of *Clostridium botulinum*, which is a soil saprophyte and common inhabitant of digestive tract of animals. The disease is transmitted by ingestion of contaminated food and water. The organism produces toxin in the decaying organic matter or soil. The organism does not produce toxin within the animal's body but preformed toxins are the main source of disease transmission.

In animals, the disease may occur in peracute, acute, subacute or chronic form. In cattle and horses, the disease is characterized by profuse salivation, partial or complete paralysis of muscles of locomotion, Knuckling and incoordination, sternal recumbency, inability to chew and swallow and protrusion of tongue from mouth. There is no rise of body temperature. In sheep, the disease is characterized by nasal discharge, salivation, abdominal respiration, stiffness upon walking and incoordination in movement, switching of the tail on the side, limb paralysis and death. In pigs, the disease is characterized by anorexia, refusal to drink, vomiting, pupillary dilatation and muscular paralysis.

Post-mortem findings show foreign material in stomach. The carcase and parts from animal affected with botulism is totally condemned. Differential diagnosis should be made with paralytic rabies, equine encephalomyelitis, parturient paresis, ragwort poisoning in horses, and other plant poisoning. In sheep, the differential diagnosis should be made with scrapie, louping ill and hypocalcaemia.

Malignant Oedema

Malignant oedema is a bacterial disease of cattle, sheep, goats, swine, horses and poultry. The disease is caused by *Clostridium septicum*. The infection is commonly soil-borne. Deep wounds associated with trauma provide ideal condition for the growth of this agent. The disease is characterized by wound infection, fever, muscle tremor and lameness, and swelling and erythema around the site of infection.

Post-mortem findings show gangrene of the skin in area of infection site, foul putrid odour, gelatinous exudates from tissues, subserosal haemorrhage, accumulation of sero-sanguineous fluid in body cavities, and dark-red muscle tissues with little or no gas. Carcases of animals affected with malignant oedema are condemned. Differential diagnosis should be made with blackleg.

Haemorrhagic Septicaemia (HS)

Haemorrhagic septicaemia is a systemic disease of cattle, buffalo, pigs, yaks and camels. The disease is caused by *Pasteurella multocida* type B of Carter. The predisposing factors of the disease are wet chilly weather, overworked and exhausted animals. Haemorrhagic septicaemia is a specific type of pasteurellosis, distinct from other forms of *Pasteurella* infections. The disease is transmitted by ingestion of contaminated feedstuff. The disease is more severe in buffalo than in cattle. The disease is characterized by high fever, salivation, difficulties in swallowing, cough, difficult breathing, pneumonia in later stage of the disease, oedematous swelling of throat, dewlap, brisket and peritoneum, and diarrhoea.

Post-mortem findings show subcutaneous swellings with yellowish gelatinous fluid around the throat region, brisket and perineum; pneumonia; haemorrhagic organs, petechial haemorrhage in the serous membranes; and enlarged and haemorrhagic lymph nodes. Carcase of an animal affected with haemorrhagic septicaemia is condemned. Cases diagnosed on ante-mortem inspection should not be allowed to enter the slaughterhouse due to reason of spread of the disease. Differential diagnosis should be made with other pasteurellosis, acute leptospirosis, anthrax, blackleg, rinderpest, snake bite and lighting stroke.

Pyelonephritis

Pyelonephritis is also known as contagious bovine pyelonephritis. This is a purulent and inflammatory disease of the kidney pelvis and parenchyma. The disease is caused by *Corynebacterium renale*. The disease occurs mainly in adult cows and sows. The predisposing factor for developing the disease is trauma to the bladder and urethra during parturition. The disease may be transmitted from clinically normal "carrier cows". The organism can enter via vulva from bedding contaminated with urine, tail swishing by carrier cows, venereal transmission by infected bulls, and non-sterilized obstetrical instruments. The disease is characterized by fever, anorexia, frequent and painful urination, colic, ceased rumen contraction, and ammoniac odour from animal.

Post-mortem findings show enlarged, pale and greyish coloured kidney; enlarged renal lymph nodes; and purulent lesion in the medulla, pelvis and ureters. Carcase of an animal affected with pyelonephritis or nephritis is condemned if renal insufficiency is associated with uraemia; and acute infection of the kidney accompanied with systemic changes in the organs and lymph nodes. Borderline cases with uraemic odours should be kept in the chiller for 24 hours, followed by a boiling test. If a urinary odour is not present after detention, the carcase may be approved. Carcase from subacute or chronic kidney infections with no systemic changes is approved. Only the affected parts are condemned. Pyelonephritis associated with kidney stones often has a favourable judgement of the carcase. Differential diagnosis should be made with cystitis, urinary obstruction, infarcts, neoplasms, hydronephrosis, white spotted kidneys of calves, peritonitis, reticulitis, metritis, post-parturient haemoglobinuria, enzootic haematuria, leptospirosis, and Johne's disease.

Endocarditis

Endocarditis is inflammation of the endocardium which is most commonly seen in the valves. In cattle, the disease is caused mainly by *Actinomyces pyogenes, Streptococcus spp*. and *Escherichia coli*. The disease is characterized by fever, breathing with accompanied grunt, pale mucous membranes, muscle weakness, jaundice and death.

Post-mortem findings show wart-like and verrucose lesions in the endocardium; and embolic lesions in other organs including the lungs, spleen, kidneys *etc*. Carcase of a debilitated animal is condemned for verrucose endocarditis if it is associated with lesions in lungs, liver or kidneys. Carcase affected with ulcerative or verrucose endocarditis with no systemic changes may be approved after heat treatment. The carcase is approved in case of endocarditis showing scar tissue. The heart is condemned. Differential diagnosis should be made with pericarditis, congenial heart disease, congenital valvular heart cysts, pneumonia, pulmonary oedema, emphysema, pleuritis, and lymphoma.

Traumatic Reticuloperitonitis

This is also known as hardware disease, traumatic gastritis or traumatic reticulitis. The disease is caused due to perforation of the reticulum by a metallic foreign body. The disease is mostly seen in adult cattle. The disease in acute form is characterized by audible "grunt" in early stages, depression, anorexia, weight loss, mild rumen bloat, scant and hard faeces, reluctance to walk, arched back and tucked up abdomen. Mild septicaemic cases manifest fever and increased heart rate. In chronic localized peritonitis, acute signs and pain lessen, temperature declines and stomach reticulo-rumen motility may return to normal.

Post-mortem findings show traumatic pericarditis; peritonitis; adhesions of rumen, reticulum and peritoneum; abscessation of rumen, reticulum and peritoneum, spleen and lungs; and pneumonia, septic pleuritis and oedema of the chest. In acute and septicaemic condition, the carcase and visceral organs are condemned. In chronic adhesive localized peritonitis and chronic pericarditis without systemic changes, the carcase is approved. The affected parts of the carcase and organs are condemned. Differential diagnosis should be made with abomasal ulceration with perforation, liver abscessation, ketosis, abomasal displacement, and "grain overload".

Mastitis

Mastitis is the inflammation of the udder. The disease may be caused by bacteria and fungi. Mastitis may occur in acute or chronic form. The acute disease is characterized by fever; swollen, hard, warm and painful udder; anorexia; depression; and abnormal gait due to painful condition. The chronic disease is characterized by purulent, bloody or watery pale exudate from teats.

Post-mortem findings show oedematous and pale yellow granular appearance of the udder parenchyma; and enlarged supramammary, iliac and lumbar lymph nodes. In acute or gangrenous mastitis with systemic changes, the carcase and viscera are condemned. In a localized condition of the udder, the carcase is approved. Differential diagnosis should be made with haematoma, rupture of the suspensory ligament and oedema in udder region.

Metritis

Metritis is inflammation of the uterus. The condition develops due to infection by pathogenic microorganisms. The disease may occur as a result of retention of placenta, abortion, twin births, abnormal labour and traumatic lesions of the uterus cervix and vagina. The disease is characterized by high fever and depression, muscular weakness, listlessness, retention of placenta, and reddish or greyish-brown foul smelling uterine exudate.

Post-mortem findings show enlarged flaccid uterus exhibiting "paint-brush" haemorrhages on the serosal surface; inflamed and enlarged uterus containing greenish-yellow purulent exudates; enlarged iliac, lumbar and sacral lymph nodes; inflamed peritoneum at the entrance of the pelvic cavity; degeneration of the liver, kidney and heart muscles; and congested carcase. Carcase of an animal affected with acute metritis along with septicaemia or toxaemia is condemned. In chronic cases without toxaemic signs, the carcase may be approved. Differential diagnosis should be made with recent calving.

Sheep and Goats

Enterotoxaemia

This is also known as pulpy kidney. Enterotoxaemia is a fatal toxaemia in sheep, goats, lambs, calves and occasionally in adult cattle. The disease is caused by *Clostridium perfringens.* The organism multiplies in abomasum and intestine and produces toxin which paralyses the vital centres in brain and damages endothelium of blood vessels. The disease occurs in sheep that have been fed heavy grain, and in animals which graze on lush growing pastures. The disease is highly prevalent in Southern Africa. The disease is characterized by green pasty diarrhoea, rapid shallow respiration, involuntary muscle contraction, staggering, recumbency, paralysis and sudden death.

Post-mortem findings show large amount of clear, straw coloured pericardial fluid; petechial haemorrhages of the heart muscle; congested abomasum, intestinal mucosa and liver; soft pulpy kidneys; haemorrhagic and oedematous sheep brain; congestion of mesenteric lymph nodes; overload of the rumen and abomasum with concentrate; and rapid decomposition of the carcase. No lesions are noted in peracute cases. The carcase of an animal affected with enterotoxaemia is condemned. Differential diagnosis should be made with sudden death in lambs (as seen in pasteurellosis, hypocalcaemia and hypomagnesaemia), polioencephalomalacia (less acute form), acute rumen impaction (long course of disease without convulsions) and other septicaemias. In adult sheep, the differential diagnosis should be made with rabies, acute pregnancy toxaemia and louping-ill and lead poisoning.

Infectious Necrotic Hepatitis

This is also known as black disease. The disease causes acute necrotic hepatitis in sheep and cattle and rarely in pigs. The disease is caused by *Clostridium novyi.* The disease is characterized by fever, rapid and shallow respiration, depression and incoordination and recumbency. Sheep may be found dead without clinical signs.

Post-mortem findings show dark brown swollen liver with necrotic areas surrounded by a zone of hyperaemia; infestation of liver flukes; darkened and cyanotic subcutaneous tissue "black disease"; and clear straw coloured fluid in the abdominal cavity. The carcase and viscera from animal affected with black disease are condemned. Differential diagnosis should be made with enterotoxaemia, blackleg, malignant oedema, anthrax and fascioliasis.

Caseous Lymphadenitis

This is also known as ulcerative cellulitis or ulcerative lymphangitis. Caseous lymphadenitis is a chronic infectious disease of sheep and goats, characterized by abscesses formation in the lymph nodes. The disease occurs worldwide and causes great economic losses to the sheep industry. The disease is caused by *Corynebacterium pseudotuberculosis*. This is a pleomorphic, Gram-positive, rod-shaped bacterium. The disease is transmitted through abraded skin. The organism then invades the lymphatic vessels and lead to the formation of abscess. The disease is sporadic in nature and spread slowly. Nodules formed on the legs, break down leading to the formation of ulcers.

The disease is transmitted through discharge from the lymph nodes via wounds caused due to shearing, docking, castration, contaminated sheep dips, and abrasions of skin or oral mucous membrane. The disease is characterized by enlarged superficial body lymph nodes, dyspnoea, purulent ocular and nasal discharge, and animal lagging behind the flock. Post-mortem findings show caseous abscesses in the superficial lymph nodes and carcase muscles; firm and dry abscess in the kidneys and other organs; and pneumonia. In goats, the content of abscess is creamy and pasty. In case of extensive involvement of many lymph nodes and tissues, the carcase is condemned. In mild cases, the carcase is approved. The affected tissues are condemned. Differential diagnosis should be made with abscesses in the organs, neoplasm, echinococcosis and other parasitic diseases.

Horses

Contagious Equine Metritis (CEM)

Contagious equine metritis is a highly contagious, acute venereal disease of horses and other equidae. The disease is caused by *Haemophilus equigenitalis*. The disease usually occurs as a result of retained placenta and injury of uterus during parturition. The disease is transmitted venereally, and also through contaminated fomites and equine handlers. The disease is characterized by high morbidity (up to 100 per cent), endometritis, vaginitis, copious thick mucopurulent discharge from vagina, and soiled hind legs with exudate.

Post-mortem findings show metritis, cervicitis, vaginitis, and mucopurulent exudate in the uterus and vagina. Carcase of an animal affected with contagious equine metritis is approved. In case of acute metritis and cervicitis with septicaemia, the carcase is condemned. In case of chronic metritis with no toxaemic signs, the carcase may be passed after thorough inspection. Differential diagnosis should be made with infections due to *Klebsiella* spp., *Pseudomonas* spp. and *Staphylococcus* spp.

Tetanus

Tetanus is an acute infectious and fatal disease of horses. The disease is caused by *Clostridium tetani*. The causative agent is found in the soil and more commonly in horse manure. It can also be demonstrated in the intestine of healthy horses. In horses, the disease is most frequently transmitted through nail wounds. In sheep, the infectious agent may enter after castration and docking. In cows, the agent may enter during puerperal infection, dehorning or castration. In swine, the disease is mostly transmitted as a result of wound infection, castration or umbilical infection in new born animals. Sheep and cattle are less susceptible. Pigs are rarely susceptible. The causative agent produces toxins and neuromuscular activity favours migration of tetanus toxins in the nervous system. The incubation period varies from 4 to 14 days which may extend to up to 4 months.

The disease is characterized by spasmodic contraction of the voluntary muscles and increased sensitivity to stimuli; stiffness of the masseter muscles and stiff stilted gait; difficulties in mastication of food "lockjaw"; erect ears and prolapse of the third eyelid; increased breathing and heart rate; congestion of mucous membranes, "sawback stance" in animal, tetany and convulsions, rise of body temperature at the end of the fatal attack, and death. Sheep, goats and swine fall to the ground. Post-mortem findings show no significant lesions. Carcase of an animal affected with tetanus is condemned. The musculature is usually greyish-yellow in colour and the carcase is inadequately bled. Differential diagnosis should be made with hypocalcaemia in mare, strychnine poisoning, lactation tetany in cattle, enterotoxaemia in lambs, meningitis, enzootic muscular dystrophy, and polioencephalomalacia.

Glanders

This is also known as farcy or malleus. Glanders is a serious zoonotic bacterial disease that primarily affects equines. Glanders results from infection by *Burkholderia mallei* (formerly known as *Pseudomonas mallei*), a Gram-negative rod in the family Burkholderiaceae. The main reservoirs of *B. mallei* are horses, mules and donkeys. Glanders is endemic in parts of the Middle East, Asia, Africa and Central and South America. It has been eradicated from Western Europe, Canada, the U.S., Australia, Japan and some other countries. *B. mallei* is readily spread on fomites including harnesses, grooming tools, and food and water troughs. Equids often become infected when they ingest *B. mallei* in contaminated food or water, and carnivores when they eat contaminated meat.

In equids, glanders is traditionally categorized into nasal, pulmonary and cutaneous forms, based on the most commonly affected sites. In the nasal form, deep ulcers and nodules develop inside the nasal passages, resulting in a thick, mucopurulent, sticky, yellowish discharge. Pulmonary involvement occurs in most clinical cases, often in combination with other forms of glanders. Affected animals develop nodules and abscesses in the lungs, or bronchopneumonia in some cases. In the cutaneous form (farcy), multiple nodules develop in the skin, along the course of lymphatic vessels. These nodules often rupture and ulcerate, discharging an oily, thick yellow exudate. Post-mortem findings show enlarged, fibrotic and abscessed lymph nodes; ulcers in the nasal mucosa, larynx and trachea; greyish centred nodules in lungs; nodules on the skin and in the subcutis; and necrosis in the internal organs and testicles. Carcase of an animal affected with glanders is condemned. The infected animal should not be allowed to enter the abattoir. Differential diagnosis should be made with strangles, dourine and melioidosis, and ulcerative lymphangitis.

Strangles

Strangles is an important contagious disease of equines. The disease occurs in stressed young horses (generally 1 - 5 years old). The disease is caused by *Streptococcus equi*. The source of infection is nasal discharge from infected animals and contaminated food and water. The disease is transmitted through ingestion of contaminated food and water or inhalation of infected droplets. The incubation period varies from 4 to 8 days. The disease is characterized by fever, inflammation of the upper respiratory tract, severe pharyngitis and laryngitis, pneumonia, dyspnoea, cough, purulent lesions in the regional lymph nodes (submaxillary, pharyngeal, parotid lymph nodes *etc.*), oedema of lower limbs, lameness, necrosis of skin and eruption of abscesses, empyema of the guttural pouch, and purulent inflammation of the brain, nuchal rigidity, terminal paralysis, and pericarditis.

Post-mortem findings show abscesses in the internal organs including the liver, spleen, lungs, pleura, peritoneum, and mesenteric lymph nodes. Multiple abscessations in the vital organs and generalized systemic infection are denoted as "Bastard strangles". Carcase of an animal affected with strangles accompanied with fever, acute infection and systemic generalized lesions is condemned. Carcase of an animal affected with mild form of the disease is approved. Differential diagnosis should be made with dourine, epizootic lymphangitis, ulcerative lymphangitis, melioidosis, equine influenza, equine viral enteritis, and equine viral rhinopneumonitis.

Pigs

Swine Valvular Endocarditis

Bacterial endocarditis (inflammation of the endocardium of the heart) is one of the significant bacterial infections in pigs and other domestic animals. The disease in swine is caused by *Erysipelothrix rhusiopathiae*, *Streptococcus* spp., *Actinomyces pyogenes*, and *Escherichia coli*. Bacteraemia may lead to infection and cause lesions in the endocardium. The valves are most frequently affected. Emboli may detach from friable growth on the valves and pass through the blood stream to organs and cause infarcts. Emboli from the right heart are a frequent cause of pulmonary thrombosis or pulmonary abscessation. The disease is characterized by fever, pallor of mucosae, arthritis, tenosynovitis and loss of condition.

Post-mortem findings show valvular lesion in the heart (valvular endocarditis), myocarditis, and embolic lesions in lungs, kidneys, spleen *etc.* Carcase of an animal affected with endocarditis accompanied with fever, loss of condition, and embolic lesions in organs is condemned. Ulcerative or verrucose endocarditis with no signs of systemic changes may be conditionally approved, pending heat treatment. The affected organs are condemned. Differential diagnosis should be made with congenital heart disease, congenital valvular heart cysts, pericarditis, pneumonia, pulmonary oedema, emphysema, pleuritis, and lymphoma.

Porcine Chronic Pericarditis

Pericarditis is one of the frequent conditions found in swine at slaughter. Pericarditis may be caused by bacteria, mycoplasma, viruses, fungi and other microorganisms. Enzootic pneumonia may be a common complication due to secondary invasion with *E. coli*. In swine, fibrinous pericarditis is associated with erysipelas, hog cholera, and Glasser's disease. Pericarditis is also observed in pasteurellosis, porcine enzootic pneumonia and streptococcal infection in suckling pigs. Primary pericarditis occurs rarely and is of viral origin. The disease is characterized by reluctance to move, pleurisy and/or pneumonia, shallow respiration, and cardiac problems.

Post-mortem findings show fibrinous deposits on the pericardium, adhered pericardium to the epicardium (heart surface), adhesions of pericardium with lungs and/or pleura, and purulent pericarditis. Carcases of animal affected with acute, diffuse or purulent pericarditis accompanied with fever and systemic changes in other body systems should be condemned. A carcase in good condition without systemic changes can be passed even though the heart showed evidence of chronic infectious pericarditis. The organs and viscera are condemned. Differential diagnosis should be made with abscess, lymphomatosis *etc.*

Swine Erysipelas

Swine erysipelas is an important bacterial infection. The disease is caused by *Erysipelothrix rhusiopathiae*. The source of infection is healthy carrier pigs that shed the infectious agent in manure, where they may survive for 5 months. In the acute form, the disease is characterized by high morbidity, fever in acute stages, diamond shaped skin lesions and their sloughing, conjunctivitis, letharginess, anorexia, and sudden death in excited animals. In the chronic form, the disease is characterized by vegetative endocarditis, arthritis and lameness. An animal affected with an acute disease of erysipelas with erythema or diffuse cutaneous erysipelas with erythema is rejected on ante-mortem inspection because of occupational hazards.

Post-mortem findings show diamond shaped skin lesions, vegetative endocarditis, arthritis, enlarged and reddened spleen and congestion of the other organs, inflamed and haemorrhagic mucosa of the stomach (paint brush effect) and intestine, cloudy swelling and ecchymotic haemorrhage of the kidney, and oedematous and haemorrhagic lymph nodes. Carcase showing skin lesions or arthritis complicated by necrosis or signs of systemic effects is condemned. A localized skin lesion requires only the removal of skin and the rest of the carcase is approved. Localized endocardial lesions of erysipelas without systemic changes or localized chronic arthritis require conditional approval of the carcase with heat treatment. Differential diagnosis should be made with dermatitis, allergies, external parasites, septicaemia, hog cholera, African swine fever, vesicular exanthema, arthritis, superficial bruises, and salmonellosis.

Melioidosis

This is also known as pseudoglanders, pseudomalleus, rodent glanders or Whitmore's disease. Melioidosis is an infectious disease of pigs, goats and occasionally of other animals. Melioidosis is a zoonotic disease and causes fatal infections in a significant proportion of infected humans, especially those who are immuno-compromised. The disease is caused by *Burkholderia pseudomallei* (formerly *Pseudomonas pseudomallei*), a saprophytic, Gram-negative, motile bacillus. Melioidosis is endemic in Southeast Asia, Papua New Guinea, much of the Indian subcontinent, southern China, Hong Kong, and Taiwan and is considered highly endemic in northeast Thailand, Malaysia, Singapore, and northern Australia.

Important sources of infection are rodents, contaminated swamps or muddy water. The organism can live up to three months in shaded soil. The disease is transmitted through subcutaneous inoculation, contamination of wounds or abrasions of the skin or perhaps bites of insects, ingestion and inhalation. The disease is characterized by fever, anorexia, enlargement of lymph nodes, particularly the sub-mandibular lymph nodes in pigs, posterior paresis, nervous signs, orchitis in boars, abortion and still births. In pigs, the disease occurs mostly in chronic form, and they may not show clinical signs. In sheep, the disease is characterized by high fever, nasal and ocular discharge and gradual emaciation.

Post-mortem findings show multiple abscesses in most organs especially in the liver and spleen, and regional lymph nodes. The abscesses contain thick, caseous greenish yellow or off white turgid pus. There is usually no calcification. In acute cases, post-mortem findings show pneumonic changes in the lungs, suppurative polyarthritis (the joint capsules contain fluid and large masses of greenish yellow pus), and meningoencephalitis. In sheep, the post-mortem findings include abscesses and suppuration in the nasal mucosa. Carcase from animal affected with melioidosis should be condemned and properly disposed of. Persons handling the suspect carcase or material should disinfect their hands, forearms, knives and other contaminated equipment. Differential diagnosis should be made with tuberculosis, non-specific purulent conditions, caseous lymphadenitis in sheep and goats and actinobacillosis in sheep.

Poultry

Fowl Typhoid

Fowl typhoid is an infectious disease in chicken and turkeys and rarely in other poultry, game birds and wild birds. The disease is caused by *Salmonella* Gallinarum. The disease is more prevalent in adult birds. Clinical signs of fowl typhoid and pullorum disease are similar in birds. Post-mortem findings show enlarged, bronzed liver and enlarged spleen in a turkey, enlarged kidneys, enteritis, and pale cadaver. Carcase and viscera affected with fowl typhoid are condemned.

Paratyphoid Infection

Paratyphoid infection is an acute and chronic infection of poultry and mammals. Several species of *Salmonella* are associated with most outbreaks in poults. The most commonly involved is *S.* Typhimurium in birds less than one month old. The disease is characterized by drooping wings, shivering and huddling near a heat source, and muscular incoordination and trembling. Post-mortem findings show enteritis, button type lesion in the intestine, nodular lesions in the pancreas, unabsorbed yolk material and omphalitis, dehydration, swollen eyelids and blindness. Carcase and viscera affected with paratyphoid infection are condemned.

Arizona Infection

This is also known as Arizoonosis or paracolon infection. Arizoonosis occurs mainly in young turkey poults. The disease is caused by *Arizona hinshawii* (*Salmonella* Arizona). *Arizona* infection is transmitted through egg. The disease is characterized by listlessness, trembling, twisted head and neck, pasting of vent area with faeces, huddling near a light source, cloudiness and enlargement of the eye, and blindness. Ante-mortem findings in young birds are similar to those of paratyphoid. Post-mortem findings show enlarged yellow liver, congested duodenum, cheesy plugs in the intestine or caecum, inflamed oviduct and peritoneum, unabsorbed yolk material, eye lesions, and purulent exudate in the brain. Carcase and viscera affected with arizoonosis are condemned.

Fowl Cholera

This is also known as pasteurellosis. Fowl cholera is an infectious disease, affecting almost all classes of fowl and other poultries. The disease is more prevalent in turkeys than in chicken. The disease occurs more frequently in stressed birds associated with parasitism, malnutrition, and poor sanitation. The disease is caused by *Pasteurella multocida*. The organism can survive several days in a humid environment and can persist for months in decaying carcases and in moist soil. The disease in the flock may be transmitted through infected animal (rodents and carnivores) bite. Contaminated feed, water, soil and equipment are also considered as potential source in the spreading of the disease.

The acute form of the disease is characterized by high morbidity and mortality, sudden death, depression, cyanosis, anorexia, nasal and oral discharge, greenish diarrhoea, swelling and discolouration of comb. The chronic form of the disease is characterized by swelling of wattles, sinuses, joints, foot pad and tendon sheaths; twisting of head and neck in some birds, loss of balance, cheesy exudate in the conjunctival sac, and infection of ears.

Post-mortem findings may be lacking in the very acute stages of the disease. Post-mortem findings in acute cases show caseous exudate in wattles, sinuses, the nasal turbinates, middle ear, joints or tendon sheaths; haemorrhages on the heart, serous and mucous membranes, on the gizzard and abdominal fat; swollen liver streaked with white areas and associated small grey areas of necrosis (corn meal liver); and free yolk in the peritoneal cavity in breeder hens and layers. Post-mortem findings in chronic cases show darkened breast muscle but haemorrhagic lesions are often missing. Carcase from birds affected with localized lesions of pasteurellosis such as infection of wattles, joints or tendon sheaths require the condemnation of the affected parts; the rest of the carcase is approved. Carcases are condemned in septicaemic cases. Differential diagnosis should be made with salmonellosis, tuberculosis, listeriosis, acute colibacillosis and erysipelas in turkeys.

Air Sac Disease

Air sac disease is a respiratory syndrome in the birds. The disease is caused by viral or mycoplasmal agents. *E. coli* is usually a secondary invader. The disease is transmitted through inhalation of contaminated air. The disease is characterized by airsacculitis, pneumonia, perihepatitis and pericarditis in broiler chickens between 4 - 8 weeks of age. Post-mortem findings show inflamed trachea, frothy exudate in the air sacs; inflamed and opaque airsacs in the early stages of infection; thickened and caseous yellow exudate in the air sacs in the later stages; and yellow fibrinous deposits on the pericardium and liver.

Chronic Respiratory Disease (CRD)

Chronic respiratory disease is a respiratory infection of the upper respiratory tract of chicken. The disease is caused by *Mycoplasma gallisepticum*. The disease affects turkeys more severely and causes infectious sinusitis. Post-mortem findings show cloudy appearance of the abdominal airsacs; and pericarditis and perihepatitis. In poultry inspection, it is important to differentiate between the inflammation of the air sacs (airsacculitis) and peritonitis, as well as pathologic changes in the air sacs and bones. On post-mortem examination affected parts with localized lesions are condemned and the rest of the carcase is approved if in good condition and no spread of infection into bones is noted. Generalized and extensive lesions of airsacculitis require total condemnation of the carcase.

RICKETTSIAL DISEASES

Anaplasmosis

This is also known as gall sickness. Anaplasmosis is a rickettsial disease. The disease is caused by *Anaplasma* spp. which is an obligate intracellular parasite. In cattle and wild ruminants, the disease is caused by *Anaplasma marginale*. The disease is transmitted through tick vector (*Boophilus* spp.) Mosquitoes and the horsefly act as mechanical transmitters. Transmission of the disease is also possible through contaminated injection needles. Acute form of the disease is characterized by high fever, severe debility, emaciation, anaemia and jaundice. Chronic form of the disease is characterized by frequent urination and constipation.

Post-mortem findings show enlarged, congested and soft spleen; enlarged and icteric liver; distended bile ducts and gall bladder with dark tarry bile; and thin, watery blood, which clots poorly. Carcase of an animal affected with acute infection of anaplasmosis is condemned. Recovered and suspect animals manifesting inconclusive signs of anaplasmosis are approved. Differential diagnosis should be made with babesiosis, leptospirosis, anthrax, icterus and anaemia of different causes, and emaciation caused by parasitism and malignant lymphoma.

Heartwater (Hydropericardium)

This is also known as cowdriosis, "black dung" when affecting African cattle and buffalo or "sheep fever" when seen in sheep. Heartwater is an acute, non-contagious disease of cattle, sheep, goats, antelopes and wild ruminants. The disease is caused by the rickettsial organism *Cowdria (Rickettsia) ruminantium*. Heartwater is transmitted by tick vectors (*Amblyomma* spp.). Transstadial transmission of the organism occurs in vector ticks. The incubation period varies from 14 to 28 days. Peracute form of the disease is characterized by fever, diarrhoea, convulsions and death. Acute form of the disease is characterized by fever, rapid breathing, anorexia, depression and listlessness. Nervous signs include walking in circles, champing of the jaw, protrusion of the tongue, paddling with legs in recumbent animals, opisthotonos and convulsions, and twitching of the eyelids.

Post-mortem findings show hydropericardium, hydrothorax, pulmonary oedema, haemorrhagic gastroenteritis, ascites, haemorrhagic abomasum and intestine, haemorrhagic and oedematous brain, and enlarged liver, spleen and lymph nodes. Carcase of an animal affected with the acute stage of the disease is condemned. In a chronic case, the carcase may be approved if adequately bled and muscles are normal in colour and texture. The affected organs are condemned. Differential diagnosis of peracute form of heartwater should be made with anthrax. The acute nervous form of the disease should be differentiated from, rabies, tetanus, cerebral trypanosomiasis, theileriosis, piroplasmosis, parasitism, strychnine poisoning, lead and organophosphate poisoning, arsenical poisoning and certain plants poisoning.

Q Fever

This is also known as query fever, abattoir fever, Queensland fever, Balkan fever, Nine mile fever, American Q fever, Australian Q fever or Burnet's rickettsiosis. Q fever is a rickettsial disease of cattle, sheep, goats, donkeys, camels, fowl, dogs, cats, pigeons and humans. Q fever is an occupational disease of livestock personnel, farmers and laboratory personnel. The disease is caused by *Coxiella burnetii*. It is a Gram-negative, short rod and intracellular organism. Domestic animals harbor the *Coxiella* organisms without showing clinical manifestations. The disease in domestic animals is transmitted through grazing of pastures contaminated with excreta of infected ticks or placenta and amniotic fluid of infected animals.

The overt disease is rare in the animals. The disease may cause abortion in some animals like cattle and sheep. The organism has also been associated with metritis, retention of placenta and infertility. Vaccination reduces the shedding of organisms in milk. Post-mortem findings show no gross lesions in the cattle. The disease in humans is characterized by sudden onset, pneumonia, anorexia, weakness and generalized malaise lasting from 1- 2 weeks. Complication of the disease may cause endocarditis in older people which may lead to death.

CHLAMYDIAL DISEASES

Poultry

Ornithosis

This is also known as psittacosis or avian chlamydiosis. Ornithosis is an acute or chronic chlamydial disease of turkeys, ducks, chicken, pheasants and pigeons.

The disease is caused by *Chlamydia psittaci*. In psittacine birds such as parrots, parakeets, cockatoos, macaws *etc.* and humans, the disease is called psittacosis. Wild and cage birds act as carrier of the infectious agent. Carrier birds shed *Chlamydia* in their secretions and excretions. The disease is transmitted through inhalation of dust or ingestion of food contaminated with faeces of infected birds. Mild cases may not show clinical manifestations or may show mild respiratory symptoms and diarrhoea. In turkeys, the disease is characterized by nasal discharge, greenish-yellow diarrhoea, depression, weakness and weight loss. In ducks and geese, the disease is clinically manifested by watery diarrhoea. In pigeons, the clinical manifestations are abnormal respiratory sounds (rales), watery diarrhoea, anorexia, conjunctivitis, and depression.

Post-mortem findings show inflamed lungs, airsacs, liver, heart, spleen, kidneys and peritoneum. In turkeys, the post-mortem findings show wasting; inflamed pericardium, airsacs, lungs; and enlarged and congested spleen. In pigeons, the post-mortem findings show swollen eyelids and conjunctivitis. In cage birds, the post-mortem findings show enlarged spleen and liver, inflamed air sacs and pericardium, and intestinal congestion. If the disease is suspected on ante-mortem, birds are treated as suspects and samples should be sent to the laboratory. Positive laboratory finding necessitate condemnation of the bird or carcase. Differential diagnosis should be made with salmonellosis, pasteurellosis and *Mycoplasma gallisepticum* infection.

MYCOPLASMAL DISEASES

Contagious Bovine Pleuropneumonia (CBPP)

CBPP is an acute, subacute or chronic, highly contagious disease of cattle. The disease is caused by *Mycoplasma mycoides* var, *mycoides*. The disease is transmitted through aerosol and droplet infection from the infected animals. The recovered animals are called "lungers" act as carriers and shedders, especially under stress. The incubation period varies from 10 to 14 days in acute form of the disease, while 3 - 6 months in chronic disease. The disease is characterized by high morbidity (up to 90 per cent in susceptible cattle), high mortality (10 - 50 per cent), fever, depression, anorexia, loss of weight, shallow rapid respiration, cough, grunting, gurgling, extended neck, lowered head, open mouth, arched back and outward rotated elbow, and arthritis in young animals.

Post-mortem findings show fibrinous inflammation of the pleura; straw coloured fluid in the thorax; lobar pneumonia with red hepatisation; enlarged mediastinal lymph nodes; haemorrhagic heart; and arthritis and tenosynovitis. Carcase of an animal affected with contagious bovine pleuropneumonia is condemned if the disease is associated with fever, serous infiltration of the brisket and emaciation and inadequate bleeding of carcase. Recovered animals without generalized signs of the disease are approved and the affected organs are condemned. Differential diagnosis should be made with pasteurellosis, tuberculosis, East Coast fever, foreign body pneumonia, infectious bovine rhinotracheïtis, chlamidial infections and lungworms disease.

Contagious Caprine Pleuropneumonia (CCPP)

CCPP is a contagious disease of goats. The disease is caused by *Mycoplasma mycoides* subspecies *capri* (*Mycoplasma* biotype F 38). The disease resembles contagious bovine pleuropneumonia; however it is not transmissible to cattle. The disease is transmitted through aerosol and droplet infection from the infected animals. Carrier or infected animals may also bring the infection into the flock. The incubation period varies from 6 to 10 days. The disease is characterized by very high morbidity (up to 100 per cent), very high mortality (60 -70 per cent), fever, cough, frothy salivation, sticking out of tongue, dyspnoea, mouth breathing in terminal stage, and death in few days.

Post-mortem findings show fibrinous inflammation of the pleura, slight interlobular pulmonary reaction, pleural adhesions and enlarged mediastinal lymph nodes. Carcase of an animal affected with contagious caprine pleuropneumonia without systemic involvement is approved. The affected organs are condemned. Carcase of an animal affected with septicaemic form of the disease requires carcase condemnation. Differential diagnosis should be made with foot and mouth disease, vesicular stomatitis, pasteurellosis, tuberculosis, East Coast fever, foreign body pneumonia, infectious bovine rhinotracheitis, chlamydial infections and lungworms disease.

VIRAL DISEASES

Cattle

Foot and Mouth Disease (FMD)

This is also known as aphthous fever. FMD is an acute viral and extremely contagious disease of cloven footed animals. There are three main strains of the virus, *viz.*, A, O and C. The disease is transmitted by direct and indirect contact with infected animals and their secretions. The disease is characterized by formation of vesicles and erosions in the muzzle, nares, mouth, feet, teats, udder and pillar of the rumen. Other symptoms include high fever, extensive salivation and drooling, extensive erosion on epithelium of bovine tongue, smacking and quivering of lips, shaking of feet and lameness, clicking sound on walking of animal.

Post-mortem findings show eroded epithelium on a bovine tongue, and necrosis of heart muscle (tiger heart), usually only in acutely infected young animals. Other findings include ulcerative lesions on gums, palate, tongue, rumen pillars and feet. Total condemnation of carcase and organs is recommended. Differential diagnosis in bovine and ovine species should be made with rinderpest, vesicular stomatitis, allergic stomatitis, bluetongue, malignant catarrhal fever, bovine papular stomatitis, bovine viral diarrhoea, feedlot glossitis, photosensitization, infectious bovine rhinotracheitis, pseudocowpox, ovine pox, contagious ecthyma, footrot, and mycotoxicosis.

Rinderpest (RP)

Rinderpest is an acute, highly contagious, fatal viral disease of cattle, buffalo and wild ruminants. The disease is transmitted by direct contact with infected animals or their excretions and secretions and fomites. The disease is characterized by high fever, nasal discharge and excessive salivation, punched out erosions in the mouth, abdominal pain (grunting, arched back), constipation followed by bloody diarrhoea and straining, inflammation, haemorrhage, erosions of the digestive tract, wasting and often bloody diarrhoea.

Post-mortem findings show punched out erosions in the oesophagus; oedema or emphysema of the lungs; eroded dental pad and the hard palate; haemorrhage in the spleen, gallbladder and urinary bladder; haemorrhagic or ulcerative lesions in the omasum; congestion and ulceration in abomasum filled with bloody fluid; severe congestion and haemorrhage in the intestine; enlarged and necrosed Peyer's patches; haemorrhage (showing "tiger stripping" of longitudinal folds) in the large intestine and rectum; and enlarged and oedematous lymph nodes. The carcase derived from a feverish and debilitated animal showing the sign of acute disease on ante-mortem inspection should be condemned. Differential diagnosis should be made with foot and mouth disease, vesicular and necrotic stomatitis and bovine papular stomatitis, bovine viral diarrhoea, malignant catarrhal fever, infectious bovine rhinotracheitis, bluetongue, coccidiosis, and vesicular diseases (do not have accompanying haemorrhage and blisters).

Vesicular Stomatitis (VS)

Vesicular stomatitis is a viral disease of ruminants, horses and swine. The disease is characterized by vesicular lesions of the mouth, feet and teats. This is caused by a vesicular stomatitis virus which has two immunologically distinct serotypes, *viz.*, Indiana and New Jersey. The disease is mainly transmitted through contact of abraded skin with saliva of infected animal. The disease is characterized by fever, mouth lesions, profuse salivation, lameness and teat lesions. In cattle, foot lesions are noticed in about 50 per cent cases.

Post-mortem findings show the skin and mucous membrane lesions which resemble the lesions of other vesicular diseases. A carcase showing acute changes and systemic lesions is condemned, while is approved if the disease is not in the acute stage and secondary changes are not present. Differential diagnosis should be made with foot and mouth disease, vesicular disease, bovine papular stomatitis, swine vesicular exanthema, the mouth and muzzle lesions (rinderpest, bovine viral diarrhoea, mycotic stomatitis, Potomac valley fever in horses, and photosensitization), and teat lesions (cowpox, pseudo-cowpox, pseudo-lumpy skin disease and bovine herpes mammillitis).

Malignant Catarrhal Fever (MCF)

MCF is an acute viral disease of cattle, buffalo, bison and deer. The disease is transmitted through close contact between cattle and wildebeest, and through drinking of water contaminated by use of common drinking troughs or through placenta of parturient dams. The incubation period of disease varies from 9 to 44 days. The disease is characterized by low morbidity, high mortality, fever, inflammation of nasal mucous membrane, profuse nasal discharge, dyspnoea, cyanosis, ocular discharge, corneal opacity and conjunctivitis, reddening of the eye lids, photophobia, blindness, encrustation of muzzle, eczema of the perineum, scrotum and udder, erosions on the lips, tongue, gums, palate, reluctance to swallow due to oesophageal erosions, drooling, and enlargement of lymph nodes.

Post-mortem findings show crater like erosions of the nose, mouth, conjunctiva, oesophagus and gastrointestinal tract; oedema and petechial haemorrhage in the intestine, swollen and reddened abomasal folds, "tiger striping" in the distal colon; congested, swollen or emphysematous lungs; white areas in the kidneys; enlarged and reddened lymph nodes, and dehydrated and emaciated carcase. In case of fever, emaciation and systemic signs, the carcase and viscera are condemned. Differential diagnosis should be made with foot and mouth disease, rinderpest, vesicular stomatitis, bovine viral diarrhoea and bluetongue.

Infectious Bovine Rhinotracheitis (IBR)

This is also known as red nose. IBR is a highly infectious viral respiratory disease of cattle, goats and pigs. The disease generally occurs in four forms, *viz.*, the respiratory form, the genital form, the enteric form and the encephalitic form. The incubation period varies from 5 to 14 days. Respiratory form is characterized by fever, nasal and ocular discharge, breathing through the mouth and salivation, hyperaemia of the nasal mucosa, necrotic areas on the nasal septum, secondary bronchopneumonia, red and swollen conjunctiva, and abortion. Genital form is characterized by frequent urination, tail elevation, oedematous swelling of the vulva, pustule formation on reddened vaginal mucosa, and exudate in the vagina. Enteric form is characterized by severe oral and stomach necrosis in new born animals, and high mortality. The encephalitic form is characterized by excitement and high mortality in calves. The respiratory form is transmitted by respiratory droplet and nasal exudate, while, the genital form is transmitted by obstetrical operations, coitus and licking of genitalia of infected animals.

Post-mortem findings show acute inflammation of the larynx, trachea and bronchi; profuse fibrino-purulent exudate in the upper respiratory tract, chronic ulcerative gastro-enteritis, lung emphysema and secondary bronchopneumonia. Carcase of an animal affected with IBR is approved if signs of acute infection are not present and the animal is in good body condition. Differential diagnosis should be made with malignant catarrhal fever, calf diphtheria, pneumonic pasteurellosis, and bovine viral diarrhoea.

Lumpy Skin Disease

Lumpy skin disease is an acute pox viral disease of cattle. The disease is characterized by sudden appearance of nodules on the skin. The disease is transmitted by insect vectors. The disease is characterized by fluctuating fever, diarrhoea, nasal discharge, salivation, appearance of the first lesion in the perineum, varying size of cutaneous nodules, swelling of superficial lymph nodes and limbs, lameness *etc.* Post-mortem findings show ulcerative lesions in the mucosa of the respiratory and digestive tract, haemorrhagic and oedematous lesions and nodules in the lungs and sloughing of skin. Carcase showing mild cutaneous lesions and no fever is conditionally approved pending heat treatment. The affected parts of the carcase and organs are condemned. Carcase of an animal showing generalized acute infection accompanied with fever on ante-mortem inspection, is condemned.

Rift Valley Fever (RVF)

Rift valley fever is an acute viral disease of cattle, sheep, goats and humans. The disease is prevalent mainly in Africa. RVF resembles influenza in humans. The disease is transmitted through biting insects and mosquitoes. In humans, the disease is transmitted by handling infected tissues. The incubation period varies from 12 to 48 hours. In sheep, the disease is characterized by fever, photosensitization, photophobia, encrustation around the muzzle, hepatitis, abdominal pain, abortion in ewes, and high morbidity and mortality in lambs. In cattle, the disease is characterized by oedematous unpigmented skin with cracking and sloughing due to photosensitization, inflammation in the mouth and salivation, haemorrhagic gastroenteritis, diarrhoea, abdominal pain, lameness, and abortion.

Post-mortem findings show cyanotic mucosae; necrotic liver in lambs; haemorrhagic internal organs and lymph nodes; haemorrhagic and oedematous gall bladder; and haemorrhages in the foetus. Carcase of an animal from clinical disease is condemned. Reactors and recovered animals are approved. Affected parts of the carcase, liver and the blood must be condemned. Differential diagnosis should be made with defect in porphyrin metabolism, fungal infections, acute viraemias/toxaemias including enterotoxaemia, bluetongue, bovine ephemeral fever, rinderpest, heartwater, East Coast fever; Nairobi sheep disease, abortions caused by *Brucella*, *Vibrio* and *Trichomonas*, and ovine enzootic abortion.

Bovine Spongiform Encephalopathy (BSE)

This is also known as "Mad cow disease". BSE is a highly contagious and fatal disease of bovine. The disease is of great economic importance. It is caused by an infectious protein particle "prion". The infectious agent is extremely resistant to physical and chemical conditions. High-temperature autoclaving is necessary to totally destroy the agent. Cattle, sheep and goats are susceptible to a group of brain diseases known as transmissible spongiform encephalopathies (TSEs). The best known of these diseases is bovine spongiform encephalopathy (BSE) in cattle. The disease is transmitted through ingestion of contaminated protein feed supplements prepared from sheep meat or sheep. BSE is a highly contagious and fatal disease of bovine, and characterized by a progressive degeneration of the central nervous system. BSE affected animals are not allowed for slaughter. In addition to this, animal's trades are not permitted in BSE outbreak regions.

BSE affected carcases are totally condemned and not allowed for manufacturing of any product through rendering. Differential diagnosis should be made with rabies, listeriosis, bovine pseudorabies, other brain infections in cattle, acetonaemia, hypocalcaemia, hypophosphataemia and hypomagnesaemic tetany. The key food safety control is the removal of specified risk material (SRM); however there are also controls on animal feed and a requirement to test certain categories of animal for BSE. In addition to these controls, cattle with BSE or suspected of having BSE and the offspring and cohorts of BSE cases are removed from the food chain.

Bovine Herpes Dermophatic Disease (BHD)

Bovine herpes dermophatic disease is a herpes virus infection of cattle and sometimes sheep and goats. The disease is transmitted by biting insects, and mechanical milking. The incubation period varies from 3 to 7 days. The disease is characterized by high morbidity, fever, cutaneous lesions, cutaneous nodules (initially round, then later become flattened and covered with dry scabs), and erosions between the digits. Post-mortem findings show dried scabs on the skin of the neck, and ulcerative lesions of the teats and udder. The judgement is same as lumpy skin disease. Carcase showing mild cutaneous lesions and no fever is conditionally approved pending heat treatment. The affected parts of the carcase and organs are condemned. Carcase of an animal showing, on ante-mortem inspection, generalized acute infection accompanied with fever, is condemned. Differential diagnosis should be made with cowpox, pseudocowpox, dermatophilis infection, vesicular stomatitis and lumpy skin disease.

Bovine Viral Diarrhoea (BVD)

This is also known as mucosal disease complex. This is an infectious viral disease of cattle. The disease is transmitted by direct contact with clinically sick or carrier animals, indirect contact with feedstuffs or fomites contaminated with urine, nasal and oral secretions or faeces and contact with aborted foetuses. The disease transmission is also possible through aerosol dispersion or by insect vector. The incubation period varies from 1 to 3 days. The disease is characterized by an active erosive stomatitis, gastroenteritis, diarrhoea, fever, cough, polypnoea, salivation, abortion in pregnant cows and laminitis.

Post-mortem findings show shallow erosions on the entrance of the nostrils, mouth, pharynx, larynx, oesophagus, rumen, omasum, abomasum, caecum and less frequently in payer's patches in the small intestine. Other findings include erythema of the mucosa with submucosal haemorrhage in the abomasum, small intestine, caecum and colon, stripped appearance on the caecal and colon mucosa and cerebral hypoplasia in calves. Carcase and viscera of acute and generalized infection with fever and/or emaciation are condemned. Differential diagnosis should be made with rinderpest, vesicular diseases, malignant catarrhal fever and blue tongue.

Bovine Leukosis

This is also known as bovine viral leucosis, bovine lymphosarcoma, lymphocytoma, malignant lymphoma or leucosis. Bovine leukosis is a persistent and malignant viral disease of the lymphoreticular system. The disease occurs in two forms, *viz.*, the sporadic and the enzootic form. The sporadic form is rare and occurs in cattle under three years of age, while the enzootic form is most commonly found in adult cattle.

The disease is transmitted from infected blood through infected needles and dehorning. Vertical transmission from the dam to the calf and by colostrum or milk is also possible. The prevalence of the disease is higher in the summer. The disease is characterized by dyspnoea, persistent diarrhoea, enlargement of superficial lymph nodes, oedema of the brisket and the intermandibular region, emaciation, pale mucous membranes, paralysis of the hind legs, swelling of the neck region, and cutaneous nodules in the terminal stage of the disease.

Post-mortem findings show enlarged spleen and lymph nodes; thin watery blood; neoplastic lesions in the heart, intestine and other organs; and ventral oedema. Carcase of an animal affected with leukosis is condemned. Differential diagnosis should be made with hyperplastic haemolymph nodes, lymphoid hyperplasia, lymphadenitis, pericarditis, enlarged spleen in septicaemic conditions, and parasitism.

Rabies

This is also known as hydrophobia, lyssa or tollwut. Rabies is an acute, highly fatal viral disease of central nervous system of mammals. The causative agent is *Lyssavirus* type 1. It is bullet-shaped neurotropic RNA containing virus belong to the family Rhabdoviridae. Rabies is enzootic/endemic and epizootic disease of worldwide importance. Infected animal bite is the most common mode of transmission of rabies. The disease is transmitted usually through saliva by a bite from a rabid animal, commonly the dog or jackal. The incubation period varies from 2 weeks to 6 months or longer. The disease occurs in two forms *viz.*, furious form and paralytic form. Furious form is characterized by restlessness, aggressiveness, sexual excitement, paralysis and death. Paralytic form is characterized by sagging and swaying of the hind quarters, holding of the tail to one side, tenesmus or paralysis of the anus, drooling, salivation, paralysis, falling of the animal to the ground, and death after 48 hours of decubitus.

Post-mortem findings may show inflammation of gastrointestinal mucosa. In endemic areas, carcases may be approved if the animal was bitten eight days before slaughter and within 48 hours of slaughter. The bite area and surrounding tissue must be condemned, and measures taken to prevent occupational hazards. Differential diagnosis should be made with poisoning, indigestion, milk fever or acetonaemia, and foreign body in the mouth.

Sheep and Goats

Contagious Ecthyma

This is also known as contagious pustular dermatitis, sore mouth, scabby mouth or orf. Contagious ecthyma is a highly contagious skin disease commonly seen in goats and sheep. The disease is found worldwide and is caused by a *Parapoxvirus*. The disease is often more severe in goats than in sheep. The disease is transmitted through direct contact between animals or indirect contact with dry scabs in pens. The incubation period varies from 2 to 3 days. The disease is characterized by inflammation and ulcerations on the lips, mouth, vulva and around the margins of hoof, udder and teats. The disease may be severe which involves cauliflower-like tumors of the mouth and mucous membrane, deep ulcerative stomatitis, pharyngitis, oesophagitis and loss of hoof.

Post-mortem findings show pustular and scabby lesions on the head, feet and udder; ulcerations in the nasal cavity and erosions in the mucosa of the oesophagus and upper respiratory tract; inflammation of the reticulum, omasum and intestine; and necrotic lesions in the lungs, pleura and liver. The carcase from animal affected with contagious ecthyma accompanied with inflammation of the stomachs and intestines, and with bronchopneumonia is condemned. Differential diagnosis should be made with sheep and goat pox, ulcerative dermatosis, cutaneous anthrax, vesicular diseases and bluetongue.

Bluetongue

This is also known as catarrhal fever of sheep or "soremuzzle disease". Bluetongue is a highly contagious viral disease of sheep. The disease is prevalent mainly in Africa, but also in Asia and the Pacific and in the Western hemisphere. The disease is transmitted by biting insects, especially Culicoides gnats and mosquitoes. Vertical transmission of the disease occurs in utero. The disease can also be transmitted through semen of infected bulls and infected blood by needles. The incubation period varies from 6 to 8 days. In sheep, the disease is characterized by fever, dyspnoea, oral lesions, excessive salivation, anorexia, weakness, emaciation, nasal discharge, oedema of the face, lips and jaw, cyanosis of the tongue and mucous membranes (bluetongue), erosion and sloughing of the oral mucosa, inflammation of the coronary band, sore feet, lameness and abortion. In cattle, the disease resembles the infection in sheep, but the clinical signs may vary from inapparent to mild.

Post-mortem findings show vesicles or ulcers in the mouth; oedematous and haemorrhagic subcutaneous tissues and muscles; enlarged spleen; congested lungs; excessive mucus in the trachea; generalized lymphadenitis; and necrosed cardiac and skeletal muscles. Carcase of an animal affected with acute infection along with generalized post-mortem lesions is condemned. The reactor animals are approved. In sheep, the differential diagnosis should be made with photosensitization, contagious ecthyma, vesicular stomatitis, sheep pox, polyarthritis, laminitis, footrot, foot abscesses, white muscle disease, muscular dystrophy in lambs, lungworm infestation and pneumonia. In bovine, the differential diagnosis should be made with foot and mouth disease, stomatitis, bovine viral diarrhoea, malignant catarrhal fever, infectious bovine rhinotracheitis, and laminitis.

Sheep and Goat Pox

Sheep and goat pox is a highly contagious viral disease of sheep and goats. The disease is transmitted through direct contact with infected animals and infected aerosols, and indirectly through fomites and contaminated transportation vehicles. The incubation period varies from 6 to 8 days. The disease is characterized by fever, laboured breathing, papular and pustular eruptions on the skin, haemorrhagic inflammation of the respiratory tract, lacrimation and salivation. Clinical signs of goat pox are less severe than in sheep pox. The benign form of sheep pox is commonly found in adult sheep, while the malignant form in lambs.

Post-mortem findings show reddish to whitish firm nodules in the mucosa of the pharynx, trachea and lungs; and inflammation of the respiratory and digestive tract in malignant form of the disease. Carcase of an animal showing the clinical disease without secondary complications is conditionally approved pending heat treatment. The recovered animals are approved. The carcase is condemned if the acute febrile or pustular stage of the disease is associated with secondary bacterial infections. Differential diagnosis should be made with contagious ecthyma, eczema, peste des petits ruminants, scabies, and ulcerative dermatitis.

Scrapie

Scrapie is a chronic disease of the central nervous system in sheep and occasionally goats. The disease is caused by a viral agent called "viroid" or "prion", which has some of the characteristics of the virus, a "slow" virus like BSE and Maedi. The disease is transmitted through abraded skin and mucous membranes. The disease can also be transmitted from sick animals to healthy animals through contaminated pasture. Vertical transmission of the disease is also possible from the dam and possibly the sire in sheep. The disease is characterized by dry wool, loss of wool, rough skin, biting of legs, flanks and belly due to severe itchiness (pruritus), grinding the teeth, smacking, curling of the lips, wagging of the tail, twitching of muscles, restlessness, continuously laying down and getting up, excitability and wild expression of the eyes, incoordination in gait, tendency to run and fall down, behavioural changes, and convulsions.

Post-mortem findings show no gross lesions. Microscopic examination reveals the presence of large vacuoles in the cytoplasm of neurons (a diagnostic lesion). Carcase and viscera affected with the clinical disease are condemned. Differential diagnosis should be made with louping ill, pseudorabies, pregnancy toxaemia, scabies, thallium poisoning, cobalt deficiency, external parasitism and photosensitive dermatitis.

Pulmonary Adenomatosis

This is also known as Jaagsiekte or driving sickness. Pulmonary adenomatosis is a chronic progressive pneumonia of sheep. The disease is caused by a *Retrovirus* and a herpesvirus acts in a secondary role. The prevalence of the disease is higher in old ewes than in the lambs. The disease is transmitted through infected aerosols. Vertical transmission of the disease may occur from pregnant ewes to foetus. The incubation period varies from 2 months to 2 years. The disease is characterized by the development of a primary lung neoplasm, dyspnoea, lacrimation, loss of weight, emaciation, running of fluid from the nose on lifting the rear of a sheep (wheel barrow test).

Post-mortem findings show enlarged and bluish-grey consolidation of the ventral part of the lung; interspersed focal lung lesions with areas of emphysema; and sometimes metastatic neoplasm into the bronchial and mediastinal lymph nodes. The carcase is condemned in case of extensive lung lesions with metastasis and loss of musculature. Differential diagnosis should be made with verminous pneumonia, Maedi/Visna, caseous lymphadenitis and other debilitating diseases.

Ovine Progressive Interstitial Pneumonia

This is also known as Maedi or Maedi-visna. Maedi/visna is a highly fatal viral disease of sheep and goats. The disease is caused by a *Lentivirus*. The disease is transmitted through colostrum to newborn lambs and less often by respiratory route. The disease is characterized by listlessness, dyspnoea, frequent cough, nasal discharge, emaciation, lameness and anaemia in chronic cases.

Post-mortem findings show enlarged greyish-yellow non-collapsible lungs of rubbery consistency; and enlarged and firm mediastinal lymph nodes. Carcase in good flesh with slight to moderate pulmonary involvement is approved. An emaciated carcase with extensive pulmonary lesions or secondary bacterial infection is condemned. Differential diagnosis should be made with pulmonary adenomatosis, pseudoglanders and parasitic pneumonia.

Nairobi Sheep Disease (NSD)

Nairobi sheep disease is a tick borne viral disease. The disease is transmitted through infected tick (*Rhipicephalus* spp.) bite. The incubation period varies from 4 to 15 days. The disease is characterized by acute haemorrhagic gastroenteritis, fever, dyspnoea, dullness, depression, mucopurulent nasal discharge, bright to dark green faeces, swollen vulva and external genitalia, abortion, collapse and death.

Post-mortem findings show excess fluid in the pericardium; ecchymotic and petechial haemorrhage in the cardiac muscles; haemorrhagic stomach and intestine; distended gall bladder containing thick syrupy bile; enlarged and oedematous lymph nodes; and hyperaemic genital tract. Carcase of an animal affected with the acute disease accompanied with fever and acute gastrointestinal lesions is condemned. Carcase of recovered animals and of animals without systemic or generalized lesions is approved. The affected organs are condemned. Differential diagnosis should be made with Rift Valley fever (diarrhoea may show blood tinged watery faeces, but is not green in colour as in NSD), rinderpest (ulcerative lesions with bloody and not green faeces), anthrax, heartwater, and plant poisoning.

Horses

African Horse Sickness (AHS)

African horse sickness is a highly fatal vector-borne febrile viral disease of equidae. The disease is transmitted by various *Culicoides* spp. and several species of mosquitoes. Mechanical transmission by biting flies is also possible. Dogs get infected by eating infected horse meat. The incubation period varies from 3 to 5 days. The disease may occur in cardiac or pulmonary form or both the forms in some animals. The cardiac form is characterized by high mortality (up to 50 per cent), fever, conjunctivitis, oedema in the supraorbital fossa, and subcutaneous oedema of the head, neck, brisket, thorax and ventral abdomen. The pulmonary form is characterized by very high mortality (over 90 per cent), sudden onset, fever, oedema of the subcutaneous tissue and lungs, dyspnoea, cough, haemorrhage and serous fluid in the body cavities and frothy exudate from the nares. In some animals both forms may occur which is characterized by fever, dyspnoea and anorexia. The mild form is known as horse sickness fever.

Post-mortem findings show pulmonary oedema and pleural exudates; froth filled trachea and bronchi; carditis, haemorrhagic pericardium, and excessive pericardial fluid; haemorrhagic stomach, intestine, kidneys; intermuscular and subcutaneous oedema and haemorrhage; gelatinous oedema in the periorbital tissue, neck muscles and ligamentum nuchae; and haemorrhagic ventral surface of the tongue. Carcase of an animal affected with acute African horse sickness with generalized clinical signs and post-mortem lesions is condemned. The carcase of recovered and reactor animals is approved for limited distribution. Differential diagnosis should be made with equine infectious anaemia, equine viral arteritis, equine influenza, equine piroplasmosis, colic and anthrax.

Equine Infectious Anaemia (EIA)

This is also known as swamp fever. Equine infectious anaemia is a viral disease of solipeds. The disease is transmitted through close contact with susceptible animals and biting insects mainly Tabanidae and mosquitoes. The disease can also be transmitted through contaminated surgical instruments, needles, and contaminated feed and bedding. Horse may die of anaemia during early viraemia or may recover and have recurrent episodes of viraemia.

The disease occurs in acute, subacute, chronic and latent forms. The acute form is characterized by sudden onset, fever, dyspnoea, depression, weakness, jaundice, oedema of the ventral abdomen, legs and prepuce, serosanguineous nasal discharge, droopy ears, half closed eyes and abortion. The subacute form is characterized by sudden onset and subsequent recovery, fever, increased pulse, sweating, and swollen icteric conjunctivae. Chronic form is characterized by anaemia, weakness, icteric conjunctivae and mucous membranes, diarrhoea and recurrence of disease in one to three months. Latent form is characterized by absence of overt symptoms but an acute attack may be caused by hard work, poor diet and parasitic infections.

Post-mortem findings in acute case show subcutaneous oedema, jaundice and subserosal haemorrhage. Post-mortem findings in chronic case show emaciation, anaemia; hydrothorax, ascites, splenomegaly, hepatomegaly, haemorrhagic organs, enlarged, oedematous and haemorrhagic lymph nodes, and replaced bone marrow fat with dark red haemopoietic tissue. Carcase of an animal affected with the clinical disease of EIA is condemned. Carcase without systemic lesions from reactor animals may be approved for limited distribution. Differential diagnosis should be made with emaciation, purulent infections causing anaemia, septicaemias due to other diseases, anthrax, glanders, leptospirosis, tuberculosis, virus encephalomyelitis, purpura haemorrhagica, babesiosis, piroplasmosis, strongylosis and fascioliasis.

Viral Encephalomyelitis of Horses

The disease is characterized by disturbed consciousness, motor irritation and high mortality. The disease has also been reported in cattle, sheep and humans. The disease is transmitted mainly from birds through insects to horses and humans. Mosquitoes of the genera *Culex*, *Aedes* and *Mansonia* are vectors of the disease. Wild birds act as reservoirs of infection. The incubation period varies from 1 to 3 weeks. The disease is characterized by fever, depression, anorexia, hypersensitivity to sounds and touch, facial muscle twitch, walking blindly into objects or in circles, paralysis, collapse and death between 2-4 days of the onset of symptoms.

Post-mortem findings show haemorrhagic visceral organs and lesions in the lungs. Carcase and viscera of the animal showing clinical signs of this disease are condemned. The carcase of reactor animal may have a limited distribution through specially licensed and closely supervised commercial channels. The brain and medulla must be condemned. Differential diagnosis should be made with equine infectious anaemia, the dumb form of rabies, botulism and plant poisoning.

Pigs

African Swine Fever (ASF)

African swine fever is a highly contagious viral disease of domestic pigs. The ASF virus is naturally transmitted between bush pigs, warthogs and giant forest hogs and some tick species (*Ornithodorus*) in which the virus replicates. The disease is transmitted by contact with infected pigs and infected fomites, ingestion of contaminated uncooked pork garbage, and tick bites. The incubation period varies from 3 to 15 days. The disease may occur in acute, subacute or chronic forms. The acute form of the disease is characterized by fever, dyspnoea, cough, nasal and ocular discharge, anorexia, vomiting, diarrhoea, incoordination, blotchy skin, cyanosis and haemorrhage, and haemorrhages of lymph nodes, and internal organs including gastrointestinal tract. The chronic form of the disease is characterized by emaciation and oedematous swelling under the mandible and over leg joints and recumbency.

Post-mortem findings show blotchy skin, cyanosis and haemorrhage; splenomegaly; haemorrhagic kidneys; enlarged and haemorrhagic gastrohepatic and renal lymph nodes; haemorrhagic heart; haemorrhagic serous membranes; hydrothorax, hydropericardium and ascites; and pericarditis and emaciated carcase in chronic case. Carcase of an animal affected with ASF is condemned. The animal is prohibited from entering the abattoir. Differential diagnosis should be made with hog cholera, salmonellosis, erysipelas, Glasser's disease (*Haemophilus suis* infection).

Hog Cholera

This is also known as swine fever or classical swine fever. Hog cholera is a highly infectious viral disease of swine. The disease is transmitted through direct contact with infected pigs and ingestion of raw and contaminated food wastes containing infected pork scraps. The incubation period varies from 5 to 10 days. The disease is characterized by high morbidity (40 - 100 per cent), high mortality (0 - 100 per cent), fever, depression, vomiting, constipation, redness of skin, goose stepping, paddling, incoordination, staggering gait, and dog-sitting posture. Septicaemia and generalized haemorrhage are noticed in acute, subacute and chronic forms of the disease.

Post-mortem findings show button ulcers in the large intestine and intestinal necrosis; tonsillar necrosis; splenic infarcts; haemorrhagic gall bladder, urinary bladder, kidneys and lymph nodes; and pneumonia in chronic infection. Carcase of an animal affected with hog cholera is condemned if kidney lesions are associated with lesions in the lymph nodes and other organs. Differential diagnosis should be made with septicaemic conditions, erysipelas, streptococcosis, pneumonia and salt poisoning.

Vesicular Exanthema of Swine (VES)

Vesicular exanthema is an acute, contagious, viral disease of swine. The disease is transmitted through direct contact with infected animals and ingestion of contaminated uncooked garbage containing infected pork scraps. The incubation period varies from 2 to 4 days. The disease is characterized by the formation of vesicles on the snout and in the mouth followed by rupture of vesicles resulting in erosion, blotchy rash in unpigmented skin (exanthema), high morbidity in hogs and high mortality in suckling pigs, loss of weight and lameness.

Post-mortem findings show vesicles on mucous membranes and skin. Carcase of an animal affected with vesicular exanthema is condemned, if the animal is suffering from fever and has generalized lesions throughout the body. In uncomplicated cases, the carcase is conditionally approved pending heat treatment. Differential diagnosis should be made with FMD (the lesions in the heart and skeletal muscles present in FMD, while not found in vesicular exanthema), vesicular stomatitis, swine vesicular stomatitis, lameness and leg injuries.

Swine Vesicular Disease (SVD)

Swine vesicular disease is a contagious viral disease of swine. The disease is transmitted through contact with excretions of infected swine, ingestion of contaminated uncooked pork wastes, and minor skin wounds. The shedding of the virus begins before the appearance of clinical symptoms and may continue for up to 3 months. The incubation period varies from 2 to 4 days. The disease is characterized by fever, vesicular lesions on snout, oral mucosa and feet, and lameness.

Post-mortem findings show skin lesions on snout, oral mucosa and feet. Histopathological examination show diffuse inflammation of brain. Carcase of an animal affected with swine vesicular disease is disposed according to national animal health regulations. In countries with an eradication programme, the carcase is condemned. The carcase is also condemned in a country free or nearly free of this disease. The infected animal should not enter the abattoir. Differential diagnosis should be made with vesicular stomatitis, vesicular exanthema, foot and mouth disease, swine pox, foot rot, and chemical and traumatic injuries.

Poultry

Avian Influenza

This is also known as fowl plague or highly pathogenic avian influenza (HPAI)). Avian influenza is a highly contagious viral disease of several avian species. The disease occurs worldwide. The disease can range from asymptomatic and mild to hyperacute and fatal. The disease occurs infrequently in humans. The disease may cause occupational health hazard. Workers in the poultry industry; employees in abattoirs, vaccinators, laboratory staff and other personnel are at higher risks of acquiring of infection. In most cases, the clinical picture is that of conjunctivitis with rare systemic reactions. Avian influenza is a notifiable disease in many countries which can be confirmed by virus isolation.

The disease is transmitted through contact with secretions from infected birds, by wild birds and contaminated feed, equipment and people. Seabirds and migratory waterfowl comprise the main reservoirs for avian influenza virus. The incubation period varies from a few hours to about one week. The disease is characterized by very high morbidity and mortality rates (up to 100 per cent), coughing, sneezing, rales, excessive lacrimation, depression, anorexia, watery diarrhoea, conjunctivitis, swollen combs with cyanotic tips and haemorrhagic surface, oedematous wattles, oedema around the eyes, head and neck, ruffled feathers, dark red skin, diffuse haemorrhages between the hocks and feet, and blood in the cloaca.

In peracute cases, the post-mortem findings show no significant gross lesions, dehydration, and fibrinous exudate in airsacs, oviduct, peritoneum and pericardial sacs. In mild to moderate infection, the post-mortem findings show inflamed trachea, airsacs and conjunctivae; congested muscles; oedematous head; haemorrhagic and cyanotic combs, wattles and sinuses; vesicles and ulceration of the comb; haemorrhagic serosal and mucosal surfaces, heart, gizzards, proventriculus and small intestine; and oedematous with haemorrhagic feet. Carcases affected with avian influenza should be condemned. Differential diagnosis should be made with fowl cholera, Newcastle disease, chlamydiosis, and mycoplasmosis.

Newcastle Disease (NCD)

Newcastle disease is a highly contagious viral disease of birds and cause high mortality (50 - 80 per cent) in chicken. Velogenic Viscerotropic Newcastle disease (VVND) is caused by the most virulent strain of the Newcastle disease virus. The disease is transmitted through direct contact, fomites, and by aerosols through coughing, gasping and respiratory fluids from infected birds. The incubation period varies from 2 to 15 days.

The chronic form of the disease in domestic fowl is characterized by anorexia, listlessness, abnormal breathing, ocular discharge and greenish diarrhoea. VVND is an acute, fatal infection of birds of all age group which is characterized predominantly with haemorrhagic lesions of the gastrointestinal tract, severe depression, and death prior to clinical manifestations. Other symptoms include oedema of the head, swelling of the lower eyelid, conjunctivitis, dark ring around the eye (black eye), excessive fluid secretion from the respiratory tract, paralysis of wings, and twisting of the head and neck (torticollis).

In peracute form, the post-mortem findings show no discernible lesions in some of the first birds dying in an outbreak. In acute form, the post-mortem findings show haemorrhagic and eroded oesophagus; oedematous head and neck; haemorrhagic tracheal mucosa; haemorrhagic, ulcerative and necrotic gastrointestinal tract; and oedematous or haemorrhagic ovaries. In chronic form, the post-mortem findings show catarrhal inflammation of the respiratory system; and oedematous surrounding connective tissues.

Carcase from bird affected with VVND or NCD is condemned and premises with equipment should be disinfected. Birds infected with VVND or NCD should not be admitted to the abattoir. Differential diagnosis should be made with highly pathogenic avian influenza, infectious bronchitis, laryngotracheitis, fowl cholera, fowl pox, Marek's disease, psittacosis, acute *Mycoplasma gallisepticum* infection, avian encephalomyelitis, and vitamin E deficiency.

Infectious Bronchitis (IB)

Infectious bronchitis is an acute, highly contagious viral disease of birds. The disease is rapidly transmitted in a flock through contaminated air. Some infected birds become carriers and shedders of the virus through secretions and discharges for many months. IB virus can persist in contaminated chicken houses for approximately four weeks. The disease is characterized by depression, weakness, sneezing, coughing, gasping, rales, nasal discharge, inflammation of the air sacs, reduced production and poor quality of egg in laying birds, renal disease and death.

Post-mortem findings show serous, catarrhal and caseous exudate in respiratory tract; cloudy airsacs; yellowish caseous exudates in abdominal airsacs; swollen and pale kidneys; and small cystic oviducts. Carcase from bird affected with acute signs of clinical disease accompanied with emaciation is condemned. A carcase from bird without systemic involvement is approved. The affected parts are condemned. Differential diagnosis should be made with laryngotracheitis, Newcastle disease, and infectious coryza. Laryngotracheitis spreads slowly in a flock but respiratory signs are more severe than in infectious bronchitis. Laryngotracheitis is not seen in young chicken.

Laryngotracheitis (LT)

Laryngotracheitis is an acute viral disease of chicken. LT virus is transmitted through the respiratory route and the intraocular route. Oral infection may occur. Mechanical transmission via fomites is also possible. The incubation period varies from 6 to 12 days. The disease is characterized by high morbidity and moderate mortality (10 – 20 per cent), dyspnoea, loud gasping or wheezing sounds, coughing up of bloody exudates, lacrimation, nasal discharge, conjunctivitis and sinusitis. Most chicken usually recover within a month.

Post-mortem findings show inflamed larynx and trachea leading to necrosis and haemorrhage of mucosa; inflamed trachea containing cheesy plugs; and extension of infection to the bronchi, lungs and airsacs. Carcase from birds affected with an acute condition and general systemic changes, is condemned. Mild form of disease and recovered birds can be approved, if carcase is in good flesh. Differential diagnosis should be made with infectious bronchitis, infectious coryza and Newcastle disease.

Fowl Pox

This is also known as pox or avian pox. Fowl pox is a viral disease of chicken, turkeys and other birds. The disease occurs worldwide and affects birds of all age groups, except the recently hatched. Fowl pox virus is present in lesions and in desquamated scabs. The virus is resistant to environmental factors and persists in the environment for many months. The disease is usually transmitted through minor abrasions or skin injuries. Mechanical transmission of the disease may occur by cannibalism. The disease is characterized by cutaneous lesions on the head, neck, legs and feet. The disease occurs in two forms, *viz.*, the cutaneous (dry form) and the diphtheric (wet form). The cutaneous form is characterized by low mortality, lack of flock vigour and weight loss, loss in egg production and scabby lesions on the head, neck and unfeathered parts of the skin. The diphtheric form is characterized by low to moderate mortality, dyspnoea, and nasal and ocular discharges.

In cutaneous form, the post-mortem findings show light coloured nodules like papules, raised and yellow vesicles and pustules. In diphtheric form, the post-mortem findings show buff to yellow plaques on oral mucosa, oesophagus and upper respiratory tract; and occlusion of trachea. Carcase affected with fowl pox accompanied with generalized lesions and emaciations is condemned. Differential diagnosis should be made with infectious laryngotracheitis and other respiratory diseases in poultry, deficiency of pantothenic acid, biotin, and vitamin E, injuries caused by external parasites and cannibalism.

Lymphoid Leucosis

This is also known as big liver disease or visceral lymphomatosis. Lymphoid leucosis is a B cell tumour in the birds which starts in the bursa and may spread to other organs before sexual maturity of the bird. The disease occurs mainly in 14 - 30 weeks old birds. The prevalence of the disease is higher in mature birds than young birds. Male birds are affected in lesser numbers than female due to the earlier regression of bursa in male birds. The disease is characterized by pale and shrivelled comb, anorexia, diarrhoea, green scant faeces, dehydration, emaciation, hepatomegaly, splenomegaly, enlargement of bursa of fabricius, and distended abdomen due to enlargement of liver.

Post-mortem findings show grey tumour lesions in the liver, spleen and bursa; sometimes lesions in lung, heart, proventriculus, mesentery, bone marrow and gonads; and haemorrhages around the skin follicles of the wing. The carcase of a bird affected with lymphoid leucosis is condemned. Differential diagnosis should be made between lymphoid leucosis and Marek's disease.

Marek's Disease

This is also known as polyneuritis, fowl paralysis or neurolymphomatosis gallinarum. Marek's disease is caused by the herpes virus, which is a DNA virus. The disease is transmitted by airborne infection involving follicle cells called chicken dander. At room temperature the virus remains viable for 16 weeks and in litter for 6 weeks. Birds are more susceptible during the first few weeks of life. Infected birds start to shed the virus in the second or third week after infection and continue shedding the virus throughout their life. Marek's disease is commonly associated with coccidiosis in the field, probably due to lack of immunity against coccidiosis in birds affected with Marek's disease. The disease occurs in six forms. The peracute form occurs in 3-5 weeks old chicks which are manifested with sudden death. The anaemic form occurs in in 3-6 weeks old chicken which is manifested with anaemia. The classical Marek's disease (range paralysis) is characterized by paralysis of two wings and legs due to peripheral nerve involvement. The acute Marek's disease occurs mainly in 6-12 weeks old birds which are manifested with tumours in the liver, heart, spleen, kidneys, brain, spinal cord and dorsal root ganglia and with sudden death. Skin leucosis form occurs in broiler chicken which is characterized by enlargement of feather follicles and associated lymphoid infiltrations. Transient paralysis form occurs in 12-18 weeks old pullets which is characterized by paralysis and may recover with 24 hours if moved to a quiet place.

Post-mortem findings show skin neoplasia; enlarged spleen; and diffuse proliferation of white pulp. Carcase of a chicken affected with Marek's disease in the extensive cutaneous form and in the visceral form is condemned. Localized skin lesions require the condemnation of the affected part and the rest of the carcase may be approved. Differential diagnosis for skin lesions should be made with lymphoid leucosis, erythema, dermatitis, pigmentation and normal large follicles. In case of involvement of the central nervous system, the lesions of Marek's disease should be differentiated from those of Newcastle disease and of avian encephalomyelitis.

FUNGAL DISEASES

Fungal diseases are caused by filamentous fungi (moulds) as well as yeasts. Moulds reproduce by the process of sporulation, while yeasts by budding. Fungal diseases may also be caused by dimorphic fungi (can change their growth to a mould or a yeast depending on the environmental conditions). On the basis of involvement of body parts, fungal diseases may be categorized into three groups, *viz.*, superficial, subcutaneous and systemic mycoses. Superficial mycoses are the most common fungal zoonoses and affect skin, hair, nail and mucous membrane, *e.g.*, dermatophytosis infection and yeast infections particularly *Candida* spp. Subcutaneous mycoses involves skin, subcutaneous tissues and bones, *e.g.*, maduromycosis, and sporotrichosis. Systemic mycoses are deep sheeted infections involving internal organs which generally result from inhalation of air contaminated with fungi, *e.g.*, blastomycosis, coccidioidomycosis and histoplasmosis. Some important fungal diseases of domestic animals are discussed as follows:

Aspergillosis

Aspergillosis is an acute or chronic granulomatous disease of many animal species, birds and humans. Aspergillosis most frequently affects the lungs, but infection may occur at other sites such as the nasal sinuses and superficial tissues. The disease is caused by *Aspergillus fumigatus, A. flavus, A. nidulans, A. niger* and *A terreus. A. fumigatus* is the most common species in causing aspergillosis, particularly invasive form. Aspergillosis occurs worldwide. The disease is transmitted through inhalation of air contaminated with spores of the causative agent. Spores of *Aspergillus* spp. are widely distributed in the soil of dairy and poultry farms and zoological gardens.

In birds, the disease is primarily bronchopulmonary, which is characterized by dyspnea, gasping, and polypnoea accompanied by somnolence, anorexia, and emaciation. Mycotic tracheitis has also been described. Torticollis and disturbances of equilibrium are seen when infection disseminates to the brain. Yellow nodules of varying size and consistency or plaque lesions are found in the respiratory passages, lungs, air sacs, or membranes of body cavities. Fur-like growth of fungus may be found on the thickened walls of air sacs. In other species, the bronchopulmonary aspergillosis may have nodular lesions in the lungs or an acute pneumonia accompanied by serosanguineous fluid in the pleural cavity and a fibrinous pleuritis.

In ruminants, the disease may be asymptomatic, appear in a bronchopulmonary form, cause mastitis, or cause placentitis and abortion. Mycotic pneumonia may be rapidly fatal. The acute form of the disease is characterized by pyrexia; rapid, shallow, stertorous respiration; nasal discharge; and a moist cough. The lesions include firm, heavy, and mottled lungs which do not collapse. In subacute to chronic mycotic pneumonia, the lungs contain multiple discrete granulomas, and the disease grossly resembles tuberculosis.

In horses, the infection of guttural pouch is characterized by a necrotizing inflammation and is thickened, hemorrhagic, and covered by a friable pseudomembrane. Mycotic rhinitis is characterized by dyspnea and nasal discharge. Epistaxis and dysphagia are common complications of gutturomycosis. Aspergillosis can be a rapidly fatal disease associated with diffuse pulmonary invasion. The disease can also cause locomotor and visual disturbances, including blindness, may occur when the infection spreads to the brain and optic nerve.

In cattle, the post-mortem findings show yellowish, caseous nodules or plaques or granulomatous lesions in the lungs (resembling tuberculosis); dark red or brownish, thickened and leathery uterine wall in case of abortion; and retention of foetal membrane. In horses, the post-mortem findings show inflamed, necrosed and thickened guttural pouches. The carcase and viscera are approved as fit for human consumption, provided condition is satisfactory and no secondary lesions are present.

Candidiasis

This is also known as candidosis, candidamycosis or thrush. Candidiasis is a localized mucocutaneous disease caused by species of the yeast-like fungus *Candida*, most commonly *C albicans*. The disease occurs worldwide in animals and humans. *C albicans* is a normal inhabitant of the nasopharynx, gastrointestinal tract, and external genitalia of many species of animals and is opportunistic in causing disease. The disease is transmitted by contact with infected tissues. The disease can also be transmitted through contaminated meat at slaughterhouse.

Superficial infections limited to the mucous membranes of the intestinal tract have been described in pigs and foals. Systemic candidiasis has also been described in cattle, calves, sheep, and foals secondary to prolonged antibiotic or corticosteroid therapy. In cats, candidiasis is rare but has been associated with oral and upper respiratory disease, pyothorax, ocular lesions, intestinal disease, and urocystitis.

Infections are rare in dogs and horses. In calves, the infection in forestomach is associated with watery diarrhea, anorexia, and dehydration, with gradual progression to prostration and death. Affected chicks are listless and have reduced feed intake and growth rate. Porcine candidiasis affects the oral, oesophageal, and gastric mucosa, with diarrhea and emaciation.

Post-mortem findings show single or multiple, raised, circular, white masses covered with scabs on the skin and mucosae. Other findings include marked keratinous thickening of the mucosae of the tongue, esophagus, and rumen. In birds, the post-mortem findings show white, circular ulcers with raised surface, and thickened mucosa of the crop and oesophagus in which an easily removed pseudomembrane is common. Depending on extent of lesions and carcase condition, the carcase and viscera are totally or partially condemned or approved (after heat treatment) as fit for human consumption.

Coccidioidomycosis

This is also known as Valley fever, Desert fever or California disease. Coccidioidomycosis is primarily a disease of domestic animals such as cattle, sheep, pig, horse, dog and cats, and humans, which affects mainly the lungs. The disease varies from inapparent (cattle, sheep, pigs, dogs, and cats) to progressive, disseminated, and fatal (dogs, nonhuman primates, cats, and humans). Coccidioidomycosis is a dust-borne, noncontagious infection caused by the dimorphic fungus *Coccidioides immitis*. Soil, dust, many species of domestic and wild animals such as coyotes, desert rodents, chinchillas *etc.* are important reservoirs of the organism. Inhalation of fungal spores is the only established mode of infection, and spores may be carried on dust particles. Cattle and dogs are most commonly affected, while sheep, horses and pigs are less frequently affected. The incubation period varies from 2-4 weeks. The disease is clinically characterized by fluctuating body temperature, dyspnoea, coughing, wheezing, colic, leg oedema and superficial abscesses. Placental infection leading to abortion and osteomyelitis has been described in horses.

The lesions are usually confined to lungs but may spread to intestine, liver and spleen. Post-mortem findings show granulomatous nodules of variable-sized containing yellowish pus or calcified material (often mistaken for tuberculosis) in the lungs. Associated lymph nodes such as retropharyngeal, bronchial and mediastinal lymph nodes are also affected. Depending on extent of lesions and carcase condition, the carcase and viscera are totally or partially condemned or approved as fit for human consumption.

Cryptococcosis

This is also known as European blastomycosis or Torulosis. Cryptococcosis is a systemic fungal disease that may affect the respiratory tract (especially the nasal cavity), central nervous system, eyes, and skin (particularly of the face and neck of cats). The disease is caused by *Cryptococcus neoformans* and *C gattii*, exist in the environment and in tissues in a yeast form. The disease occurs worldwide in sporadic form. The causative agents are found in soil and fowl manure, especially in pigeon droppings. The disease is transmitted through inhalation of spores or contamination of wounds.

The disease occurs most commonly in cats but also is seen in dogs, cattle, horses, sheep, goats, birds, and wild animals. In bovine, the disease has been associated only with cases of mastitis, and many cows in a herd may be infected. In cows, the disease is manifested by anorexia, decreased milk flow, swelling and firmness of affected quarters, and enlarged supramammary lymph nodes. The milk becomes viscid, mucoid, and gray-white, or it may be watery with flakes. Involvement of central nervous system is characterized by ataxia, stiffness, hyperaesthesia, circling and sometimes blindness. The disease in horses almost invariably is a respiratory ailment, which is characterized by obstructive growths in the nasal cavities.

Post-mortem findings show granulomas, abscess and ulcers in the mouth, upper respiratory tract and thoracic cavity. Other lesions include swollen lymph nodes, granulomatous lesions in the mammary tissues and thickened meninges. The lesions in ruminants resemble tuberculosis. Depending on extent of lesions and carcase condition, the carcase and viscera are totally or partially condemned, or conditionally approved as fit for human consumption.

Epizootic Lymphangitis

Epizootic lymphangitis is a chronic granulomatous disease of the skin, lymph vessels, and lymph nodes of the limbs and neck of horses, mules, donkeys *etc.*, caused by the dimorphic fungus *Histoplasma farciminosum*. The disease is prevalent in Asian and Mediterranean areas. The organism forms mycelia in nature and yeast forms in tissues and has a saprophytic phase in soil. The disease is transmitted through wound infection or by bloodsucking insects.

The disease is characterized by freely movable cutaneous nodules, which originate from infected superficial lymph vessels and nodes and tend to ulcerate and undergo alternating periods of discharge and closure. Affected lymph nodes are enlarged and hard. The skin covering the nodules may become thick, indurated, and fused to the underlying tissues. Lesions may also be present in the lungs, conjunctiva, cornea, nasal mucosa, and other organs. The nodules are pyogranulomas with a thick, fibrous capsule and contain thick, creamy exudate. Depending on extent of lesions and carcase condition, the carcase and viscera are totally or partially condemned, or conditionally approved (after heat treatment) as fit for human consumption.

Histoplasmosis

This is also known as Darling's disease or reticuloendothelial cytomycosis. Histoplasmosis is a chronic, noncontagious, disseminated, granulomatous disease of animals and humans, caused by the dimorphic fungus *Histoplasma capsulatum*. The organism is commonly found in soil that contains bird and bat manure. Histoplasmosis occurs worldwide. The disease usually occurs in dog, cat, bat, rodents, carnivores, humans *etc.* Infections have also been reported from horses and birds. The disease is transmitted through inhalation of spores of the causative agent from soil contaminated with bat guano or bird droppings. Histoplasmosis is latent in most of the domestic animals. In dogs, the disease is manifested by intermittent fever, cough, lethargy, lymphadenitis and diarrhea. In cats, the disease is characterized by the development of granulomatous nodules in the lungs and involvement of spleen and heart. In horses, it causes fever, pulmonary symptoms and abortion.

Post-mortem findings show enlarged liver, spleen, and mesenteric lymph nodes; ascites; yellow-white, variable-sized nodules in the lungs; and enlarged bronchial lymph nodes. The enlarged liver may have multiple, scattered, irregular-shaped, pale yellow foci of granulomatous inflammation. Pale foci may be present in the myocardium, and the small intestine may have thickened, gray walls and ulceration of the mucosa. Depending on extent of lesions and carcase condition, the carcase and viscera are totally or partially condemned or approved (after heat treatment) as fit for human consumption.

Rhinosporidiosis

Rhinosporidiosis is a chronic, pyogranulomatous infection, primarily of the nasal mucosa and occasionally of the skin of horses, cattle, dogs, cats, and aquatic birds. The disease is caused by *Rhinosporidium seeberi*. Humans and animals such as cattle, buffaloes, dogs and horses are most commonly affected. The disease occurs mainly in India, Sri Lanka, Argentina and Brazil. The exact mode of transmission of disease is unknown. However, the disease transmission is believed to be by contaminated soil, dust and stagnant water.

The disease is characterized by development of large polyps on the nasal mucous membrane which may cause obstruction in breathing. The polyps may be soft, pink, friable, and lobulated with roughened surfaces. There may be purulent nasal discharge. The cutaneous lesions may be single or multiple, sessile or pedunculated. Depending on extent of lesions and carcase condition, the carcase and viscera are totally or partially condemned or approved (after heat treatment) as fit for human consumption.

Ringworm

This is also known as dermatophytosis or dermatomycosis. Ringworm is a common disease of the stratum corneum of the skin, hair and nail. In animals, the disease is caused by *Microsporum canis*, *M. gallinae*, *Trichophyton equinum* and *T. verrucosum*. The disease caused by *Trichophyton* spp. is worldwide. The disease is transmitted through direct contact with infected animals. The disease is also transmitted through indirect contact with contaminated bedding, harness, saddlery, grooming kit, clothing *etc.* The cattle is the most commonly affected animal species, the young cattle are more commonly affected.

In cattle, the hair becomes brittle and skin on the face appears as circular plaque-like lesions. In equines, the skin lesions are characterized by patches of alopecia. In dogs and cats, the disease is manifested as scaly lesions with crusty center on the skin. Differential diagnosis should be made with mycotic dermatitis and dermatophilosis in cattle, and pityriasis and exudative dermatitis in pigs. The carcase and viscera are approved, provided condition is satisfactory.

Sporotrichosis

Sporotrichosis is a sporadic, chronic contagious and granulomatous disease of animals and humans, caused by *Sporothrix schenckii*. The organism is dimorphic and forms mycelia on vegetation. It is ubiquitous in soil, vegetation, and timber. The disease occurs worldwide, mainly in moist tropical and subtropical zones. The disease is transmitted through wound infection via contact with plants or soil or penetrating foreign bodies. Disseminated disease caused by inhalation of spores is rare. Sporotrichosis has been reported in dogs, cats, horses, cows, camels, goats, mules, birds, and pigs, including humans. Zoonotic infections can occur. The cat may be the species with the greatest zoonotic potential, and transmission from cats to humans has been reported without evidence of trauma.

Sporotrichosis may occur in three forms, *viz.*, lymphocutaneous, cutaneous, and disseminated. The lymphocutaneous form is the most common. Small, firm dermal to subcutaneous nodules, 1-3 cm in diameter, develop at the site of inoculation. Lesions ulcerate and discharge a serohaemorrhagic exudate. The cutaneous form tends to remain localized to the site of inoculation, although lesions may be multicentric. Disseminated sporotrichosis is rare but potentially fatal and may develop with neglect of cutaneous and lymphocutaneous forms or if the animal is inappropriately treated with corticosteroids. Infection develops via hematogenous or tissue spread from the initial site of inoculation to the bone, lungs, liver, spleen, testes, gastrointestinal tract, or central nervous system. Depending on extent of lesions and carcase condition, the carcase and viscera are totally or partially condemned or approved (after heat treatment) as fit for human consumption.

PROTOZOAN DISEASES

Protozoan parasites are microscopic, single-celled organisms that can be free-living or parasitic in nature. Pathogenic protozoan parasites are conveniently dealt with by placing them in four groups, *viz.*, sporozoa (*e.g.*, *Cryptosporidium parvum, Toxoplasma gondii etc.*), amoebae (*e.g.*, *Entamoeba hystolytica*), flagellates (*e.g.*, *Giardia lamblia*, *Trypanosoma* spp., *Leishmania* spp. *etc.*) and miscellaneous group (*e.g.*, *Babesia* spp., *Balantidium* spp. *etc.*). The mechanism of pathogenesis varies according to the type of parasite, *e.g.*, *Babesia* and *Plasmodium* affect the red blood cells; which are ruptured releasing the contents to cause haemoglobinuria, whereas, *Coccidia* invades and destroys the mucosal cells of intestine and produces intense diarrhoea, emaciation and often death.

Babesiosis

This is also known as piroplasmosis, Texas fever, red water fever, or tick fever. Babesiosis is a febrile, tick-borne disease of cattle, horses, sheep and swine. The disease is caused by various species of the protozoan parasite of the genus *Babesia*. The disease is transmitted through tick vectors of the family Ixodidae. The *Babesia* spp. can be transmitted transstadially and transovarially within a tick species. The incubation period varies from 7 to 10 days. The disease is characterized by high mortality (up to 50 per cent or over), high fever, dark reddish-brown urine in the terminal stage, and reddened mucous membranes at the early stage and anaemic mucous membranes at the later stage of the disease. In cerebral form of babesiosis, the clinical manifestations may resemble rabies.

Post-mortem findings show oedematous and congested lungs; enlarged and yellow coloured liver and distended gall bladder with thick dark green bile; splenomegaly; anaemia, jaundice and pale muscles; haemorrhagic brain in bovine; and oedematous and haemorrhagic lymph nodes. Carcase of an animal affected with acute form of the disease accompanied with icterus is condemned. An emaciated, jaundiced carcase showing yellow gelatinous fat is also condemned. Carcase from animal affected with mild form of the disease without icterus, may be approved. Differential diagnosis should be made with anaplasmosis, theileriosis, leptospirosis, bacillary haemoglobinuria and trypanosomiasis.

Coccidiosis

Coccidiosis is a major disease problem in poultry industries and the most common cause of enteritis. Coccidiosis causes severe losses in intensive broiler production and in layers and causes a drop in egg production. The disease is caused by various species of the protozoan genus *Eimeria*. The important species of *Eimeria* are *E. maxima, E. acervulina, E. brunetti, E. necatrix* and *E. tenella*. Coccidia are host specific which means that coccidia in chicken will not be found in turkeys and vice versa. Birds affected with Marek's disease and infectious bursal disease is more susceptible to coccidiosis.

The disease is transmitted through ingestion of oocysts. Oocysts damage the intestinal lining and later the oocysts are shed in the faeces. Moist environment and a temperature range of 20 - 30°C are favourable for sporulation of oocysts within 12 - 30 hours and they become infective. Birds raised on the range in the spring and summer are more susceptible in comparison to birds raised in total confinement of barnyards. The disease is characterized by bloody diarrhoea, dehydration, weight loss, anaemia, ruffled feathers, and retracted head and neck.

Post-mortem findings show fibrinous enteritis with bloody mucoid exudates; and whitish blood streaked faeces. The carcase from birds affected with severe coccidiosis accompanied with emaciation or anaemia are condemned. Otherwise, the carcase is approved, after the condemnation of affected tissue. Differential diagnosis should be made with salmonellosis, necrotic enteritis, ulcerative enteritis, ascariasis, capillariasis, haemorrhagic disease, leucosis, blackhead, bluecomb and haemorrhagic enteritis in turkeys.

Histomoniasis

This is also known as enterohepatitis or blackhead. Histomoniasis is a protozoal disease of turkeys, chicken and game birds. The disease often occurs in turkeys run with or after chicken. The disease is caused by *Histomonas meleagridis*. The agent is transmitted in the ova of caecal worms (*Heterakis gallinarum*) and by earth worms. Well-developed hepatic and caecal lesions are pathognomonic for histomoniasis. In young turkeys, the disease is characterized by very high mortality and morbidity (up to 100 per cent), blood in faeces, anorexia, emaciation, depression, drooping wings, closed eyes, cyanotic (black) head, and yellow sulphur coloured faeces.

Post-mortem findings show thickened wall, ulcerated mucosa and caseous exudates of the caeca; inflamed peritoneum due to perforation of caecal wall; and circular yellow necrotic lesions in liver surrounded with white rings. The carcase of bird affected with histomoniasis accompanied with emaciation or septicaemia is condemned. The carcase of bird affected with histomoniasis is approved if the flesh is in a good condition. The liver and intestine are condemned. Differential diagnosis should be made with coccidiosis, trichomoniasis, salmonellosis and aspergillosis.

Sarcocystosis

This is also known as sarcosporidiosis. This is one of the most common parasitic conditions in domestic food animals. The disease is highly prevalent in cattle in various parts of the world. Sarcocystosis is caused by the various species of the protozoan parasite of the genus *Sarcocystis*. The causative agent is host specific. Herbivore acts as an intermediate host, while carnivore or omnivore as a definitive host. In cattle, three species have been recognized. *S. cruzi* is most pathogenic species in cattle and occurs worldwide. The definitive hosts are dog, coyote, red fox, raccoon and wolf. It forms microscopic cyst, less than 0.5mm long. It can cause fever, anaemia, abortion, neurologic signs and even death. *S. hirsute* occurs worldwide. Its definitive host is cat. It forms macroscopic cysts, up to 8mm long and 1mm wide, fusiform in shape. It is mildly pathogenic. *S. hominis* occurs in Europe. It forms microscopic cysts. It is mildly pathogenic to cattle.

In pigs, sacocystosis is caused by *S. miescheriana, S. suihominis*, and *S. porcifelis*. *S. suihominis* is a zoonotic species in which it uses humans and non-human primates as definitive hosts. *S. hominis* infects cattle and *S. suihominis* infects pigs. These both species are zoonotic, in which human acts as a definitive host. In cattle, the disease is transmitted through ingestion of feed, pasture or water contaminated with sporocysts. The incubation period varies from 5 to 11 weeks. All *Sarcocystis* spp. forms cysts in the muscles of the intermediate hosts (the food animals). The symptoms are fever, anorexia, excessive salivation, anaemia, abortion and loss of hair especially at the tip of the tail. In pigs, the disease is characterized by dyspnoea, muscle tremors, weight loss, purpura of the skin especially of the legs and buttocks, and abortion.

Judgement should be made on the basis of macroscopic presence of cysts. In heavy and widespread infestations with the visible cysts, the whole carcase is condemned. In lighter infestations those parts of the carcase which are not affected are passed for human consumption. Differential diagnosis should be made with cysticercosis, trichinosis, toxoplasmosis, neurofibromatosis and eosinophilic myositis.

Theileriosis

This is also known as East Coast fever. Theileriosis is a haemoprotozoan disease of cattle. The disease is caused by *Theileria parva*. The disease is transmitted through infected ixodid tick vectors of the genus *Rhipicephalus*. The disease is characterized by high mortality (up to 90 per cent), high fever, dyspnoea, cough, nasal discharge, salivation, watery eyes, enlargement of lymph nodes, circling to one side, convulsions and death. Chronic cases of theileriosis may manifest loss of condition, emaciation, diarrhoea and blindness.

Post-mortem findings show froth in nostrils and bronchi due to pulmonary oedema and emphysema; swollen, oedematous lungs and interstitial pneumonia; enlarged and haemorrhagic lymph nodes; splenic lymphoid hypertrophy; enlarged and mottled liver; infarcts, thrombosis and lymphoid hypertrophy in spleen; haemorrhagic and rarely ulcerative enteritis; white spots of lymphoid aggregates in a kidneys; and brownish coloured fat. Carcase and viscera of an animal affected with febrile chronic theileriosis without systemic lesions are approved. Carcase and viscera of an animal affected with acute febrile theileriosis accompanied with fever and generalized lesions are condemned. Differential diagnosis should be made with babesiosis, trypanosomiasis, bovine leucosis, Rift Valley fever, heartwater, haemorrhagic septicaemia and malignant catarrhal fever.

Toxoplasmosis

Toxoplasmosis is a contagious disease of sheep, swine and other animal species. The disease is zoonotic, which is caused by protozoon *Toxoplasma gondii* in animals and humans. The disease occurs worldwide. Toxoplasmosis most frequently affects the pigs and sheep, while cattle are rarely affected with clinical toxoplasmosis. Old animals are more susceptible to toxoplasmosis than the young animals. The potential source of infection for the final host (cat) and intermediate host (pigs and man) is sporulated cysts in the environment or tissue cyst in meat.

In humans, the disease is transmitted by ingestion of sporulated cysts or cysts containing raw or undercooked meat. Tachyzoites may be transmitted to the foetus from infected pregnant women. In humans, the disease is characterized by fever, malaise, skin rash, pneumonia, myocarditis, lymphadenopathy, encephalitis, chorioretinitis, abortion and stillbirth. In animals, the disease is characterized by neonatal mortality, fever, pneumonia in young pigs, dyspnoea, cough, diarrhoea, weakness, wasting, incoordination, trembling and abortion.

Post-mortem findings show pneumonia, hydrothorax, ascites, ulcerated intestine, necrosed liver, spleen and kidneys, granulomatous lesion in the brain and inflamed lymph nodes. Carcase of an animals affected with acute disease is condemned. Carcase from recovered and reactor animals is approved. Differential diagnosis should be made with abortion in pigs due to brucellosis, leptospirosis, porcine parvovirus infection, hog cholera and Aujeszky's disease. Differential diagnosis should also be made with encephalitis due to salt poisoning, chlorinated hydrocarbons, lead and mercury poisoning, hypoglycaemia, vitamin A deficiency, encephalomalacia, meningitis, scrapie and rabies.

Trypanosomiasis

Trypanosomiasis in horse is also known as dourine, or mal du coit. This is a protozoan disease of animals and humans. The disease is caused by protozoan parasites of the genus *Trypanosoma*, which are found in blood plasma, various body tissues and fluids. This disease in horse is caused by flagellated protozoan called *Trypanosoma equiperdum*. The disease is transmitted primarily by *Glossina* spp., *Stomoxys*, tsetse fly, tabanid and reduviid bugs, and by venereal contact. In horses, the disease is transmitted mainly by coitus and rarely by bloodsucking flies (Tabanidae and *Stomoxys*). *Trypanosoma* spp. in the insect vector undergo one or two cycles of development. The disease is characterized by intermittent fever, anaemia, weight loss, weakness, oedema in the face and legs, haemorrhage, corneal opacity, keratitis, photophobia and enlargement of body lymph nodes. Contagious trypanosomiasis in horses is characterized by oedematous swelling and inflammation of the genitalia, cutaneous lesions and paralysis.

Post-mortem findings show enlarged spleen, liver, kidney and lymph nodes; oedematous and emaciated carcase; and mild icterus. The carcase affected with trypanosomiasis or any other protozoan diseases is condemned if an acute condition is associated with systemic body changes. The carcase of recovered and reactor animals may be approved if generalized lesions are lacking. The affected parts of the carcase and organs are condemned. Differential diagnosis should be made with equine infectious anaemia, heartwater, babesiosis, anaplasmosis, helminthiasis, malnutrition and other chronic wasting diseases.

9

Parasitic Diseases

A parasite is an organism that lives on another organism (the host) derives nutrients from it, but confers no benefit in return. The most common use of the term 'parasite' refers to the multicellular helminthes (worms), arachnids (ticks, mites, lice *etc.*) and arthropods. A parasite can harm to the host in various ways such as sucking blood and fluid, competing for food, mechanical injury and release of toxic products. In meat inspection, the important parasites may be categorized into trematodes, cestodes, nematodes and arthropods.

Diseases Caused by Trematodes

Trematode parasites are commonly known as flukes. They are flatworms belonging to class Trematoda of the phylum Platyhelminthes. They have short unsegmented flattened bodies, covered by a protective cuticle, and bear hooks or suckers for attachment to their host. Trematodes are classified into two main groups, *viz.*, the Monogenea, which are mostly parasite of the skin and gills of fish, and Digenea. They parasitize the intestine, biliary system, vascular system and urogenital tract of vertebrates, and include several important pathogens of domestic animals and humans. There are several species of trematodes of zoonotic importance. In most of the zoonotic trematode infections, human acts as a definitive host. Transmission of infections occurs mostly through intermediate hosts. Some important diseases caused by trematodes are discussed as follows:

Fascioliasis

Fascioliasis is caused by different liver flukes. *Fasciola hepatica* occurs worldwide. It occurs most frequently in sheep and cattle and less often in goats and swine. *Fasciola gigantica* occurs in Africa and some parts of Southeast Asia and *Fasciola magna* in North America including Canada and Europe. *Fasciola* spp. lives in the bile ducts of ruminants and other mammals. Moist conditions and high rainfall favour the development of intermediate host (amphibious snail of the genus *Lymnaea*). The final hosts are ruminants, especially sheep. In animas, the disease is transmitted through ingestion of metacercariae along with vegetation. Larvae penetrate the small intestine wall and peritoneal space to the liver. The disease is characterized by weight loss, emaciation, anaemia, chronic diarrhoea, and swelling in the mandibular region.

Post-mortem findings show emaciated, anaemic or oedematous carcase, presence of flukes in bile ducts and in the liver parenchyma, hepatic abscesses, and icterus, and haemorrhagic tracts of migratory immature flukes in the liver. In severe infestation with associated emaciation or oedema, the carcase is condemned. In mild, moderate and heavy infestation without emaciation, the case may be approved. Differential diagnosis should be made with *Dicrocoelium dendriticum* and *Gigantocotyle explanatum* infections, melanosis and melanoma.

Dicrocoelium dendriticum Infection

Dicrocoelium dendriticum is a small fluke occurs in the liver. It is also known as the lancet fluke. Two intermediate hosts are required for completion of its life cycle. The eggs excreted with faeces by the final host (ruminants) are ingested by the first intermediate host (land snail) followed by further development of the agent in the second intermediate host (ant of the genus *Formica).* In ruminants, the disease is transmitted through grazing of vegetation adhered with infected ants. *Dicrocoelium dendriticum* is only slightly pathogenic and does not produce clinical symptoms in the animal. Post-mortem findings show thickened bile ducts with damaged liver parenchyma and presence of the parasites in the bile ducts. Judgement and differential diagnosis are same as fascioliasis.

Diseases Caused by Cestodes

Cestodes are also known as tapeworms. They are flatworms belong to the class Cestoda and the family Platyhelminthes. The adult stage of cestodes parasites are found usually in intestine and their larval stages in the body tissues of the hosts. Several genera are of zoonotic significance, including *Dipylidium* in dogs and *Taenia* in cattle, pigs and humans. Most of the infections require intermediate hosts. Adult parasites cause little harm to the hosts. The larvae are usually more damaging than the adults, causing diseases such as coenurosis, cysticercosis, sparganosis and hydatid disease in their intermediate hosts. Some important diseases caused by cestodes are discussed as follows:

Cysticercosis

Cysticerci are the larval stage of *Taenia* parasites. *Cysticercus bovis* (the larval stage of *Taenia saginata*), *Cysticercus ovis* (the larval stage of *Taenia ovis*), *Cysticercus cellulosae* (the larval stage of *Taenia solium*), and *Cysticercus tenuicollis* (the larval stage of *Taenia hydatigena*) cause cysticercosis in cattle, sheep, pig and dogs/cats, respectively. The disease in animals (intermediate host) is transmitted through grazing of pastures or ingestion of foods and water contaminated with faeces of infected human (definitive host). The disease is characterized by muscle stiffness, fever, anorexia, weakness and depression. In sheep, usually no clinical signs are recognized.

Post-mortem findings show small white lesions in muscle tissue; clear transparent bladders of 5 × 10 mm size; opaque and pearl like lesions in later stage of the disease (over 15 weeks of infection); degeneration, caseation and calcification (after 12 months or more after infection); and degenerative myocarditis. Carcase and viscera of heavily infested animals are condemned, while light infestation can be treated either by boiling or freezing (10 days at - 10° C). An animal is considered as "heavily infected", if lesions are discovered in two of the usual inspection sites including the masseter muscles, tongue, oesophagus, heart, diaphragm *etc.* Differential diagnosis should be made with infectious myositis, abscess, granuloma, and *Hypoderma* spp. migration to heart.

Coenurosis

This is also known as gid or sturdy. Coenurosis is a parasitic disease of the nervous system in sheep. The disease is caused by *Coenurus cerebralis* (the intermediate stage of *Taenia multiceps).* The causative agent inhabits the intestine of dogs, cats and wild carnivores. The disease in intermediate host (sheep) is transmitted through ingestion of pastures contaminated with eggs from faeces of infected dog (definitive host). After ingestion the *larvae* hatch in the intestine and pass with the blood stream towards different organs. The larvae which reach the brain and spinal cord grow to the coenurid stage. *Coenurus cerebralis* further mature in the brain and spinal cord. During migration of larval stage, the disease is characterized by muscular tremor, incoordination, blindness, excitability and collapse. Infection with the fully developed larval stage is characterized by salivation, wild expressions, frenzied running, convulsion, head pressing, incomplete paralysis and inability to rise.

Post-mortem findings show thin walled cyst in the brain, and lesion in the lumbar region and in the cervical area of the spine. Carcase affected with coenurosis is approved. Affected brain and organs are condemned. Differential diagnosis should be made with encephalitis, rabies, abscess, haemorrhage and brain tumours.

Hydatid Disease

This is also known as hydatidosis or echinococcosis. Hydatid disease is an important parasitic zoonotic disease. The disease is caused by a cestode parasite "*Echinococcus* spp.", particularly *Echinococcus granulosus* and *Echinococcus multilocularis*. The disease occurs primarily in herbivorous animals and incidentally transmissible to humans. The disease in animals (intermediate host) and human (incidental intermediate host) is transmitted through ingestion of food and water contaminated with faeces of infected dog (definitive host). The disease can also be transmitted through inhalation of air contaminated with dog faeces. In humans, hand to mouth transmission of infection is also possible through transfer of eggs by handling or playing with infected dogs. Hydatid cysts are found mainly in liver and lungs. Animals show no clinical evidence of infection. However, in humans, hydatid cysts can cause serious disease.

Post-mortem findings show hydatid cysts in liver, heart, lungs, spleen, kidneys, muscle, brain, and other tissue including bone. Carcase of hydatid disease affected animal with emaciation, oedema and muscular involvement is condemned and destroyed. Otherwise the carcase is approved. Affected viscera and any other tissue are also condemned and destroyed. Differential diagnosis should be made with retention cysts in kidneys, cysts in liver, granulomatous lesions, *Cysticercus tenuicolis* and tuberculosis.

Sparganosis

Sparganosis is a tape worm disease in pigs. It occurs in the Asia-Pacific region and some other parts of the world. The disease is caused by spargana, the larval (plerocercoid) stages of the tape worm *Spirometra erinacei*. The adult tape worm *Spirometra erinacei* lives in the small intestine of the cat, fox and dog (definitive host). The egg passed in the faeces develops into a ciliated coracidium in water, which when ingested by cyclops (the first intermediate host), develops into a procercoid. If the cyclops with the procercoid is eaten by a frog (the second intermediate host), the procercoid develops into a plerocercoid which resembles the adult tape worm in miniature but without the genitalia. When these frogs are eaten by a cat, fox and dog the plerocercoid develops into mature tape worm *Spirometra erinacei*. However, if a frog is eaten by a pig or other animals such as snakes the plerocercoid migrates to certain tissues, particularly the skeletal muscles where they encysted (appears as cyst up to 6 mm long) or as ribbon like structures about 5 cm long with a miniature scolex. These are known as spargana. Humans can be infected with spargana. No significant clinical signs are noticed.

Post-mortem findings show caseated cysts or cysts up to 6 mm long with spargana in the skeletal muscle or other tissues. Carcases with heavy infestations are condemned. In moderate to light infestations the lesions and parasites are removed and the carcase frozen at -12°C or below for 5 or more days before being passed for human consumption, but such carcases may not be acceptable for export. Differential diagnosis should be made with cysticercosis, sarcocystosis, trichinellosis, myositis, and nerve fibres.

Diseases Caused by Nematodes (Roundworms)

Nematodes are invertebrate animals belonging to the phylum Nematoda. They typically have a smooth cylindrical unsegmented body tapered at both ends and protected by an outer cuticle. Nematode possesses a mouth with two or three lips and an alimentary canal which runs the length of the body, and a terminal anus. Mostly, they are found as separate sexes. The shape and size of eggs vary in different species of nematodes. The life cycle may be direct (*e.g., Ancylostoma* spp. and *Ascaris* spp.) or indirect (*e.g., Habronema* spp.) involving intermediate hosts. Alternatively, the eggs may be taken up by a transport host (*e.g.*, earthworm, as in *Syngamus* spp.). They are found virtually in all habitats and include both free-living and parasitic forms. Parasitic forms cause diseases in animals and humans. The adult parasites occur in various sites in the body. The adult forms of nematodes can reside in the gastrointestinal tract, blood, lymphatic system or subcutaneous tissues. Alternatively, the immature (larval) states can cause disease through their infection of various body tissues. Some important diseases caused by nematodes are discussed as follows:

Ascariasis

Ascariasis is an important parasitic disease of swine which accounts for significant losses to the swine industry due to reduction in growth rate, stunting of young pigs and liver condemnations. The disease is caused by a nematode parasite "*Ascaris suum*". Young pigs are more commonly affected. Adult worms live in the small intestine of pigs where it lays a large number of eggs. After ingestion by a host, the eggs become infective within a few weeks and larvae are released in the small intestine. The larvae migrate through the intestinal wall and portal vein to the liver, further larvae enter the blood stream and reach the lungs. The larvae, during this migration, damage the liver and lungs and sometimes the kidneys. Larvae reach the pharynx through the bronchi and trachea, after they are swallowed by the host; they mature in the intestine and lay eggs. The disease is characterized by haemorrhage and verminous pneumonia due to larvae, rarely cough, dyspnoea, rarely vomiting up the adult worms and poor growth. In young animals, there may be marked respiratory signs called "thumps".

Post-mortem findings show large congested liver in early stages; "white spots or milk spots" lesions in the liver; degeneration and cirrhosis of liver; obstruction of intestine by adult worms; mild enteritis and rarely obstruction of the bile ducts caused by adult worms; oedematous and haemorrhagic lungs with parasitic pneumonia; and jaundiced carcase of poor quality flesh. Severely "white spotted" and cirrhotic livers are condemned. If mild isolated lesions disappear after holding the liver overnight in the offal cooler, it can be released for human consumption. Differential diagnosis should be made with enzootic pneumonia, chronic enteritis caused by *Salmonella* and *Treponema* spp.

Lung Worms Disease

Lung worms disease in cattle and sheep is caused by *Dictyocaulus viviparous* and *Dictyocaulus filaria,* respectively. The disease is transmitted through ingestion of feed and water contaminated with larvae present in the faeces. Larvae migrate through the intestinal wall to the mesenteric lymph nodes, and they pass via the lymphatics to the venous circulation and to the heart. Further, from the heart they reach the lung alveoli, and migrate to bronchi where they mature and lay eggs. The disease is characterized by fever, verminous pneumonia or bronchitis, husk or hoose, presence of mature lung worms in the bronchi, dyspnoea, grunting, nasal discharge, cyanosis and recumbency.

Post-mortem findings show haemorrhagic and inflamed bronchi with froth; oedematous and emphysematous lungs with lung worms; and enlarged associated lymph nodes. Carcase of animal affected with lung worms is approved if infestation is slight and no secondary changes are observed. The lungs are condemned. Carcase of animal affected with lung worms accompanied with pneumonia, emaciation or anaemia is condemned. Differential diagnosis should be made with bacterial bronchopneumonia, tuberculosis, actinobacillosis, necrobacillosis, hydatid disease, and abscess.

Oesophagostomiasis

This is also known as pimply gut or nodular worm disease. Oesophagostomiasis is a parasitic disease of ruminants and swine. *Oesophagostomum radiatum*, *Oesophagostomum columbianum* and *Oesophagostomum dentatum* are found in cattle, sheep and swine, respectively. The larvae develop to the infective stage on pasture. The disease is transmitted through grazing of pastures contaminated with infected larvae. The larvae penetrate into deeper mucosal layers provoke an inflammatory reaction and nodules of "pimply gut". Further, they develop in the intestinal wall and reach the colon. They become adult in the colon and attach themselves to the colonic mucosa where they lay eggs. The disease is characterized by diarrhoea with black-green faeces which may be mixed with mucus and blood, stiff gait, emaciation and anaemia.

Post-mortem findings show greyish-white nodules contain a greenish pasty material in younger lesions or a yellowish-brown crumbly material in older lesions; thickened intestinal wall, peritonitis and enteritis. The carcase and intestine from animals affected with severe infestation of nodular worm disease with emaciation and oedema are condemned. In mild, moderate and heavy infestation without emaciation, the carcase can be approved with condemnation of intestines.

Onchocercosis

In cattle, onchocercosis is caused by nematodes of the genus *Onchocerca*. Several species are involved, but the most important species is *Onchocerca gibsoni*. It causes sub-cutaneous nodules or "worm nests" in cattle in some countries of the Asia-Pacific region and Southern Africa. The adult worm lives in the nodules and the fertilised females liberate microfilariae into the tissue lymph spaces from where they are taken up by an insect vector (intermediate host). The common vectors are the midges of the genus *Culicoides*. Other biting flies can act as intermediate hosts. The larvae develop to the infective stage in these insect vectors. In cattle, the disease is transmitted through biting flies with the infective larvae feed on them. The disease is characterized by sub-cutaneous nodules in the brisket and buttock regions.

Post-mortem findings show firm fibrous nodules (0.5 cm - 5 cm in diameter) singly or in clusters in the regions of brisket, buttocks and thighs; tightly coiled worms in the nodules; and dead or calcified worms in older nodules. The carcase of animal affected with onchocercosis is passed after removal of the nodules. In heavy infestations the affected briskets are removed. Differential diagnosis should be made with cysticercosis, eosinophilic myositis, abscesses, and neurofibromatosis.

Trichinellosis

Trichinellosis is also known as trichiniasis or trichinosis. The disease is caused by a nematode parasite mainly "*Trichinella spiralis*. The adult form of the parasite lives in the intestine of mammals and the larval form lives in the muscular tissues of the same host. *Trichinella spiralis* is found in the intestine and muscular tissues of the pig, wild boar, rat, mouse, dog, cat, bear, human and other mammals. The main source of infection is flesh infected with the parasite. The disease is transmitted through consumption of trichinosed flesh. Trichinellosis in the pig is most likely to be acquired from the flesh of another pig via raw or unboiled garbage. Human acquires the infection from infected pig meat.

Larvae penetrate the epithelial lining of the small intestine, undergo four moults and become sexually mature adults. The adult worms may be 1-4 mm long. The newborn larvae pass to the striated muscles by the lymphatics and the blood stream. In the muscle, they grow, curl up in a spiral coil and are encysted. The predilection sites for these larvae are the tongue, diaphragm, eye, masticatory and intercostal muscles. The most characteristic symptoms of human trichinosis are high fever, weakness, arthralgia, myalgia, abdominal pain, diarrhoea, oedema of the face and eyelids, and hives. Neurological symptoms include dizziness and paresis. The carcase affected with *Trichinella spiralis* should be condemned. Differential diagnosis should be made with sarcosporidiosis and *Cysticercus cellulosae* infection and tyrosine crystals in muscles.

Diseases Caused by Arthropods

Arthropods are invertebrate animals possessing segmented body and a hard exoskeleton. The phylum Arthropoda includes several classes, *viz.*, Insecta, Archnida, Diplopoda and Crustacea. Many are of great zoonotic importance. Arthropods like flies, mosquitoes and midges may cause public health problems such as erythema, swelling, urticaria, pruritus and pain. The members of the order Diptra are also of zoonotic importance. They have a single pair of functional membranous forewings. The hindwings are reduced to so-called halters. Some important arthropod diseases are discussed as follows:

Hypoderma spp. Infestation

In cattle, the disease is caused by warble flies *viz.*, *Hypoderma bovis* and *Hypoderma lineatum*. The disease occurs mainly in summer season. The adult fly lays its eggs on the leg hair and occasionally on the body of cattle. Within a week the larvae hatch and burrow into the skin followed by travel through the body for several months. *Hypoderma bovis* migrates into the thoracic and abdominal cavities towards the spinal canal before moving under the skin of the back, while *Hypoderma lineatum* migrates to the oesophageal area before reaching the dorsal area of the animal. The larvae reach the area of the back and burrow a breathing hole and increase in size to approximately 8 mm × 25 mm. Larvae are visible for a month and thereafter maggots fall to the ground where they develop into flies and start the whole cycle once again.

The disease is characterized by swelling or erosion on skin of the back, protrusion of larvae from the skin, violent rushing and kicking the abdomen with hind legs, erected tail and paralysis of the lower body and legs in case of involvement of spinal cord. Post-mortem findings show inflamed subcutaneous tissue with red, green or yellow colour around the maggot or at the site of lodging of maggot;

inflamed oesophagus; and presence of *Hypoderma bovis* larvae. Carcase of an animal affected with *Hypoderma bovis* is approved. Subcutaneous lesions are removed.

Screwworm Myiasis

Screwworm myiasis is caused by larvae of the flies *Cochlliomyia hominironux* (New World Screwworm) and *Chrysomya bezziana* (Old World Screwworm). The New World Screwworm occurs in Central and South America including the Caribbean region, while the Old World Screwworm occurs in India, Southeast Asia, and Tropical Africa and in the Persian Gulf area. A temperature range of 20 - 30°C is favourable for the flies. After mating, the female lays one or more batches of up to 300 eggs at the edge of wound or break in the skin in warm blooded animals. The larvae develop within 24 hours, and burrow into the living flesh, creating large, deep, open wounds. The disease is characterized by weight loss, small openings in the skin which may be with extensive pockets of screwworm larvae beneath; and discharge of serosanguineous exudes from the infested wounds. In dogs, screwworm larvae commonly tunnel under the skin. Post-mortem findings show wound with larvae. The carcase affected with screwworm myiasis can be passed after removal of the wounds tissues. Differential diagnosis should be made with other blow flies such as *C. macellaria* and *Sarcophagidae* spp.

10

Metabolic Diseases

Metabolic diseases are associated with production in the animals which develop due to imbalance or deficiency in uptake of nutrients. Some metabolic diseases such as imperfect osteogenesis and mannosidosis in certain cattle, ceroid lipofuscinosis and photosensitization of sheep, and porcine syndrome in pig are associated with hereditary factors. Mannosidosis occurs due to deficiency of enzyme mannosidase which leads to nervous symptom, ataxia, abortion and neonatal death. Some important metabolic diseases associated with meat inspection are discussed as follows:

Milk Fever (Parturient Paresis)

This is a non-febrile metabolic disease of dairy cows. The disease occurs due to acute hypocalcaemia in cows after parturition and in pregnant and lactating ewes, goats and mares. The disease is characterized by muscle weakness, incoordination, recumbency, turning of head to one side, loss of appetite, coma and death due to lack of treatment. Post-mortem examination does not show any lesion.

Hypomagnesaemic Tetany

This occurs in cows and ewes. The disease occurs in high productive dairy cows and ewes grazing lush pastures. The disease occurs due to intake of magnesium deficient feed or fodder and deficiency in uptake of magnesium. The acute form of the disease is characterized by twitching of muscles and ears, incoordination, salivation, convulsion and death within 2 hours, if not treated. Chronic form of disease is characterized by dullness, anorexia, ataxia and nervous symptoms. Sudden death may occur in chronic form of the disease. Post-mortem findings show extravasation of blood in the pericardial sac and haemorrhages in the subendocardium, pleura, peritoneum and intestinal mucosa.

Acetonaemia (Ketonaemia)

This is an energy metabolic disorder, usually affecting dairy cows 4-6 weeks after calving. High demand of glucose in heavy lactation and insufficient glycogen in the liver causes breakdown of fat into volatile fatty acids and ketones. The disease may evident in two forms, *viz.*, wasting form and nervous form. The wasting form is characterized by reduced milk yield, anorexia, letharginess and odour of ketones (sweetish smell) in the breath and milk. The nervous form is characterized by muscle tremors, staggering gait, circling, hyperesthesia, head pressing *etc.*

Pregnancy Toxaemia (Twin-Lamb Disease)

This is a highly fatal metabolic disease of ewes, occurring towards end of pregnancy. The disease is characterized by twitching of ears, periorbital muscles, head and leg muscles, staggering, apparent blindness, ataxia, teeth grinding, coma and death. The disease results from deficiency of glucose (hypoglycaemia) due to increased need for glucose in ewes carrying multiple lambs. Post-mortem reveals fatty degeneration of liver, which is pale yellow or greyish in colour.

Post-Parturient Haemoglobinuria

This is a highly fatal disease of high milk yielding cows. The disease is associated with hypophosphataemia resulting from feeding on phosphate-deficient diets. The disease occurs in 2-4 weeks after parturition, which is characterized by haemoglobinuria and anaemia. The later stage of the disease is manifested by anorexia, weakness and reduced milk yield, jaundice and sometimes haemoglobinuria. Post-mortem findings show dark reddish coloured urine, anemia, icterus and fatty infiltration of liver.

Azoturia (Paralytic Myoglobinuria)

This is a metabolic disorder of horses. The disease occurs due to excessive exercise of animal after a period of inactivity on full feed. The predisposing factor for the disease occurrence is excessive feeding of grain to animal having little or no exercise. The disease is characterized by stiff gait, posterior paralysis and recumbency and passing of dark-coloured urine containing the muscle pigment myoglobin. Post-mortem findings show dark, moist and pale areas on the muscles of hind quarters especially gluteal and quadriceps muscles. Other findings are swollen kidneys with brownish streaks in their medulla.

Photosensitization

Phtosensitization is local or generalized reaction to bright sunlight resulting in inflammation, and itching, which can be severe. It occurs due to the presence of photodynamic substances such as hypericin, saponin, phagopyrin, phylloerythrin *etc.* in the blood. The photodynamic substances are mainly found in plants (*Senecio jacobea, Pithomeces charterum etc.*), but certain chemicals like tetracyclines, phenothiazine sulphoxide, corticosteroids *etc.* can also directly act as photodynamic substances. The disease is characterized by hypersensitivity of skin to sunlight, which results in dermatitis and oedema. The disease is commonly encountered in cattle and sheep. The skin lesions occur as white unpigmented areas on the face and ears. Later on, the lesions become red and oedematous with exudation of serum.

Judgement of Metabolic Diseases

Depending on overall condition, the carcase and viscera are totally or partially condemned or approved as fit for human consumption.

11

Nutritional Deficiency Diseases

Nutritional deficiency diseases may be due to deficiency of minerals and vitamins. The most frequently encountered nutritional deficiencies are energy and protein starvation, which occurs mainly during winter and drought time.

Mineral Deficiency Diseases

Minerals are important nutrients for growth, repair, production and other metabolic processes. Hence, the deficiency of minerals leads to development of various diseases in the animals. Some important diseases caused due to deficiency of minerals are discussed as follows:

Calcium

Calcium deficiency leads to poor growth and ossification of bone, brittleness of bone which may lead to its fracture, rickets in young and osteomalacia in adult or old animal, and impaired functions of muscles and heart. Other problems associated with calcium deficiency include hypocalcaemia, milk fever, lactation tetany *etc.*

Phosphorus

Phosphorus deficiency leads to retarded growth, osteodistrophy, osteomalacia and distortion of bones with enlargement of joints. Hypophosphataemic animals may show pica (deprived appetite) and will chew bones.

Sodium Chloride

Sodium chloride deficiency leads to pica, incoordination, cardiac problems and death.

Potassium

Potassium deficiency leads to hypokalaemia, anaemia, anorexia and poor growth. Hypokalaemia causes severe muscular weakness and exhaustion, sometimes muscular tremors and irregular heartbeat. Potassium deficiency can also affect transmission of nerve impulse and muscle function.

Magnesium

Dietary deficiency of magnesium leads to hypomagnesaemic tetany (grass staggers) in adult ruminants. The signs are incoordination, muscle tremors and excitability, with animals often having a "wild" look. Post-mortem findings show haemorrhages in peritoneum, pericardium, and endocardium. Subcutaneous haemorrhages may be seen in some cases.

Manganese

Manganese deficiency leads to bone deformities, joints enlargement, retarded growth and infertility in cows and ewes.

Iron

Iron deficiency leads to anaemia, particularly in newborn piglets and sometimes calves and lambs. Post-mortem lesions reveal anemia, pale mucous membranes, subcutaneous oedema of carcase, hypertrophy of heart and liver.

Copper

Copper deficiency leads to anaemia, retarded growth, alopecia, wool pigmentation, myocardial fibrosis, bone fragility, incoordination and infertility. Post-mortem findings show anaemia, emaciation, osteoporosis, and sometimes enlarged joints.

Cobalt

Dietary deficiency of cobalt leads to anorexia, emaciation, retarded growth, and reduced reproductive performance and milk yield. Post-mortem findings show anaemia, emaciation, osteoporosis and bone fragility, enlarged joints, hydrocephalus with an increased cerebrospinal fluid.

Selenium

Selenium deficiency leads to liver necrosis, mulberry heart disease and iron hypersensitivity in pigs. Selenium deficiency causes white muscle disease and paralytic myoglobinuria in cattle. Selenium deficiency also causes exudative diathesis in poultry and crazy chick disease.

Iodine

Dietary deficiency of iodine leads to goitre, dwarfism and myxoedema. Iodine deficiency causes infertility in cattle, sheep and pigs, and muscular dystrophy and encephalomalacia in poultries. Post-mortem findings show enlarged thyroid gland and myxoedema.

Vitamins Deficiency Diseases

Different animal species require varying amount of vitamins, *e.g.*, vitamins B, C and K are synthesized by ruminants, and therefore they are not required in their diet. Almost all animal species require vitamin A to maintain the integrity of epithelial tissues. The deficiency of vitamin A in cattle and sheep leads to conjunctivitis, diarrhoea, lacrimation, coughing and nasal discharges.

Judgement of Nutritional Deficiency Diseases

Total condemnation is recommended in case of emaciation and generalized oedema of the carcase. Carcases with less severe and satisfactory condition are approved as fit for consumption.

12

Environmental Pollutants Causing Diseases

The problems of environmental pollution are being increased in recent years. The reasons of the problems include urbanization, industrialization, intensive farming, animal husbandry practices *etc.* The environmental pollutants may be organic and inorganic chemicals, pesticides and antimicrobials. These pollutants directly or indirectly may reach in the foods which cause health hazards in the consumers.

Sudden development of clinical signs in the healthy animals may be suspected as case of poisoning. Poisoning may occur due to spraying or dipping for destroying ectoparasites, dressing for destruction or removal of warble flies using organophosphorous compounds, disposed paint tins, lead containing items and grazing of certain pastures containing poisonous plants.

Differential diagnosis of poisoning at ante-mortem is a difficult task for the veterinarian, *e.g.*, lead toxicity produces nervous symptoms which may be confused with some infectious diseases like listeriosis. Animals suspected of having chemical residues must not be certified for slaughter for human food. In all cases the liver should be collected for diagnosis of poisoning. The kidneys should be collected in cases suspected of arsenic, lead, mercury and copper poisoning. Blood samples should be collected for diagnosis of suspected cases of poisoning of lead, copper and strychnine. Some important diseases caused by environmental pollutants are discussed as follows:

Arsenic

The common sources of arsenic are dips against ectoparasites, weedicides, rodenticides, wood preservatives *etc.* Post-mortem findings show haemorrhagic gastroenteritis, degenerative changes in liver, heart, kidneys and adrenal glands. Chronic toxicity may show fatty degeneration in liver and kidneys.

Lead

Lead is used in solders, paints, petrol, water pipes, insecticides *etc.* Animals get toxic lead through licking and chewing of different sources. Lead poisoning is characterized by sudden death, bellowing, nervous symptoms, excitability, head pressing, muscular spasm, apparent blindness, diarrhoea *etc.* Post-mortem findings in acute poisoning show gastroenteritis, liver and kidneys degeneration, pulmonary congestion and epicardial and endocardial haemorrhages. In chronic toxicity, the lesions include softening and cavitation of central nervous system with yellowish discolouration and anemia.

Mercury

The common sources of mercury are antiseptic (*e.g.*, mercuric chloride), counter irritant ointments (*e.g.*, red mercuric iodide, yellow mercuric oxide *etc.*), parasiticides, diuretics and purgatives. Acute poisoning may occur through licking of skin dressings and ointments or ingestion of mercurial seed dressings and contaminated feeds, inhalation of vapours of mercury compounds or via the skin. Post-mortem findings in acute poisoning show gastroenteritis with petechial hemorrhages in the mucosa. Chronic toxicity may cause enlargement and congestion of liver and kidneys, pulmonary congestion with haemorrhages, and petechiation of liver, kidneys and nasal mucosa.

Cadmium

The common sources of cadmium are fungicides, zinc mines, sewage, sludge *etc.* Cadmium poisoning occurs rarely, but cases occur mostly from aerial pollution of pastures near mining operations and by accidental ingestion of farm chemicals. Acute poisoning causes tubular necrosis and renal failure, mainly in horses and pigs. In poultry, it causes enteritis, nephritis, hydropericardium and myocardial haemorrhages.

Selenium

Soluble form of organic and inorganic selenium compounds is taken up by certain plants, *e.g., Acasia, Astragalus* spp., *Aster* spp. *etc.* which are consumed by cattle and sheep. Acute selenosis causes sudden death of animal. Post-mortem lesions in acute poisoning show generalized haemorrhages and ascites, while in subacute and chronic selenosis cause necrotic and cirrhotic changes, especially in the liver and spleen, congestion, haemorrhages, ascites and cerebral oedema.

Copper

Copper is used in fungicides, parasiticides, seed dressings, mineral mixture and growth promoters. Ingestion of plants such as subterranean clover (*Trifolium subterraneaum*), Heliotrope (*H. europaeum*) and ragwort (*Senecio* spp.) containing alkaloids leads to retention of copper in the liver and causes toxicity. Acute poisoning with copper may occur in farm animals which are characterized by haemorrhagic gastroenteritis with ulceration and rupture of abomasum in some cases. Chronic poisoning with copper is characterized by enlargement and yellowishness of liver, icterus, enlarged and gun-metal coloured kidneys, and haemoglobinuria.

Zinc

Zinc is used in fungicides, rodenticides, paints, plumbing and galvanization of metals. Zinc poisoning occurs rarely, but may occur accidentally due to grazing of pastures contaminated from factories dealing with zinc or its compounds. Acute poisoning with zinc causes pulmonary emphysema, kidney haemorrhages, liver degeneration and pale flabby myocardium. Chronic toxicity causes gastroenteritis and internal haemorrhages in axillary spaces.

Molybdenum

Molybdenum is used in industries dealing with alloys and in the hardening of steel. Molybdenum rich soil causes its accumulation in certain plants, which serve as a source of high concentration of molybdenum. Therefore, molybdenum poisoning may occur in animals through ingestion of plants containing high concentration of molybdenum from molybdenum rich soil or industrial contamination of pastures. Acute toxicity with molybdenum causes gastroenteritis, with ulceration and icterus, while, chronic toxicity (molybdenosis or tert) causes enlargement of liver with yellow colouration and enlargement of kidneys and spleen.

Phosphorus

Phosphorus is used in rodenticides. Poisoning with phosphorus is uncommon, but may occur due to accidental ingestion rodenticides. Phosphorus poisoning causes severe haemorrhagic gastroenteritis, and enlargement and yellowishness of liver with necrosis and haemorrhages.

Fluorine

The main sources of fluorine are soil and water. Acute poisoning of fluorine (fluorosis) is rarely encountered. Chronic toxicity with fluorine may occur through grazing of pastures and forages contaminated with fluorine dusts and hydrofluoric acid from factories dealing with aluminium and calcine ironstone, phosphate fertilizer *etc.* Acute poisoning causes haemorrhagic gastroenteritis, and congestion of liver and kidneys in pigs. Chronic toxicity includes mottling, staining and excessive wear of teeth, and softening and thickening of bones.

Sodium Chloride

Death due to acute poisoning with sodium chloride has been reported in many animal species including poultry. Acute poisoning may occur due to accidental consumption of large amount of sodium chloride by animal and consumption of feed and drinking water contaminated with sodium chloride. The disease is characterized by gastroenteritis, blood-stained faeces, watery blood and cerebral oedema in ruminants and pigs. In poultry, associated lesions include ascites, anasarca, hydropericardium, myocardial haemorrhages, cardiac hypertrophy and nephritis.

Pesticides

Pesticides include insecticides, herbicides and fungicides. The insecticides include carbamates compounds (*e.g.*, carbaryl, propoxur *etc.*), chlorinated hydrocarbons (*e.g.*, aldrin, benzene hexachloride, dichlorodiphenyltrichloroethane *etc.*) and organophosphates (*e.g.*, dichlorvos, malathion, fenthion *etc.*). The poisoning with the carbamate and organophosphate compounds cause salivation, bronchial constriction, vomiting, diarrhoea, muscular twitching and paralysis. Organophosphate poisoning may occur through consumption of contaminated feed, feed crops and water or swallowing of dip.

Rodenticides

Substances such as alpha-naphthylthiourea, thallium acetate, strychnine, warfarin, red squill and sweet clover may cause poisoning in the animals. Alpha-naphthylthiourea causes pulmonary oedema, hydrothorax and congestion of trachea and kidneys. Thallium acetate causes gastroenteritis, dermatitis and alopecia. Gastroenteritis may be haemorrhagic, ulcerative and necrotic extending to mouth and gums. Strychnine (*Strychinos nuxvomica* and *S. ignatii*) is an alkaloid which is used as rodenticide. This causes muscle stiffness, tremor and tetanic convulsion. Post-mortem lesions show venous congestion of lungs and meninges. Warfarin is an anticoagulant rodenticide which acts as competitive antagonist of vitamin K which is necessary for the formation of blood coagulation factor I, II, VII, IX and X in the liver. Red squill (*Urginea maritime*) is a plant containing a glycoside which causes gastroenteritis. Sweet clover poisoning may occur due to feeding of moldy sweet clover hay or silage that contains dicumarol. It is the dicumarol that causes the sweet clover poisoning, preventing normal blood clotting and causing extensive haemorrhage.

Acids

Acids like sulphuric acid, hydrochloric acid, nitric acid, and phenol, cresol *etc.* are used as disinfectants. These may come in contact with skin and cause burns. Inhalation of vapours of acids causes corrosion in the respiratory tract mucosa and lungs.

Alkalis

Alkalis like caustic soda (sodium hydroxide), hydrated lime (calcium hydroxide) and sodium carbonate are used for cleaning and other purposes. Poisoning with alkalis causes corrosive action on the body tissues which leads to stomatitis, oesophagitis, and gastroenteritis with congestion, oedema and thickening of mucosa.

Ammonia Compounds and Urea

Ammonia compounds such as ammonium sulphate and nitrate are used as fertilizers. Urea is added in an optimum concentration in animal feeds to enhance the nutritive value, but accidentally excessive intake of urea causes poisoning which is characterized by salivation, nystagmus, excitement and diarrhoea.

Tetrachlorodibenzodioxin (TCDD)

Dioxins are teratogenic which may occur as contaminants during the manufacture of certain oxin herbicides and fungicides. Post-mortem shows lesions in the liver and kidneys. In poultry, the lesions include accumulation of yellowish fluid in body cavities and subcutaneous tissues.

Feed Additives

Various types of feed additives like antimicrobials, anthelmintics, hormones *etc.* are added to livestock feeds to improve the growth and enhance the productivity. Addition of these substances in higher dosage may cause toxicities and death of animals. The use of antimicrobial in higher or improper dosage may cause diarrhoea, gastrointestinal haemorrhages and neuromuscular damage. Hormone like oestrogen in higher doses may cause prepucial, vaginal and rectal prolapse, raising the tail, head, and nymphomaniac behaviour in female animals. Post-mortem lesions due to higher doses of oestrogen includes chronic cervicitis and cystic endometritis in cattle and sheep, while in pigs the lesions include inflammation, necrosis and prolapse of rectum, nephritis, ureteritis, prostitis and enlargement of seminal vesicles. Urea addition in higher dose in ruminant's feeds can lead to sudden death, severe colic, bloat, convulsion and death.

Hydrocyanic Acid (HCN) and Cyanides

Hydrocyanic acid is a highly toxic cyanogenic glycoside which is found in certain plants, *e.g.*, *Sorghum, Acacia, Eucalyptus, Lotus, Prunes, Panic* spp. *etc.* Poisoning can also occur due to ingestion of contaminated effluent from gold mining operations, chrome and nickel plating, and hydrocyanic acid-containing rodenticides. HCN in higher doses causes anoxia in the central nervous system and sudden death of animal. Post-mortem lesions include congestion in carcase and organs, haemorrhage in the trachea and lungs and petechiation in the abomasum and small intestine, and subepicardial and subendocardial haemorrhages along with bitter almond smell on opening of the abdomen.

Nitrate and Nitrite

Excessive application of nitrate fertilizers raises the nitrate concentration in streams and wells. Many plants may, in certain circumstances, contain dangerous concentration of nitrate. Ingested nitrates are converted to nitrites in the alimentary tract, especially in ruminants. It converts haemoglobin to methaemoglobin and consequently interferes with tissue oxygenation.

Poisonous Plants

There are many plant species that cause poisoning in animals include clovers, bracken fern (*Pteridium aquiline*), *Brassica* spp. (turnip, rape, kale *etc.*), buttercup (*Ranunculus* spp.), horse tail (*Equisetum* spp.), perennial ryegrass, (*Solium perennial*), ragwort (*Senecio jacobea*), rhododendron (*Erica* spp.), *Solium* spp. (woody and garden nightshade, Japanese cherry) *etc.*

Poisonous Fungi

Mycotoxins are produced by toxigenic fungi growing on food and foodstuffs. Different types of mycotoxins are produced by different types of fungi. Aflatoxins are produced by *Aspergillus flavus* and *Aspergillus parasiticus*. Aflatoxins are of four types, *viz.*, AFB1, AFB2, AFG1 and AFG2. Among these toxins, AFG1 is the most commonly produced and the most toxic. Aflatoxins are retained for a longer periods in liver and kidneys in comparison to other tissues. Ochratoxins are produced by some *Penicillium* spp. and some *Aspergillus* spp. Ochratoxins are carcinogenic and accumulated in higher concentrations in kidneys than in liver, fat and muscles.

Poisonous Algae

Poisonous algae that have caused poisoning and deaths in animals include *Anabena, Aphanizomenon, Chlorella, Microcyctis aeruginosa, Noduleria, Oscillatoria* spp. *etc.* Algae secrete exotoxins in water that serve as a source of the poison. Algal endogenous toxins like saxitoxin, β-peptide toxin and certain polysaccharides cause severe skin irritation, oral blisters and photosensitization. The poisoning may occur due to consumption of poisonous algae or drinking of water by animals from stagnant ponds, lakes *etc.* having algal bloom (heavy growth of algae). *Anabena flos-aqua* causes most serious algal poisoning, which is characterized by paralysis of central nervous system, respiratory failure and death. The poisoning associated lesions include congestion of central nervous system, pulmonary oedema and hydrothorax in peracute cases. Lesions in chronic toxicities include hepatic necrosis, jaundice, haemorrhagic gastroenteritis, visceral congestion, widespread petechiation, ascites and photosensitization. The post-mortem findings in *Chlorella* spp. poisoning (chlorellosis) include enlarged liver and hepatic lymph nodes with delineation of the hepatic lobules, with a bright green pigments giving it a honeycomb appearance.

Judgement of Inorganic and Organic Poisoning Including Poisoning with Poisonous Plants and Algae

Total condemnation is recommended in all these cases.

Radioactive Substances

Radioactive contamination arises from both natural and man-made sources. Naturally the radioactive substances such as uranium, thorium, radium, rubidium, carbon, potassium and radon are found in soil, rocks and water. Man-made sources of radioactive substances are nuclear reactors, chemical industries, laboratories and X-rays for diagnosis. Radioactive substances may reach to the animal body through contamination of pastures, drinking water and atmosphere.

Consumption of meat from animals having radioactive substances in their body tissues can cause health hazards in consumers. Large doses of radiation cause immediate effect (necrosis in cells), while, small doses cause delayed somatic and genetic effects (metabolic disturbances in cells). The lymphoid organs are damaged first, followed by testes and ovaries, bone marrow, epithelial cells of the alimentary tract, skin, liver, lungs, kidneys, pleura, peritoneum, nerve cells, skeletal muscles and connective tissues.

13

Ante-Mortem Inspection of the Food Animals

The term ante-mortem means "before death". Professional examination of the live animal before slaughter by a competent and qualified veterinarian is known as ante-mortem inspection. Ante-mortem inspection has two important components that are the screening and segregation of animals suspected of being diseased or in an unsatisfactory condition and veterinary examination and diagnosis of the screened-out animals. All livestock presented for slaughter by the establishment must receive ante-mortem inspection. The purpose of ante-mortem inspection is to ensure that only clean, healthy, residue and stress-free animals are presented for slaughter. Ante-mortem inspection is equally important as post-mortem inspection. During ante-mortem inspection the animals should be observed for the following conditions:

1. General behaviour of the animal
2. Nutritional status of the animal
3. Cleanliness of the animal
4. Obvious sign of disease and other abnormalities in the animal.

Principle of Ante-Mortem Inspection

The basic principle of ante-mortem inspection is to determine the animal is normal or not.

Objectives of Ante-Mortem Inspection

There are several objectives of ante-mortem inspection of animals.

1. To provide clean, wholesome and disease free meat to consumers.

2. To ensure the selection of normal and rested animals for quality meat production.
3. To enable the veterinarian to make the decisions as to whether the animal can be slaughtered for human consumption.
4. To determine whether there is any sign of any condition which might adversely affect human or animal health.
5. To ensure the collection of reliable clinical information in order to conduct efficient post-mortem inspection.
6. To ensure that injured animals or those with pain and suffering receive emergency slaughter and that animals are treated humanely.
7. To prevent contamination of equipment, premises and personnel by animals suffering from communicable diseases.
8. To separate the animals suffering from rabies, tetanus *etc.* which are usually undetectable on routine post-mortem inspection.
9. To assist in establishing and promoting the meat export trade by ensuring export of clean and wholesome meat.
10. To determine whether animal welfare has been compromised.
11. To identify sick animals and those treated with antibiotics, chemotherapeutic agents, insecticides and pesticides.
12. To determine whether any test should be carried out in relation to disease diagnosis or for residues of veterinary medical products.

Facilities Required for Ante-Mortem Inspection

For efficient ante-mortem inspection the facilities of adequate lighting, adequate space for inspection and satisfactory restraining materials such as rope, travis *etc.* are required. Metal or plastic ear tags and neck and leg bands are required for identification of animals. Thermometer, stethoscope *etc.* are required to conduct clinical inspection of animals. There should be facility of isolation pen for suspected and diseased animals with separate drainage and situated as to avoid contamination of other animals.

Prerequisites for Ante-Mortem Inspection

1. The animal must be subjected to ante-mortem inspection (except in case of emergency slaughter) by a veterinarian or inspector; thereafter the animal can be preceded for slaughter.
2. Ante-mortem inspection of animal should be carried out on the day of arrival of animal at slaughterhouse.
3. Ante-mortem inspection of animal must be carried out within 24 hours of arrival of animals at slaughterhouse.
4. Ante-mortem inspection of animal must be repeated immediately before slaughter.

5. Ante-mortem inspection should be carried out according to the professional rules under suitable lighting.
6. Ante-mortem inspection of animal should be carried out in adequately lighted pens where the animals can be moved freely and inspector can observe the animals while at rest and in motion.
7. The animal should be inspected from both sides that are left and right. It should be inspected at rest and in motion because certain abnormal signs, such as labored breathing, are easier to detect while the animals are at rest, while other abnormalities, such as lameness, may not be detected until the animals are observed in motion.
8. Any sign of disease, distress or injury should be noted and appropriate action taken.
9. Tired and agitated animal must be rested for at least 24 hours unless the official veterinarian decides otherwise.
10. The ante-mortem inspection of animal must determine the followings:
    i. Whether the animals are suffering from zoonotic diseases.
    ii. Whether the animals are tired, agitated or injured.
    iii. Whether the animal shows the symptoms of generalized conditions, which is likely to make their meat unfit for human consumption.

Procedure of Ante-Mortem Inspection

1. Place the animals in properly lightened holding pens where they can move about freely so they can be properly inspected both at rest and in motion.
2. Observe the animal in-motion either from outside of the pen or inside the pen depending on class of animals and safety conditions. Take the help of establishment employee to move all the animals slowly and individually out of the pen, and observe each animal for any abnormalities by viewing the head, neck, shoulder, flank, legs *etc*. Inspection can be carried out safely inside the pen in case of small animals such as sheep, goat, calves and small pigs, but it is unsafe in case of large animals such as cattle, buffalo *etc*.
3. Observe the animals carefully to determine whether they are fatigued or rested.
4. Observe each animal for its general condition and appearance. The animals showing abnormal conditions should be isolated at this stage.
5. Observe the animal for lameness while in motion.
6. Observe the animals from both sides for the presence of abscess, wound, other lesions *etc*.
7. Observe the calves for maturity, enlargement of joints, infection of navel *etc*.

Conditions to be Recorded on Ante-Mortem Inspection

Temperature

An elevated body temperature indicates systemic disturbance.

Pulse Rate

The pulse rate of a healthy animal is fairly constant. Pulse rate of young animal is often faster. Increased or decreased pulse rates indicates diseased condition.

Respiration

Healthy animals breathe regularly and easily with noiseless respiration, while animals suffering from respiratory problems breathe heavily.

Behaviour

Behaviour of the animal may be manifested as dull, reeling, anxious and aggressiveness. Moreover, the animal may manifest other abnormal behaviour like walking in circles or show an abnormal gait or posture, pushing its head against a wall *etc.*

Gait

Gait is observed in moving/walking condition of animal. An altered gait is indicative of some abnormal conditions associated with legs, chest and abdomen. The veterinarian must observe each animal while in moving and at rest. The inspection must be sufficient to identify animals showing neurological symptoms, respiratory symptoms, alimentary tract abnormalities, change in gait, or external abnormalities. An abnormal gait in an animal is associated with pain in the legs, chest or abdomen or an indication of nervous disease, stiffness and pain in joints, chemical poisoning *etc.*

Posture

Posture is observed when the animal is in rest that is in standing or lying positions. The animal may stand with an extended head and stretched out feet. The animal may be lying and have its head turned along its side. When the animal is unable to rise, it is often called a "downer". Downer animals should be handled with caution in order to prevent further suffering.

Structure

Abnormal body structure (conformation) of animal are manifested by many conditions such as abscesses, haematoma, hernia, umbilical swelling (may be due to umbilical hernia or omphalophlebitis), bloated abdomen, enlarged joints, enlarged sensitive udder (may be due to mastitis) and enlarged jaw (lumpy jaw).

Discharge

Discharges from nose and mouth, excessive saliva from the mouth, and bloody diarrhoea are indicative of bodily abnormalities.

Protrusion

Protrusions such as prolapsed uterus and prolapsed rectum are indicative of bodily abnormalities.

Physical Exhaustion

Physical exhaustion of animal may be manifested as fatigue, stiffness, weakness and pain. Physical exhaustion of animal due to dehydration may be manifested by loss of skin elasticity, sunken eye balls and dry mucous membranes. The animals in oestrus may undergo stress. They should be isolated and given adequate rest before slaughter. Heifers in oestrus subjected to slaughter may exhibit imperfect bleeding.

Nutritional Status

Nutritional status of the animal is directly associated with the quality of meat. The animal should be adequately nourished and healthy for quality meat production.

Cleanliness

Cleanliness of the animal is directly associated with production of hygienic quality of meat. The veterinarian must verify that the animals do not have such hide, skin or fleece conditions that present an unacceptable risk of contamination of the meat.

Colour

Abnormal body colour of the animal may be due to various conditions such as black areas on horses and swine, red areas on light coloured skin (inflammation), dark blue areas on the skin or udder (gangrene).

Odour

An abnormal odour is difficult to detect on routine ante-mortem inspection. The abnormal odour may be due to abscess, a medicinal odour, stinkweed odour or an acetone odour in ketosis.

Table 13.1. Normal Temperature, Pulse Rate and Respiration Rate of Different Species of Animals

Species	*Temperature (°F)*	*Pulse Rate/min*	*Respiration Rate/min.*
Horse	100.5	30-40	8-10
Cattle	101.5	60-80	10-30
Buffalo	101.5	60-80	16-22
Sheep	102.0	70-90	10-20
Goat	103.0	70-90	25-35
Pig	102.0	75	10-20

Diseases/Abnormalities Encountered on Ante-Mortem Inspection

On ante-mortem inspection, various diseases and abnormalities are encountered that include anthrax, brucellosis, tuberculosis, Johne's disease, tetanus, black leg, haemorrhagic septicaemia, actinobacillosis, actinomycosis, caseous lymphadenitis, swine erysipelas, white scour, listeriosis, foot and mouth disease, rabies, sheep pox, goat pox, bovine spongiform encephalopathy, babesiosis, ringworm, mastitis, metritis, fever, milk fever, ketosis, abortion, still-birth, diarrhoea, dysentery, jaundice, poisoning, bloat, yoke gall, tumor, abscess, haematoma, gangrene, fracture, lameness, arthritis, malnutrition, rickets, emaciation, immaturity, oedema, traumatic reticulopericarditis *etc.*

Categorization of Animals on Ante-Mortem Inspection

The immediate purpose of ante-mortem inspection is to separate healthy and unhealthy/ diseased animals. Healthy animals are passed for slaughter. Unhealthy/ diseased animals may be categorized as follows:

1. **Unfit:** Animals under this category may contaminate the premises and meat and may pose a risk for slaughterhouse personnel. As per the BIS regulations the animals under this category include dead animals and animals affected with any condition that may result in condemnation of the carcase at post-mortem inspection. The unfit category of animals include those suffering from anthrax, haemorrhagic septicaemia, black leg, tetanus, Johne's disease, caseous lymphadenitis, sheep pox, goat pox, rabies, foot and mouth disease, acute swine erysipelas, swine fever, rickets, emaciation, malnutrition, immaturity, malignant neoplasm, acute fever, septicaemia, mastitis, metritis, generalized melanosis *etc.*
2. **Slaughter Authorized under Special Condition:** Animals showing localized conditions such as fracture, injuries, abscess, tumors *etc.* need to be segregated and subjected to a detailed examination. The animals under this category are preferred to slaughter at the end of days kill or slaughtered separately with thorough post-mortem inspection in order to prevent contamination of meat and premises.
3. **Delayed Slaughter:** The animals under this category may be fatigued, inadequately rested, excited or affected with transit sickness, and such animals require rest or treatment.
4. **Casualty Slaughter:** It refers to the situation where an animal is not in acute pain or immediate danger of death but affected with some chronic conditions such as obturator paralysis, post-partum paraplegia following milk fever in cow and benign superficial tumor.
5. **Emergency Slaughter:** It refers to the situation where an animal is in acute pain or suffering from a condition where a delay in slaughter would be contrary to animal welfare and may lead to economic losses due to death of animal. Emergency slaughter of animals is carried out with careful and extensive post-mortem inspection in order to ensure the safety of consumers and preventing contamination of meat and premises.

Emergency slaughter of animal is recommended in multiple fracture, extensive bruising, severe injuries, tympanitis, uterine prolapse and post-partum haemorrhage in bovine. The meat of emergency slaughtered animal may be used for human consumption only if it is from an otherwise healthy animal that has suffered with an accident and slaughtered for animal welfare and economic reasons. The chief causes of emergency slaughter in cattle are injuries and affections of the uterus and udder.

14
Inspection of Poultry, Eggs and Fish

Poultry meat production throughout the world has been increasing both in quantity and in sophistication. Adequate floor space for each bird is essential for its growth, health, quality and general well-being. A continuous adequate supply of clean water is essential to prevent dehydration. The raw materials present in the bird's food are carefully sourced and are examined for quality and chemical and microbiological purity. The birds are caught by hand and put directly into the compartments, thus avoiding the need for multiple handling or carrying the birds for long distances before crating.

Ante-Mortem Inspection of Poultry

All poultry farms should have a veterinarian, specializing in poultry who can advise on the health and management of the poultry flock. Ante-mortem inspection of poultry includes checking of procedures record which depending on the type of bird includes date of arrival, mortality, suppliers of feeding stuff, and expected date of slaughter. Poultry can also be examined for other reasons that are whether birds are suffering from a disease which can be transmitted to humans. Whether the birds show disturbance of general behavior or sign of sickness which may make the meat unfit for human consumption.

Ante-mortem inspection of birds may be difficult if the birds are placed in crates or liners, and hence only a superficial inspection of their general condition is carried out. The remainder of the poultry inspection should take place after the birds have been hung in shackles and before they are bled. It is essential to maintain the records of ante-mortem inspection and should include date and time of inspection, truck number, species, the total number of birds and the name of the owner. The main objectives of ante-mortem inspection of birds are as follows:

1. To determine the general condition of the birds.
2. To establish if a disease or condition requires particular handling such as segregation of diseased birds, delayed slaughter or adjustment of line speed.

Ante-mortem inspection of birds should be carried out for overall condition of the birds including the head, with attention to the eyes, legs, and the body of the birds; and whether there are any unusual swellings or other abnormalities on the birds. Poultry ante-mortem inspection must be performed on the day of slaughter. Ante-mortem inspection of birds may lead to three possible outcomes such as passed for slaughter, suspect and condemned. In birds, the symptoms of disease that may be observed on ante-mortem inspection include emaciation, swelling around the head and eyes, oedema of the wattles, torticollis or wry neck, gasping and sneezing, off-coloured diarrhoea, skin lesions, dermatitis, lameness, bone or joint enlargement *etc.*

Post-Mortem Inspection of Poultry

Post-mortem inspection in poultry refers to inspection of techniques and inspection of carcases and viscera. Post-mortem inspection of poultry should be carried out immediately after slaughter under suitable lighting (540 lux). Post-mortem inspection consists of use of sense of smell to detect any abnormal odour, visual inspection, palpation and incision if necessary. The shape, colour and consistency of organs and tissues must be observed singly or in combination. The colour of the poultry carcase depends on age, sex, nutrition and the scalding temperature during slaughter.

In post-mortem inspection of poultry the intestinal tract, liver, spleen and heart must be exposed for visual inspection and palpation. The inspector should be able to look inside the carcase and detect any pathological lesions such as peritonitis, salpingitis, air sac inflammation *etc.* Contamination of carcase and visceral organs with faeces and bile should also be observed during post-mortem inspection of poultry. Both hands should be used during post-mortem inspection of poultry carcase and viscera. External lesions on the carcase include the swelling of the sinuses, skin lesions, joint swellings *etc.*

Poultry unfit for human consumption includes generalized infectious disease and chronic localization in organs with pathogenic microorganism transmissible to humans, systemic mycosis and local lesions in organs suspected of having been caused by pathogenic agents transmissible to humans. Other local conditions include poisoning, abnormal colour and smell, general soiling or contamination, major lesions and hemorrhage, ascites *etc.* To reduce the microbial contamination in meat, it is necessary to ensure that the birds entering the slaughter or processing chain should be free from infection. Localized lesions could be disposed by an inspector; however the final judgement of the carcase should be done by a veterinarian.

Poultry carcases are usually condemned due to pathological, non-pathological and aesthetic reasons.

Inspection of Poultry Eggs

Eggs are inspected for cracks, leaks, cleanliness of the shell, freshness, soundness, size *etc.* Poultry eggs are inspected for the following conditions:

1. **Contamination of Eggs with Manure:** Eggs may get contaminated with poultry manure in laying cages due to poor hygienic conditions and lack of sanitary measures. Heavily soiled eggs are difficult to clean once the manure gets dried up on the shell.
2. **Fly Speck:** Flies in the poultry farm may result in fly speck on the eggs. It is an unhygienic condition of eggs.
3. **Cracked Shells:** The quality of egg shell is affected with storage conditions. Nutritional deficiency of calcium, phosphorus, vitamin C and vitamin D also influence the egg shell quality.
4. **Albumen Quality:** Albumen quality of eggs is affected by age of birds and handling, gathering and cooling of the eggs. Keeping of eggs in hot poultry pens (above 30°C) causes their deterioration and rapidly declines in albumen quality. However, oiling of the eggs may prevent this deterioration in quality.
5. **Diseases:** Diseases in poultry may result into transmission of pathogens in eggs. Eggs are usually rejected due to poor albumen quality. Diseases in poultry and their eggs can be controlled by proper preventive health management of the flock.
6. **Spoilage/Rotting of Eggs:** Unsound eggs may be due to growth of bacteria and moulds. Spoilage of eggs may occur due to cracking of the eggshell, improper washing and storage techniques. The most predominant spoilage of eggs is rot which is caused by Gram-negative motile rods such as *Pseudomonas, Proteus, Alcaligenes, Aeromonas,* and coliforms. Rotten egg smell of hydrogen sulphide gas may occur from the egg due to putrefaction of the sulphur containing proteins of the egg. The predominant bacteria causing spoilage in pasteurized egg products are psychrotrophic Gram-negative. Dried eggs are not susceptible to microbial spoilage due to low water activity. Unbroken egg can be examined with transmitted light using candle can identify grossly contamination of eggs or rotten egg. Spoilage of eggs can be prevented by the following methods:
   i. Maintenance of hygiene in the poultry houses.
   ii. Timely collection of eggs from the poultry houses.
   iii. Fumigation of eggs with gaseous ethylene oxide before storage of eggs against bacterial spoilage.
   iv. Pasteurization of eggs during heat treatment of shell egg.

Method of Inspection of Poultry Eggs

The quality of eggs can be determined by the following methods:

1. **Shaking:** A fresh egg makes no sound, while a stale (spoilt) egg makes a sound when shaken.
2. **Candling:** Freshness of egg may be tested by candling. This is performed by holding the egg between the eye and a light source such as a candle flame or the sun. As the shell is translucent, the internal quality and the size of the yolk can be assessed. The quality of eggs can also be determined by candling to detect the general bloodiness or translucent spots in the egg yolk. Nowadays colour additives are used in poultry feeds to impart desirable colour to the egg yolk.
3. **Floating:** Fresh eggs usually sink to the bottom of a bowl of water, whereas spoilt eggs float and can be removed. Spoilt egg floats in the water due bigger air cavity in it which makes the egg more buoyant. The disadvantage of this method is that the water may enter through the eggshell pores so it is important to use clean water, change it frequently and not to leave eggs in the water.
4. **Breaking:** In this test, around 10 eggs out of 100 are taken randomly and checked for spoilage by breaking them open to see what is inside. This is the most accurate method of testing the eggs but it is not cost-effective, therefore, this method should be used only when the other methods are not exercised (*e.g.*, in large-scale operations).

Storage of Eggs

Eggs are perishable food items; therefore, they need proper storage. They should be kept cool and dry. Maintenance of the egg's internal quality depends on the time, temperature and other conditions of storage such as the presence of tainting substances in the storage environment. Eggshells are porous and eggs can quickly absorb foreign odours which causes development of taint in the egg contents. It is therefore advise to avoid storing strong-smelling and volatile materials such as kerosene, varnish, phenol, formalin *etc.* near egg stores.

Eggshell has minute pores which permit oxygen for embryo inside but microorganisms may also gain entrance through these pores. Penetration of microbes may be prevented by dipping them in some blend oil, petroleum jelly *etc.* and storing at 0°C for 6 months. After grading, the eggs may be stored as broader end upside, providing sufficient air space.

Inspection of Fish

Fish is a highly perishable commodity and spoilt very rapidly. Newly caught fish is fresh and bright on the surface. The eyes are clear and tense and gills are bright and pink in colour. With few hours after being caught, the body of the fish becomes rigid and continues for 6-24 hours depending on the climatic conditions. Within few hours after being caught, a thick slimy grayish exudate covers the entire surface of the fish. The red colour of the gills cannot always be regarded as sign of freshness as in some cases it may continue even after putrefaction has commenced. Grayish slimy gills indicate that fish has no longer fresh. The spoilt fish are unfit for consumption which need total condemnation.

Table 14.1. Differentiation of Fresh Fish from Spoilt Fish

Sl.No.	Organs/body	Fresh Fish	Spoilt Fish
1	Eyes	Clear, bright, prominent and convex	Dull, sunken and grey
2	Gills	Pink and odourless	Pale pink, dark brown or grey, slimy and foul odour
3	Scales	Adhere closely	Loose and many scales missing
4	Skin	Clear, bright, flesh and firm to touch	Flesh soft and limp
5	Surface slime	Covered with moderate amount of surface slime which has characteristic odour	Surface slime becomes yellowish in colour emit bad odour

Factors Causing Contamination of Fish

Fish are generally considered clean and fresh, but several environmental factors can make fish unfit for human consumption.

1. **Quality of Fish Food:** Contamination of fish may occur through contaminated food which is offered to fish.
2. **Cleanliness/Safety of the Water Body:** Cleanliness of the water body is important to keep them clean, fresh and healthy. Unprotected water body can also lead to contamination from outside as well as challenge to their safety. Contamination of water bodies may occur through industrial chemical wastes which may contain heavy metals, farm chemical drainage containing pesticides which may bioaccumulate (*e.g.*, DDT accumulates in fish tissues) and domestic and commercial wastes, drainage and runoff, which may be contaminated with faeces or other pollutants.
3. **Handling of Fish:** Careful and safe handling of fish is also important to maintain the quality of fish. Careless and poor handling of fish at any stage (from being caught to being eaten) may lead to contamination of fish.

Diseases Associated with Poor Fish Hygiene

Fish is a very perishable and potentially hazardous food item if not handled properly. There are many fish-borne diseases associated with the environment in which the fish is grown, and with the way it is handled after it is brought out of the water, particularly if it is kept at room temperature. Microorganisms and autolytic enzymes are more active at higher temperatures, so deterioration proceeds faster. Autolytic enzymes are naturally occurring proteins in an animal that cause automatic breakdown of cells and tissues after its death. Water contaminated with faeces may lead to contamination of fish with fish tapeworm, *Shigella* spp., *Salmonella* spp. and other organisms. Contamination of fish with *Salmonella* also occurs through unhygienic handling by infected persons.

15

Inspection of Game Animals, Rabbits and Hares

Game Animals

Game animals refer to land mammals and birds either in the wild or farmed that are not normally considered to be domestic animals such as bovines, domestic swine, sheep and goats, domestic solipeds, domestic fowl, turkeys, guinea-fowl, ducks and geese. Therefore, these domestic animals are specifically excluded from game animals. The common game species include bear, bison, antelope, caribou, deer, elk, moose, reindeer, wild boar, snake, alligator, rabbit, squirrel, beaver and birds (pheasant, grouse, quail, wild turkey, wild geese and ducks).

Some parts of the world continue to be blessed with large, thriving populations of game animals, in Africa particularly antelopes such as impala, kudu and eland, in the Southern part of Latin America hares and some deer and antelope species and in Eastern Europe red and roe deer. Controlled cropping of these herds can provide a significant, sustainable source of supplementary protein especially in rural areas. The antelope is the most preferred and frequently hunted species of game animal in Africa for the specific purpose of providing meat for human consumption. The impala is the easiest of the antelope to cull on a large scale in Africa. Although, the management of eland is easy as management of domestic cattle.

Two basic systems of culling and preparation of carcases are used for export of game meat in some countries. The first system includes night shooting on foot using spotlights. Apparently healthy animals are shot, immediately bled and the stomachs and intestine eviscerated on the spot. After a number of carcases have been collected on an accompanying vehicle, they are then transferred to a central, permanent abattoir facility suitable for their dressing, inspection and refrigeration.

Since ante-mortem inspection is performed by hunters, they should be trained in basic ante-mortem procedures in order that they may be able to select healthy from sick animals. In second system, the animals are rounded up and herded into a temporary, funnel like structure. The animals are rested and ante-mortem inspection is carried out. Animals are then shot at point blank range, bled and eviscerated immediately and removed to a temporary butchery for dressing, inspection and refrigeration. Ante-mortem and post-mortem inspection can be objectively carried out with this system. Although hygiene could be somewhat compromised. A judicious combination of features form both these systems, which can be modified, can be used to suit a variety of circumstances in the field.

Ante-Mortem and Post-Mortem Inspection Procedures of Game Animals

The inspection procedures that are most appropriate to any particular type of game animal or carcase varies according to species and also according to whatever other information may be available about the wildlife population from which they are harvested. Minimum inspection procedures as set out in the Joint FAO/WHO Codex Alimentarius Commission's Code of Hygienic Practice for Game are useful in developing appropriate inspection procedures. On post-mortem inspection the carcase and offal of game animals can be classed into the following categories of judgement:

1. **Approved as Fit for Human Consumption:** When the inspection and any other information available has revealed no evidence of any disease or defect, and if the dressing has been implemented in accordance with hygienic requirements, the carcase and offal should be approved as fit for human consumption without restriction provided no animal health restrictions are otherwise applicable.
2. **Totally Unfit for Human Consumption:** The game animal and all offal should be condemned and properly disposed of in the following conditions:
   i. When they cause hazards to other animals, food handlers and consumers.
   ii. When they show extensive injury, swelling, oedema, emaciation, contamination and decomposition.
   iii. When they show signs consistent with natural death, death by trapping or a moribund state.
   iv. When the meat shows unacceptable deviations from normal game meat on detection by sensory means.
3. **Partially Condemned:** Where lesions are localized, affecting only part of the carcase or offal, the affected parts should be removed and the unaffected parts conditionally or unconditionally passed.

Conditions Affecting Antelope

The causes of condemnation of the carcase, meat and offal in impala to fall into two categories, namely, management related and disease related.

1. **Management Related:** The causes of condemnation of the carcase, meat and offal in impala may occur due to traumatic injury from gunshot, contamination chiefly gross dirt attained from the environment during bleeding and de-gutting or intestinal contents during careless evisceration, and spoilage and putrefaction lead to wastage. These losses can be minimized if hunting and dressing are performed during the winter months, at night and during lower ambient temperatures.
2. **Disease Related:** Different types of "measles" tapeworm cysts have been found in carcases of game animals such as impala, kudu, bushbuck, reedbuck, sable, wildebeest (gnu, antelope) and warthogs. Cysts of varying sizes are often found in the peritoneum, viscera and musculature. There are no special predilection sites of the muscular cysts. Sarcocysts are frequently seen in the skeletal muscle of impala. Carcases severely affected with sarcocysts are needed to be condemned. *Stilesia* tapeworm may be found in the liver of small antelope which can be removed by trimming. *Cooperoides hepatica* is a small brown filarial worm which occurs coiled up in a cyst in the liver, most frequently in impala. Trimming is required for removal of cysts. *Cordophillus* is a filarial worm found encysted in the heart muscle of kudu. Approximately 25 per cent of these animals are affected. This parasite is occasionally found in other muscle and may also occur in the heart muscle of domestic cattle. Affected tissues should be trimmed. Hydatid cysts are found in the lungs and livers of impala, zebra, giraffe and warthog. Hydatid cysts can be removed by trimming of affected part or removal of organs. Hydatid cysts involving multiple sites in the carcase and organs needs total condemnation.

Wild Game Meats

Wild game meats have health benefits in comparison to meats from domesticated or farm-raised animals due to the fact that wild game animals eat their natural diet and are very active in the wild contributes to the lower fat content of the meat. The most common wild game includes venison, bison, rabbit, elk *etc.* The elk (*Cervus canadensis*) or wapiti is one of the largest species within the deer family, Cervidae, in the world, and one of the largest land mammals in North America and Eastern Asia. They are raised primarily for their velvet antlers; however, there are also good markets for breeding stock, meat, and other by-products. Commercial elk farming has steady growth because of increasing demand for deer products and minimal acreage requirements for production.

Elk farming has several advantages in comparison to other livestock enterprises. Elk converts forage efficiently into protein, and with proper management they can be raised on marginal land. Elk also fit well into an existing grazing operation. Another advantage is that it has high ratio of lean meat produced per pound of live weight. The labour requirements for elk production are minimal, while the profit potential can be much greater than that of a comparable beef cow-calf operation.

Farm-raised elk venison is a fine-grained, mild, tender meat with a delicate flavor that is distinctly different from wild game venison. The meat has a taste somewhere between beef and venison and is higher in protein and lower in fat and cholesterol than beef, pork, and chicken. Elk is lean, with only 22 per cent of the meat energy derived from fat, as compared with 35 per cent to 47 per cent for lamb and 33 per cent for beef. Venison is low in cholesterol. In venison the ratio of polyunsaturated and monounsaturated fatty acids is higher than in conventional red meats. Venison also has more vitamins and minerals than beef.

Post-Mortem Inspection of Farmed Game Animals

All post-mortem inspection procedures of farmed game animals (ruminants) are similar to those described for cattle. The post-mortem inspection of wild game (reindeer, caribou and musk ox) involves the following parts:

1. **Inspection of Head:** The head must be visually examined to detect any gross lesion/abnormality. In the case where heads and/or tongues are saved for human consumption, mandibular lymph nodes should be exposed, examined visually and then incised, and the tongue palpated.
2. **Inspection of Viscera:** The lungs should be visually examined and palpated, and special attention should be given to presence of hydatid cysts. Associated lymph nodes should be incised and examined. The liver should be visually examined for the presence of both liver flukes and cysticerci; and palpated and incised if necessary. Hearts should be examined visually and then incised longitudinally for visual inspection of parasitic and other lesions. Mesenteric lymph nodes should be visually examined and palpated. Kidneys should be exposed and examined visually.
3. **Inspection of Carcase:** The internal and external surfaces of the dressed carcase should be carefully examined prior to washing of the carcase. The joints, outer muscle surface, the body cavities, the diaphragm and its pillars should be visually inspected. Pillars of the diaphragm must be incised in order to detect parasites (*e.g.*, cysticeri). The pre-femoral lymph node should be palpated to detect lesions resulting from migration of warble. The carcase exterior, cut surface of bones and muscles, pleura, peritoneum and diaphragm should be observed for cleanliness, efficiency of bleeding, colour, odours, evidence of bruising and other conditions. If necessary, the palpation and incision of parts may be carried out, *e.g.*, triceps brachii muscles for *Cysticercus bovis*. The carcase associated lymph nodes such as superficial inguinal, external and internal iliac and prepectoral lymph nodes should be observed and if necessary, palpated and incised. The thoracic and abdominal cavities should be inspected for the evidence of inflammation, abscess, actinobacillosis, mesothelioma or tuberculosis. The carcase has no such lesions are passed for food.

Inspection of Ostriches

1. **Handling:** Ostrich pens should be enclosed to create darker conditions.
2. **Transportation:** The skin and meat of ostriches is particularly valuable, therefore, careful transport by vehicle is the only suitable method of transport. For transport of slaughter animals, it is necessary to take into account numerous factors in order to ensure the health and welfare of the animals.
3. **Slaughter:** Electrical stunning should be carried out in ostriches. The tongs can be placed either at the sides of the head below and behind the eye or above and below the head. Poultry can be stunned electrically using a manually operated device or using an automatic water bath. In this process the birds are dragged through a trough of water charged with a low voltage current.

Post-Mortem Inspection of Ostriches, Rheas and Emus

The post-mortem inspection should be carried out in the following sequential steps:

1. **Inspection of Head:** The head, eyes, sinus openings and the neck should be visually inspected.
2. **Inspection of Viscera:** The abdominal and thoracic air sacs must be visually inspected. The heart is visually inspected, palpated, and incised through the interventricular septum to expose the inner surfaces for inspection. The lungs should be visually inspected and palpated on all external surfaces. The oesophagus and gizzard must be visually inspected. The liver, kidneys and spleen should be visually inspected and palpated.
3. **Inspection of Carcase:** The carcase should be inspected after removal of viscera but before washing of carcase. The inspection consists of a careful examination of the external surfaces of the carcase and the internal cavity.

16

Slaughter of the Food Animals

In developing countries, billions of animals are slaughtered for food annually, more than half of them without the stunning benefit. The slaughter animals are transported in poorly designed vehicles in overloaded conditions with little protection. They are transported for long distances without feed and water over harsh roads. In unauthorised premises the animals are stunned by repeated hammer blows to the head. They are sometimes stabbed with sharp knives until they collapse. While the animals are still conscious, their throats are cut, and they die from excessive blood loss after minutes of struggling. These brutal methods cause immense animal suffering. Such cruel practices also have significant economic impact.

Transportation of animals in poorly designed vehicles with careless handling and management causes extensive bruising in them. Bruising renders the meat unfit for human consumption; damage to the hides causes loss of product; and worker injuries result in decreased productivity. At the same time, unhygienic practices cause contamination of the meat and thereby spread of many diseases such as salmonellosis, *E. coli* infection, campylobacteriosis *etc.* which are serious public health concern. More humane transport, handling, and slaughter practices and the introduction of modern systems and equipment in the slaughter process not only decrease animal suffering but also important in quality meat production and safety to the consumers and provide economic benefits as the amount of meat and hide wastage is reduced. At the same time, the safety of slaughterhouse personnel is greatly increased.

Humane handling of animals is important in order to follow the animal welfare regulations as well as quality meat production. The responsibility for the welfare of animals from the time they arrive at the slaughterhouse until they are dead lies with the operator of the establishment. The design of the handling facility in the lairage which delivers the animals to the point of slaughter should utilize knowledge of animal behaviour to reduce the fear or apprehension felt by the animal to a minimum. The number of variable factors involved in modern stunning systems, especially electrical and gaseous methods, makes a system of approval prior to operation desirable. Accurate monitoring of stunner performance means more effective stunning, fewer second shots, greater operator efficiency and safety, and a high standard of animal welfare. For rabbits, electrical stunning is carried out using a handheld or wall-mounted V-shaped spiked electrode. Restrainer systems have been manufactured to deal with small farm animals like calves and sheep.

Principles of Slaughter

The principles in a slaughtering process are divided in "dirty" and "clean" operations. Dirty operations include lairage, slaughter and bleeding, and skinning and scalding/dehairing. Clean operations include evisceration, chilling/hanging, cutting/deboning, and freezing/delivery.

Preparing Animals for Slaughter

The animals should be healthy and physiologically normal at the time of slaughter. Slaughter animals should be adequately rested, preferably 24 hours or at least overnight, particularly if they have travelled for sometimes over long distances. However, the resting period for pigs and poultry are relatively shorter. Holding of pigs in the lairage pens for longer periods cause stress in them enhance the chances of disease transmission among them. Animals should be watered during holding and can be fed, if required. The holding period of animals in the lairage pens also provide a condition for injured and victimised animals to be identified and for sick animals to be quarantined. The animals should be moved to the stunning area in a quiet and orderly manner. Movement in the animals can be facilitated using flat canvass straps, rolled plastic or paper, and sometimes prodders can be used in case of stubborn animals. Animals should never be beaten nor have their tails twisted.

Restraining Animals for Slaughter

Restraining of animals is very important before stunning or bleeding. Restraining ensures the stability of the animal so that the stunning operation can be carried out accurately and efficiently. Animals should never be left standing for a prolonged period in a restraint device and must be stunned immediately after being secured. The operator must be adequately trained and supervised. In some countries, people who handle and stun animals have to be trained and licensed. Different types of restraints are appropriately used for different animal species as follows:

1. **Cattle:** Cattle are the most commonly restrained in the stunning box. The stunning box should be of such size and width so that the animal cannot turn around otherwise it will be difficult to stun the animal. The floor of the stunning box should be non-slippery. Sometimes a simple neck crush is used to restrain the cattle on small-scale operations.

2. **Sheep/Goats:** A properly constructed metal stunning box is suitable for restraining of sheep/goat; however, they can also be restrained manually.
3. **Pigs:** A stunning box is suitable for pigs. Putting a few pigs in a small room is suitable but only for electrical stunning. Pigs should not be restrained manually.
4. **Poultry:** Chickens are shackled by their legs onto a conveyor line. Shackling should be carried out gently to avoid injury and stress. In a small slaughterhouse, birds can be placed headfirst in cones.
5. **Ostriches:** Ostriches are temperamental animals. They must be securely restrained to avoid injuries to handler due to kicking by them. This can be done by leading them into a padded V-shaped pen, with the head facing the apex of the pen. Also the feet can be clamped immediately after electrical stunning has begun.

Method of Slaughter of Animals

Slaughtering means putting the food animal to death and thereafter prepare the carcase for human consumption. Animals arrived at slaughterhouse must be slaughtered without undue delay. The main consideration in adopting a particular method of slaughter is to ensure humane conditions during slaughter to minimize cruelty to the animal and to achieve perfect bleeding. There are various methods of slaughter:

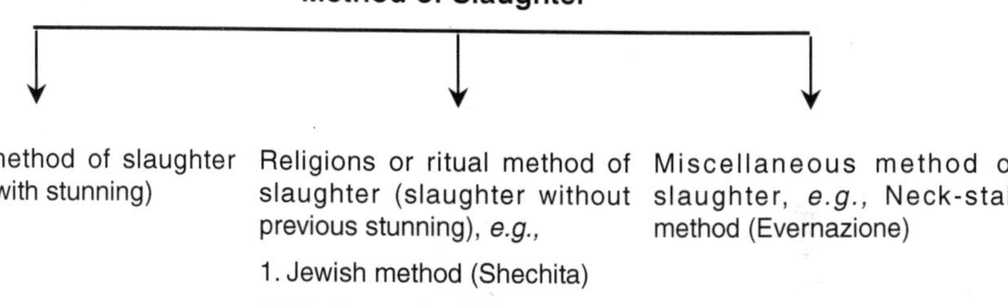

Humane Method of Slaughter

Slaughter of animal with previous stunning is known as humane slaughter.

Stunning

Stunning is the process to render the animal unconscious or insensible. Stunning is desirable to render an animal unconscious before it is slaughtered in order to eliminate pain, discomfort and stress from the procedure. Humane treatment of slaughter animals not only reduces unnecessary suffering but also reduces loss of quality and value of meat and animal by-products, thus contributing to food security and income in many countries. Most developed and many developing countries have legislation that requires pre-slaughter stunning, with the exception of authorised ritual slaughter like Kosher or Halal. In some circumstances, traditional slaughter may be exempted from pre-slaughter stunning.

Stunning raises the blood pressure and heart rate which facilitate immediate bleeding. The initial effect on the animal is immediate unconsciousness, accompanied by tonic activity. The animal collapses, stops breathing, and becomes rigid. This period of rigidity normally lasts for 10-20 seconds following stunning. The forelegs may be flexed initially and then gradually straighten out, but this depends on the species and the severity of the blow. Tonic activity is followed by a period of involuntary kicking, which gradually subsides. If an animal is stunned properly, it collapsed immediately. There is no rhythmic breathing, no blinking, no corneal reflex, and no vocalizing. The animal has a fixed, gazed expression and relaxed jaw, and the tongue is hanging out.

Objectives of Stunning

1. To perform humane slaughter of the animal.
2. To perform easy slaughter of the animal.
3. To prevent cruelty against animal.
4. To prevent struggling of the animal during slaughter.
5. To prevent the animal from severe or intense pain during slaughter.
6. To safeguard the personnel from injury by animal.
7. To facilitate immediate bleeding.

Choice of Method of Stunning

The choice of a particular method of stunning depends on the following factors:

1. Class of animal (species, age *etc.*)
2. Intended line speed
3. Number of animals to be slaughtered
4. Humane aspect
5. Capital and maintenance costs
6. Efficiency of equipment
7. Ease of operation
8. Safety of personnel
9. Effect on carcase and brain
10. Religious and legal requirements

Methods of Stunning

Various methods of stunning such as physical or mechanical (percussive), electrical and chemical (use of carbon dioxide and other gases) are adopted throughout the world. Percussive and electrical stunning methods are commonly used in developing countries.

Physical or Mechanical (Percussive) Stunning

This method produces a physical shock to the brain. Many percussive stunners are used for percussive stunning throughout the world. Percussive stunning is carried out by using percussive stunners. Percussive stunners are of two types *viz.*, penetrative type and non-penetrative type.

1. **Penetrative Type of Percussive Stunners:** Penetrative percussive stunning may be carried out by using the following types of stunners:
   i. **Captive Bolt Pistol:** Captive bolt pistol is most commonly used stunner and very effective in cattle, sheep and calves but less effective in bulls and pigs especially in sow and boars in which percussive stunner produces immediate and permanent insensibility by destroying the cortex and deeper part of the brain. A captive bolt pistol is not suitable for stunning of ostrich. Its brain is small and lobulated, and the bolt does not produce proper concussion. Captive bolt stunner is comprised of a steel bolt with a flange and a piston at one end that is held in the barrel. The piston fits tightly into the breech and the bolt is free for moving forward and backward in the barrel. Upon firing, the expansion of gases, produced by the explosion of the charge, propels the piston forward, and the bolt projects through an aperture in the front of the barrel. The bolt remains captive in the barrel, however, because the flange at the rear prevents it from passing through the hole. Penetrating stunner causes unconsciousness as a result of a concussive blow to the skull and the physical damage due to entry of the bolt into the brain. Penetrating stunner is generally preferred, as it results in more rapid unconsciousness and death. For effective stunning, the operator should be well trained in its use of the stunner. If the operator becomes fatigued, accuracy of stunning is reduced, so in large meat plants, rotation of two stunners is recommended.
   ii. **Free Bullet Pistol:** The free bullet pistol is frequently used for humane killing of horses and sometimes cattle. Great care must be taken to avoid accidents when using free projectiles. In circumstances where animals are too fractious to be handled in the normal way, such as when they cannot be loaded on the farm or led into the stunning restraint, gunshot with a free, soft-nosed bullet is effective. A 22-calibre bullet is sufficient for most animals. Shooting with a free bullet can be dangerous to operators.
   iii. **Pneumatic Stunner:** In pneumatic stunner, the bolt is activated under the pressure of 80-120 p.s.i.
2. **Non-Penetrative Type of Percussive Stunners:** Non-penetrative type of percussive stunners such as hammer, club *etc.* may be used for stunning of animals. Heavy strike on the head with these blunt objects induces sudden unconsciousness. Stunning by such stunners sometimes lead to failure of stunning if the stunner is light weight, struck with light force and applied at inappropriate site.

Head Sites for Percussive Stunning in Various Animal Species

1. **Adult Cattle:** A point at the fore head at the intersection of two lines drawn from the base of the horns to the inner canthus of eyes on the opposite side.

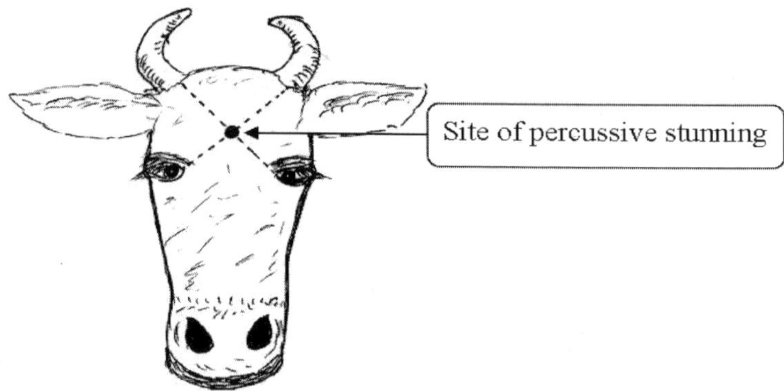

Figure 16.1. Site of Percussive Stunning in Cattle.

2. **Bulls and Old Cows:** The muzzle of the pistol is placed 1.5 cm to the side of the ridge running down the centre of the forehead.
3. **Calves:** The muzzle of the pistol is placed slightly lower on the head than for adult cattle.

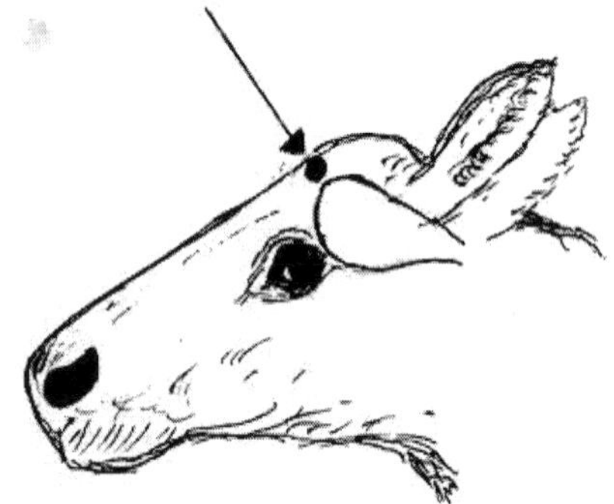

Figure 16.2. Site of Percussive Stunning in Calf.

4. **Hornless Sheep and Goat:** The muzzle of the pistol is placed on the top of the head and aimed towards gullet.
5. **Horned Sheep and Goat:** The muzzle of the pistol is placed behind the ridge which runs between the horns and aimed towards gullet.

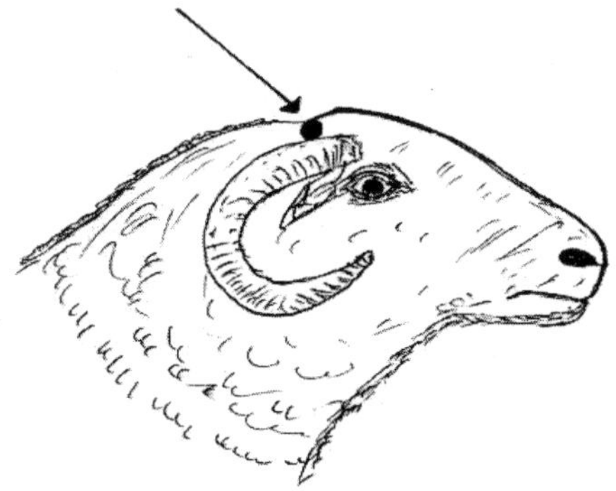

Figure 16.3. Site of Percussive Stunning in Sheep.

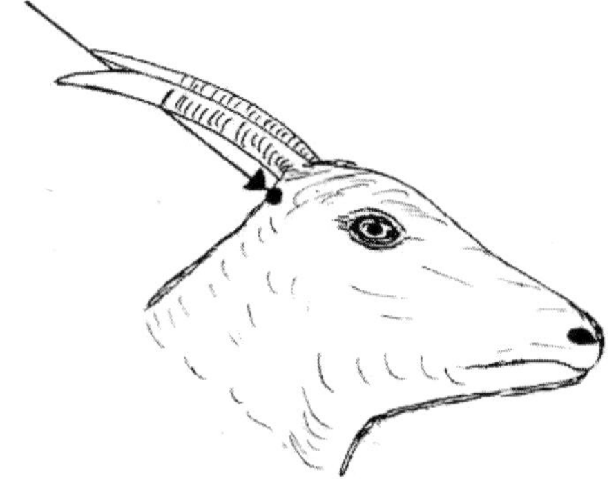

Figure 16.4. Site of Percussive Stunning in Goat.

6. **Bacon-Weight Pig:** The muzzle of the pistol is placed about 2.5 cm above the level of eyes in the mid-line of the skull.

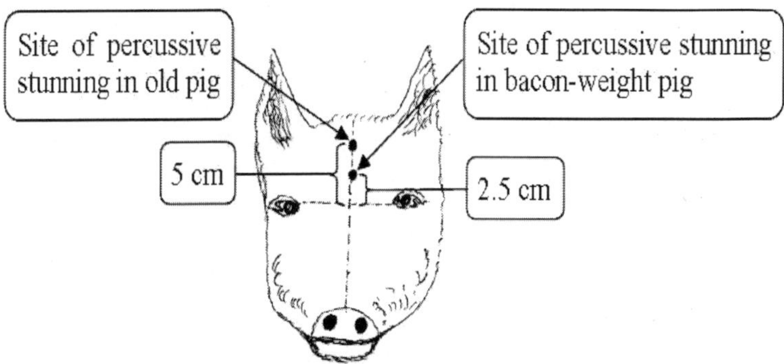

Figure 16.5. Site of Percussive Stunning in Pig.

7. **Old Pig:** The muzzle of the pistol is placed about 5 cm above the level of eyes. In older pigs, the captive bolt stunning is less reliable owing to the massive nature of the skulls and the large frontal sinuses.

8. **Horse:** The muzzle of the pistol is placed at right angles to the frontal surface 1 cm above the point where imaginary lines from eye to ear cross as the brain is present in the upper part of the head.

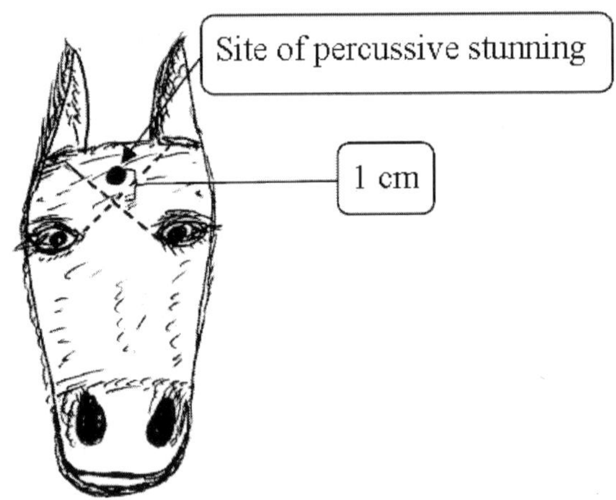

Figure 16.6. Site of Percussive Stunning in Horse.

Chemical Stunning

Chemical stunning may be carried out by exposure to carbon dioxide (CO2) and other gases. Stunning of pigs by carbon dioxide can be carried out using gas tunnel, dip lift and compact CO_2 immobilizer. Oval gas tunnel is suitable for higher rate of slaughter (600 pigs per hour). The use of carbon dioxide gas is a relatively new method of stunning suitable for pigs and poultry. However, it is applicable only at large industrial plants, as the sophisticated technical equipment is relatively costly to install. Moreover, the acceptability of this method on welfare grounds has been questioned. This method is suitable for pigs. A concentration of 80-95 per cent carbon dioxide in air is the most suitable for pre-slaughter anesthesia. Argon may be used along with carbon dioxide or oxygen in different concentration. Currently Argon gas is being tested for stunning purposes. It is assumed that Argon gas may have some advantages over CO2, but the costs may be higher. Nitrous oxide may be used as an alternative anaesthetic gas for stunning of pigs.

Optimum Concentration and Conditions for Chemical (Carbon Dioxide) Stunning

1. Concentration of carbon dioxide in air: 80-95 per cent.
2. Period of exposure: 45 seconds.
3. Bleeding time: within 30 seconds, after leaving the gas chamber.
4. Average recovery (regain consciousness) time: 90 seconds.

Factors Affecting Chemical (Carbon Dioxide) Stunning

1. If the concentration of carbon dioxide is too low, the pigs will not be properly stunned.

2. If the concentration of carbon dioxide is too high, there is a tendency for the pigs to become stiff, show reflex muscular activity and bleed poorly.
3. If the exposure of pigs to carbon dioxide is too long, superficial congestion of the skin occurs, and when pigs are scalded the skin is bluish in colour.

Electrical Stunning

This method of stunning is well suited for pigs, sheep, goats, poultry and ostriches. The use of electrical stunning in cattle or other large animal species is in development, but if not properly applied it may lead to excessive haemorrhage in the muscles or spinal fractures. In this method, an alternating current (A.C.) is passed through the brain/heart of the animal with an instrument which resembles a pair of tongs. When electric current is passed through the animal body, it causes massive polarization of neurons in the brain, resulting in an epileptic seizure.

Electrical stunning is most widely used for pigs but also satisfactory for sheep and calves. The efficiency of electrical stunning is not dependent on the individual factors of voltage, amperage or time, but on the total quantity of electrical energy supplied in the animal body. Electrical energy can be expressed in terms of 'watt-seconds' (Watt-Seconds = Voltage x Amperage x Time). If too high current is employed, carcase quality may be affected (muscle hemorrhages and broken bones). If too low voltage is used, animal may be paralyzed but still conscious of pain. There are three systems of electrical stunning depending on the application of electric current on different parts of the animal body.

1. **Head Only Electrical Stunning System:** A minimum electric current of 400 mA for pigs, 250 mA for sheep and lambs has been recommended to produce an effective stun, when the tongs are placed in ideal position. This system renders the animal unconscious in about 50 seconds. Head-only electrical stunning can be done either at low voltage or at high voltage as follows:
   - i. **Low-Voltage Electrical Stunning:** Using less than 150V is the most common type in operation in Britain. The terminal electrodes of tongs are applied to the sides of the head of the animal. In order to create better contact, the electrodes are immersed in a saline solution before use.
   - ii. **High-Voltage Electrical Stunning:** This is more effective than low-voltage electrical stunning. High voltage (300V or above) electric current is passed for at least 3 seconds.
2. **Head-to-Brisket Electrical Stunning System:** In this system, an electric current of 2.5 ampere at 550 Volt is applied between the nose and neck yoke for 3 seconds to stun the animal with an additional current applied between the neck and brisket to produce cardiac arrest.
3. **Head-to-Back/Leg Electrical Stunning System:** In this system, a high voltage head-only electrical stunning is carried out with simultaneous application of additional electricity to the head and back/leg. In this system, the brain is anesthetized and the heart put into arrest, thus cutting

the blood supply to the brain, which suffers death before the anesthesia ends. This system renders the animal unconscious in about 23 seconds.

Missed Shock

A condition produced due to application of poorly positioned electrical electrodes on the body of animal during electrical stunning. The condition is characterized by paralysis in the animal but it is fully conscious.

Assessment of Unconsciousness/Insensibility of Animals at Slaughter

The assessment of unconsciousness at slaughter is important to determine the efficiency of stunning system and to implement the animal welfare regulations, as the bleeding and dressing operations must not begin until complete stunning has been achieved. After percussive stunning the fully unconscious cattle exhibits the typical signs of immediate collapse of the animal followed by tonic (that is continuous) spasm lasting about 10-15 seconds followed by slow clonic (that is repeated) movements of hind limbs and finally vigorous hind limbs movements. In pigs the tonic phase lasts 3-5 seconds. There is a cessation in the normal rhythmic breathing, bulging of eyeballs with a fixed gaze. After carbon dioxide anesthesia the fully unconscious pig exhibits the typical signs of increased respiration rate are followed by slow respiratory movements and finally dyspnoea, absence of corneal and palpebral reflexes, extreme muscle flaccidity and relaxation of jaw and limbs.

In electrical stunning of sheep, goats, pigs and ostriches, a "grand mal" seizure (a severe form of epilepsy) is induced which causes instant unconsciousness in them. Gasping is a "grand mal" seizure, that characterized by paroxysmal transient disturbances of the electrical activity of the brain. This results in periodic recurrent convulsions of the body or "epileptic fit". After stunning of animal by electrical method, the fully unconscious animal exhibits the typical signs of immediate collapse with flexion followed by rigid extension of the limbs, opisthotonous posture like arching back of the neck and spine, downward rolling of the eyeballs with tonic muscular spasm followed by clonic violent muscular spasm and finally muscle flaccidity. The tonic phase lasts about 10-25 seconds and clonic phase about 15-45 seconds, in both pigs and sheep. The stunned animal showing vocalisation such as squeal, moo or bellow, and indicates improper stunning. Vocalising animal still feels pain. Electrically stunned poultry head should hang straight down. Improperly stunned birds show a strong righting reflex and raise their heads.

Effect of Stunning on Meat/Carcase

1. In penetrative percussive stunning especially by captive bolt, the blood splashing may occur in muscles of diaphragm, intercostal muscles, heart and abdominal wall. Blood splashing occurs mainly due to delay between stunning and sticking.
2. High-voltage head only electrical stunning in pigs sometime causes petechial hemorrhages throughout the loin.
3. Head-to-back/leg electrical stunning produces a very low incidence of blood splashing but petechiae may occur in connective tissues and fat.

4. High-voltage electrical stunning also results in the fracture of bones particularly scapula and vertebrae, with associated hemorrhage into surrounding tissues. The fracture occurs probably due to the force of the tonic convulsion induced during and immediately after stunning.

Muscular Hemorrhages (Blood Splashing)

Muscular hemorrhages are also known as splashing or blood splashing. Muscular haemorrhage is usually associated with penetrative percussive stunning (especially by captive bolt pistol) resulting from long interval between stunning and sticking and/or inadequate penetration of bolt. In head-to-back/leg electrical stunning, a very low incidence of blood splashing in the form of petechiae may occur in connective tissues and fat. Muscular hemorrhages may be seen in various locations of the animal body:

1. Large intestine of sheep
2. Endocardium of calves
3. On the neck and longisimus dorsi muscle in cattle.
4. Beneath the splenic capsule of cattle.
5. Skeletal muscles
6. Muscular portion of diaphragm
7. Inner aspect of thoracic and abdominal wall.

Causes of Splashing

1. Delay between stunning and sticking
2. Increased blood pressure arising violent muscular contraction at stunning
3. Excitement of the animal before slaughter
4. Long journey or pre-slaughter stress
5. Hot weather

Among above causes, 1 and 2 are the main causes of splashing.

Pithing

The stunned animals are sometimes pithed before bleeding by inserting a long thin rod or closely coiled wire into the hole made by the penetrating bolt of the pistol. The insertion of rod destroys the motor centres of the brain, so that reflex muscular action does not occur at sticking, thus avoiding injury to operatives and speeding carcase dressing. Pithing was used with the bed system of carcase dressing. Pithing is very unhygienic and time taking processes. It is contraindicated in those countries where bovine spongiform encephalopathy is known to exist because of risk of contamination of meat and infection of operatives/handlers.

Sticking/Bleeding

Sticking is the process of cutting the throat of slaughter animal using a very sharp knife to sever the major blood vessels in its neck/chest that supply the brain, ensuring rapid blood loss and therefore death. Animal should be bled immediately after stunning to ensure proper bleeding. Bleeding may be carried out by severance of all the neck blood vessels with one transverse cut on the ventral aspect of the neck. The rail can be used for efficient bleeding of slaughtered animals as it permits a centralized collection of blood. The head of the slaughtered animal is immediately removed from its body after complete bleeding. There are two main methods of bleeding cattle.

1. Bilateral severance of the carotid arteries and jugular veins by making an incision across the throat region caudal to the larynx as in ritual slaughter.
2. Making an incision in the jugular furrow at the base of the neck, the knife being directed towards the entrance of the chest to sever the brachiocephalic trunk and anterior vena cava.

Bleeding should continue for 6 minutes. The average yield of blood in cattle slaughter is 13.6 kg. About 58.3 per cent of blood is yielded in the first 30 seconds after sticking, 76.6 per cent after 1 minute and 90 per cent after 3 minutes of sticking. The knife use for slaughter should be sharp. A blunt knife will prolong the incision and the cut ends of the blood vessels will be damaged. This may cause premature clotting and blockage of the blood vessels, delaying bleeding out and prolonging the onset of unconsciousness. Incisions should be made swift and precise. In poultry, sheep, goats and ostriches, the throat is cut behind the jaw.

Oversticking/Back Bleeding

If the pleura are punctured during sticking, blood may be aspired into the thoracic cavity and adhere to the parietal pleura, particularly along the posterior edges of the ribs. This contamination is known a back bleeding or oversticking, if not washed immediately, may necessitate striping of the pleura. To prevent this contamination, care must be taken to avoid the passing of the knife too far towards the chest. In cattle, the blood cannot infiltrate between the pleura and contiguous chest wall, this may occasionally occur in pigs but over a small area immediately posterior to the first rib.

Electrical Stimulation

Directly after bleeding, clamps are attached to the ear, cheek or throat of the carcase and an appropriate current passed through for 50 seconds from a stimulation unit. After completion of the cycle, the cables are removed and prepared for the next animal. The main objectives of electrical stimulation are as follows:

1. To accelerate tenderness and reduce aging times.
2. To improve neck cleanliness in carcase.
3. To improve colour and flavour.
4. To allow rapid chilling.
5. To avoid cold and thaw shortening.
6. To allow for hot boning on the dressing line since cut can be chilled rapidly.

Weasand Rodding

Weasand rodding is a process of separation of oesophagus from trachea and surrounding tissues using a metal rod, and tying of the oesophagus with the rubber ring using stainless steel instrument (rodder). The process is usually applied in cattle slaughter to prevent the spread of gastrointestinal contents on the carcase and organs in an attempt to prevent the contamination.

Carcase Dressing

Carcase dressing is a process of removal of hide, visceral organs, head, tail, extremities (part of limb) *etc.* of a slaughtered animal to make it fit for human consumption.

Principles of Carcases Dressing

The important principles of carcases dressing are as follows:

1. Carful dehiding of the slaughtered animal without causing contamination of the carcase and organs.
2. Careful evisceration of the slaughtered animal to avoid piercing of the gastrointestinal tract and prevention of contamination of the carcase and organs during evisceration.
3. Careful removal of organs such as udder, uterus and urinary bladder of the slaughtered animal without causing contamination of the carcase by the contents of these organs.

Method of Carcase Dressing

The practice of air inflation in the carcases for removal of skin or other purposes is traditionally practiced with hygienic precautions in some countries. However, this practice may cause contamination of the carcases, therefore, it should be discouraged. The feet are removed before skinning the carcase. The hind feet are removed by skinning the area above and below the place where the leg is to be cut and thus removed without contacting the hide. The horns and front feet are removed in such a manner as to avoid contamination of the carcase.

During removal of the feet and the hide, the skin should be cut from the inside-out to prevent carcase contamination with hair and dirt, except for the necessary starting cuts. Avoid carrying contamination from one head to another. The knife used to begin skinning operations must be adequately rinsed and sanitized prior to reuse. Skinning should begin at the hind shanks after foot removal and proceed downward, reflecting the hide away from the carcase.

The oesophagus should be effectively tied before evisceration. The process of separation of oesophagus from trachea and surrounding tissues using a metal rod and tying of oesophagus with a thread is known as rodding. Rodding is carried out in order prevent spillage of gastrointestinal contents and thereby contamination of interior and exterior surface of the carcase. Bunging (sealing the rectum) is also important to prevent faecal contamination of carcase. In bunging, a cut is made around the rectum to free it from the carcase, and then it is tied off to prevent spillage of fecal material.

Washing of Carcase

After dressing, carcases are subjected to washing with potable water to remove the blood from the surface of the carcase; bone dusts (produced due to splitting of carcase) and other extraneous matter adhere to the carcase. Spray washing is valuable for cleaning purposes. Washing the carcase is essential to reduce the level of contamination in the carcase/meat.

Slaughter of Cattle/Buffalo

Pre-slaughter ante-mortem inspection of animal is essential in order to detect recently developed abnormalities/diseases and to prevent contamination

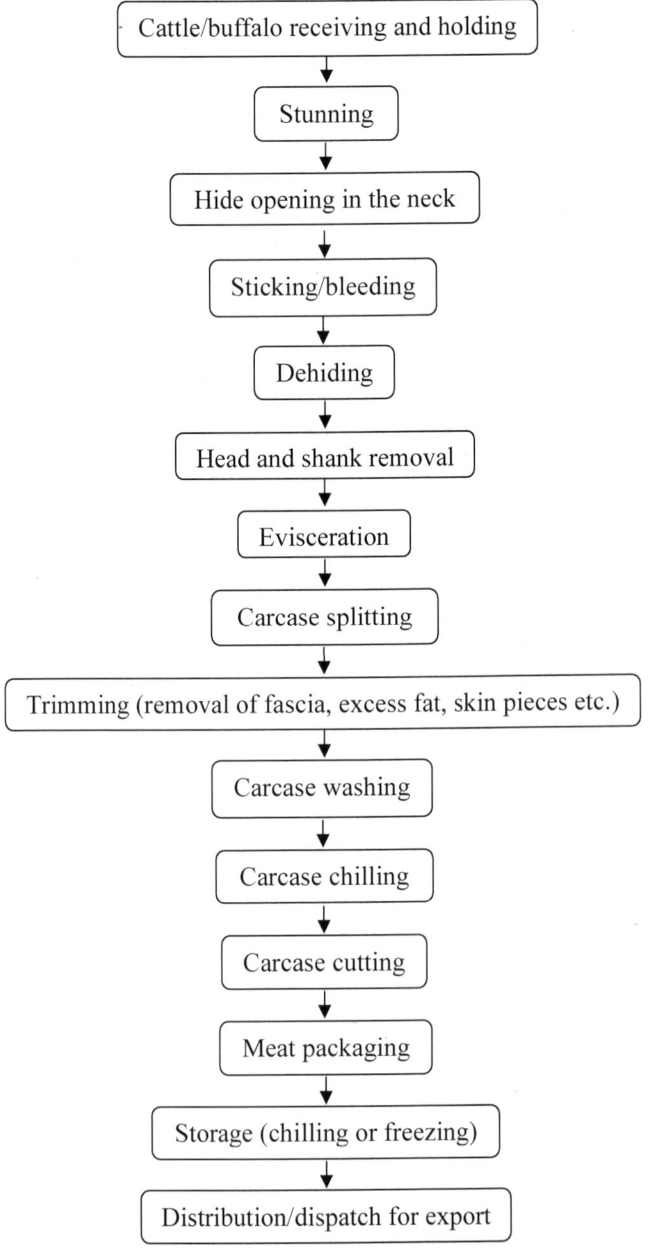

Figure 16.7. Flowchart of Cattle/Buffalo Slaughter.

in production chain. Stunning of animal is carried out in a stunning pen. After stunning, an incision is made in the neck region and hide is opened before sticking. After sticking/bleeding, the carcase can be dressed on-the-rail. After bleeding the dehiding may be carried out either by manually or mechanically using hide-puller. The neck is cut after removing the hide and placed for post-mortem inspection.

Evisceration should be carried out with great precautions to avoid puncture of gastrointestinal tract and contamination from its contents. The thoracic cavity including organs in it and eviscerated organs are subjected to post-mortem inspection. The eviscerated and dressed carcase is split into two halves along the midline of the body using carcase splitting saw. The purpose of splitting of the carcase is to facilitate its efficient cooling and easy handling. Undesired parts such as fascia, excess fat, remains of skin pieces *etc.* are trimmed from the carcase with the object of proper dressing of carcase and avoid contamination from skin.

The dressed carcase is well washed using potable water to remove blood contents and bone dusts arising from bone cutting and carcase splitting and any type of contaminants like faeces, soiled materials *etc.* After washing of the carcase the meat or offal may require chemical or microbiological examination if necessary. The carcase is kept in a chilling hall for 24 hours, during which rigor mortis develops in the carcase. This is helpful in maintaining the quality of meat and also for making desired pieces of meat.

Slaughter of Pig

Ante-mortem inspection of pigs must be repeated immediately prior to their slaughter. Thereafter, stunning of pigs is carried out in order to their humane slaughter. Stunning of pigs can be done by exposure to carbon dioxide gas or passing alternating current (AC) through the body of the animal. After stunning, the pigs are subjected to sticking/bleeding. The pig carcase is then transported to the tunnel on a rising rail so that the head is lower than the other parts of the body during the whole scalding process. Scalding is carried out for 6 minutes in a scalding tank containing water maintained at 61-64°C. Scalding may be carried out in a vertical scalding tank. This tank is double-walled tunnel in which steam generated from a water bath in its bottom is blown over the carcases and through a ventilator located over the condenser. The cooling water from the condenser in the tunnel is used to flush the pig carcases during the dehairing processes. Vertical scalding improves the bacteriological standard of the pig meat. Vertical scalding also reduce the incidence of pale soft exudative(PSE) muscle due to the fact that vertical scalding does not produce a rise in body temperature to above 41°C as in normal scalding operations.

Dehairing (hair scarping) may be carried out by manually operated equipment or mechanical dehairing machines. After scalding the carcases are placed on-the-rail with the help of gambrel. Singeing is carried out to burn out the hair stumps. A pre-evisceration wash is essential to remove burnt hair or other tissues. The next process is bunging (a technique of sealing of rectum using threads, tapes *etc.*) to prevent faecal contamination of carcase and premises.

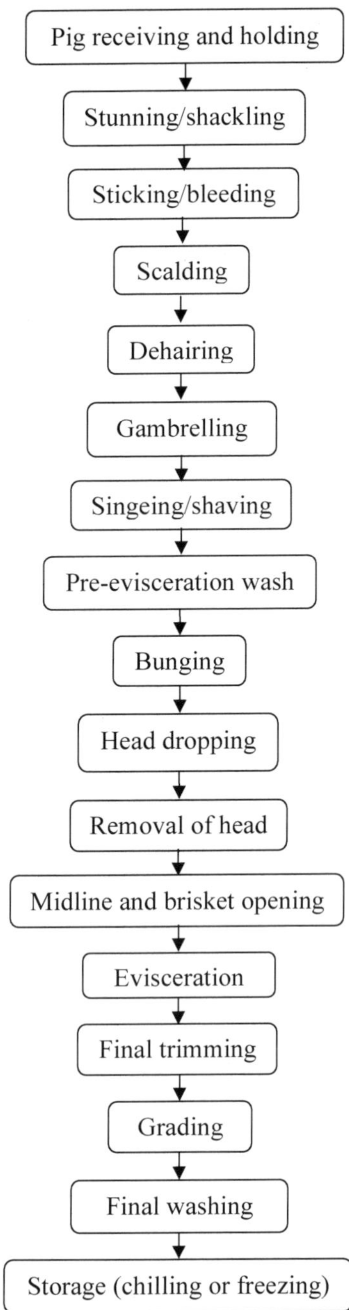

Figure 16.8. Flowchart of Pig Slaughter.

After bunging the next steps are head dropping and removal of head. An opening is made in the midline and brisket after removal of head. The evisceration is carried out very carefully to avoid unnecessary incision or puncture and contamination from gastrointestinal contents. Unwanted parts like fascia, part of skin, excess fat *etc.* from the carcase are trimmed. Thereafter, the carcases are finally washed. The carcases are kept in a chilling hall for 24 hours for development of the rigor mortis as well as setting of the carcases.

Slaughter of Poultry

A pre-slaughter inspection of poultry is essential in order to avoid their sufferings and effective implementation of Regulations for Prevention of Cruelty to Animals. Before slaughter the poultries should be subjected to ante-mortem inspection. After removal from the crate or module the birds should be checked to see whether they are suffering from a condition which may cause them pain or distress. The birds suffering from these conditions should be killed immediately either by dislocation of the neck in case of small birds or by stunning knife in case of small and large birds.

Wing flapping can be reduced by applying the head down the bird's body or to briefly hold on to its leg. The poultries are shackled and suspended from the conveyor before their stunning. A strip of smooth plastic sheeting installed parallel to the conveyor line, along which the breast of the suspended birds rub also has a quieting effect. Runts (growing/small) and diseased birds, which may have been inadvertently crated, should be killed and not shackled.

Electrical stunning is the most common method of stunning under commercial conditions. Electrical stunning is usually carried out in an electrically charged water bath by dragging the head of the birds through water in which an electrode is submerged. The shackles of the killing line simultaneously touch an earth electrode,

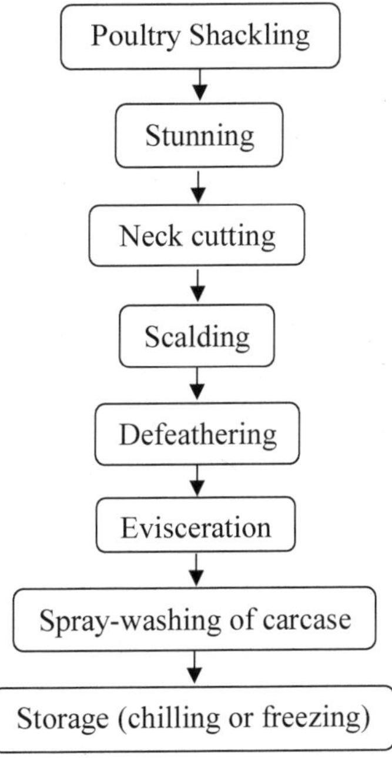

Figure 16.9. Flowchart of Poultry Slaughter.

causing an electric current run throughout the body of the bird. A minimum current of 120 mA is required to induce cardiac arrest in birds. The efficiency of stunning can be improved by adding salt in the water. Gas stunning of birds may be carried out by using 90 per cent argon or other inert gases; and a mixture of 25-30 per cent carbon dioxide and 60 per cent argon or other inert gases in air.

The neck of the bird should be cut within 15 seconds after emerging of bird from stunner. Birds are killed manually by passing a knife across the side of the neck at the base of the bird's head, which should sever jugular veins and carotid arteries. The bleeding should be allowed for 60-90 seconds.

After bleeding, the birds are scalded either by immersion in hot water or by spray scalding. The scalding may be carried out either by soft scalding or hard scalding methods. For the fresh and chilled market meat, a soft scalding is preferred which is carried out at 50-51°C for 3.5 minutes. Soft scalding does not damage skin. Soft scalding also prevents discolouration and drying (barking) on air chilling. For the frozen market, a hard scalding is preferred which is carried out at 56-58 °C for 2-2.5 minutes. Hard scalding facilitates defeathering process. However, hard scalding causes discolouration and drying (barking) on air chilling. Birds scalded by spray scalding is more hygienic but more expensive.

Defeathering can be done either by manually or by using feature plucking machine. For mechanical defeathering the line of carcases passes through machines which have a series of rotating drums fixed with rubber fingers which remove the feathers. The feathers drop into a trench below with flowing water carrying to the area of rendering plant.

The next process is evisceration, usually carried out mechanically but manual evisceration is still in practice. For mechanical evisceration the conveyor line of birds passes through an automatic venting machine, which pushes up on the abdomen slightly while a blade cuts open the carcase without disturbing the bowel. The next machine or part of the process scoops into the cavity and pulls out the unbroken viscera for inspection. The giblets (heart, liver and gizzard) are often pooled and inserted into the body of the chicken. Giblets are more frequently contaminated with *Salmonella* than other sample sites and chickens which contain them are more often contaminated than those without giblets.

Finally the carcase is subjected to spray washing using the water containing 40 ppm free chlorine. This gives visually clean carcase and also decreases carcase contamination. Thereafter, the carcases are stored under refrigeration.

Religious (Ritual) Method of Slaughter

Most developed and many developing countries of the world require by law an animal to be rendered unconscious before it is slaughtered. This is in order to ensure that the animal does not suffer pain during slaughter. However, exceptions are made for the Jewish (Kosher), Muslim (Halal) and Sikh (Jhatka) method of slaughter of livestock. Here stunning generally is not allowed and the animal is bled directly using a sharp knife to cut the throat and sever the main blood vessels. This results in sudden and massive loss of blood with loss of consciousness and death. However, many authorities consider that religious slaughter can be very unsatisfactory and that the animal may not be rendered unconscious and suffer considerable discomfort and pain in the slaughter process.

In religious method of slaughter, the animals should be securely restrained, particularly the head and neck, before cutting the throat. Movement results in a poor cut, improper bleeding, slow loss of consciousness and feeling of pain. This has serious implications for animal welfare. Animals should not be shackled and hoisted before bleeding. This causes them severe discomfort and stress, which is against the animal welfare regulations. Hoisting should be done only after the animal has lost consciousness. Restraining equipment should be comfortable for the animal.

Jewish Method of Slaughter (Shechita)

The Jewish method of slaughter is controlled in Britain by the Jewish Board of Shechita. In this method, a swift cut is made across the neck with a very sharp knife. This severs the following content of the neck:

1. Skin
2. Underlying muscles
3. Trachea
4. Oesophagus
5. Jugular veins
6. Carotid arteries.

Principles of Shechita

There are five principles of Shechita. As per these principles the neck of the animal should be cut without pause, pressure, stabbing, slanting and tearing. If the knife receives any nick during the operation, the slaughtered animal is considered unfit for Jews consumption.

Some Common Terms Associated With the Jewish Method of Slaughter

1. **Shechita:** The Jewish method of slaughter is known as Shechita.
2. **Shochet:** The person who performs Jewish slaughter is known as shochet. Shochet is also known as cutter.
3. **Shomer:** The person who assists the shochet is known as shomer. Shomer is also known as sealer. Shomer is responsible for putting the kosher mark on the brisket and on edible offal.
4. **Chalef:** It is a razor-sharp knife, used for slaughter of the animals in the Jewish method of slaughter. Shochet slaughters the fully conscious animal with single, deliberate and swift action of chalef.
5. **Searching:** It is somewhat a type of post-mortem inspection, carried out by the shochet. Besides performing act of slaughter, the shochet offers prayer and carries out a post-mortem inspection by making an incision posterior to the xephoid process and inserting the arm to detect any adhesions in the thoracic cavity.

6. **Porging:** It is the removal of larger blood vessels from meat/carcase particularly from forelimb and hind limb as it may contain blood according to Jewish law.
7. **Kosher:** The carcase/meat which fulfills the requirements of Jewish law and fit for Jews consumption is known as kosher. For production of kosher meat, the animal must be alive and healthy, have suffered no injury and not stunned at slaughter.
8. **Terefa:** The carcase/meat which does not fulfill the requirements of Jewish law or the carcase showing any disease lesion *i.e.* unfit for Jews consumption is known as terefa.

Muslim Method of Slaughter (Halal)

Unlike Shechita, the Muslim method of slaughter is not controlled by a Central Board, but is overseen by the local Islamic authority. This method is similar to the Jewish method but there is no searching or porging after slaughter of the animal.

Sikh Method of Slaughter (Jhatka)

In this method, sheep or goat is decapitated by one stroke of a special sword.

Miscellaneous Method of Slaughter

There are some uncommon methods of slaughter of animals throughout the world. Neck-stab (Evernazione) method of slaughter of animal is one of the methods.

Neck-Stab Method of Slaughter (Evernazione)

In this method, the cattle were traditionally slaughtered in Spain, part of Italy, Mexico and some South American countries, but now this technique is banned. In this method, a short double edged knife (puntilla) is plunged into the occipito-atlantal space at the nape of the neck, severing the medulla oblongata, which immobilizes the animal without inducing insensibility/unconsciousness.

Supervision of Hygiene during Carcase Dressing

During dressing, the carcase is exposed to contamination from various sources and processes:

1. **Environment:** Abattoir environment may have a variety of microorganisms like bacteria, fungi, yeasts *etc.*, if insanitary and unhygienic conditions and practices are prevailing in the abattoir. Therefore, the slaughterhouse environment should be free from such contaminants for quality meat production.
2. **Equipment:** The equipment used for slaughtering, dressing and meat inspection may be exposed to contamination. Contaminated equipment may have *Salmonella* on its surface in a range of 0-270 per cm^2 or more depending on its regular cleaning and sanitation.
3. **Animal Hygiene:** Animal hide may be a potential source of contamination. Heavily contaminated hide may contain up to 3×10^6 bacteria per cm^2 or more.

4. **Personnel Hygiene:** Slaughterhouse personnel such as animal handlers, butchers, meat handlers *etc.* should be regularly monitored for their personal hygiene as well as hygiene adopted by them during animal slaughter, meat handling and other processes.
5. **Gastrointestinal Contents:** Gastrointestinal contents have the heaviest load of microorganisms. Faeces contain up to 9.0×10^7 bacteria per gram, and various numbers of yeast and mould.
6. **Carcase Dressing Techniques:** During meat inspection, it is an important duty of the meat inspector to ensure the hygienic dressing of the carcase. During cutting into the hide and exposure of the carcase, the external surface of the hide should not contact the carcase. The viscera are not accidentally opened during the dressing procedures or during evisceration. If a carcase or part is contaminated with faeces or visceral contents, such areas should be removed by trimming.
7. **Implementation of HACCP:** The implementation of a Hazard Analysis Critical Control Point (HACCP) programme in a meat plant can be helpful to maintain high standards of slaughter and dressing hygiene based on an assessment of the risks to human and animal health.

17

Examination of Lymph Nodes

Examination of lymph nodes in meat inspection is very important. The meat inspector should be well versed with the regional lymph nodes of various species of food animals. The following parameters are considered during lymph node examination:

1. **Size:** The size of the lymph nodes greatly varies in different animal species and different age groups within the same species of animal. The size of the lymph nodes varies from pin head to 20 cm. The size of the lymph nodes is relatively larger in young animals than the adults.
2. **Shape:** The shape of the lymph nodes is variable. Lymph nodes are generally oval and somewhat compressed. The shape of the lymph nodes may also be circular, elliptical *etc.*
3. **Structure:** The structure of the lymph nodes is also variable which may be compact or lobulated. The lymph nodes of pigs are lobulated.
4. **Colour:** The colour of lymph nodes is variable which may be whitish to greyish-blue or almost back in colour. The lymph nodes of pigs are almost white in colour with the exception of those of head and neck which are reddish in colour. The mesenteric lymph nodes of ox are black in colour.
5. **Consistency:** The consistency of the lymph nodes varies in different parts of the body. The consistency of the lymph nodes is usually firm rather than soft. The lymph nodes of abdomen are generally softer than those of the thorax. Exposure of lymph nodes to an irritant causes their enlargement, congestion, softening of the tissues *etc.*

Conditions Observed in Lymph Nodes during Post-Mortem Inspection

Lymphadenitis

Lymphadenitis may be acute or chronic. It is a sequel in various infectious diseases. It must be distinguished from hyperplasia of the lymph nodes which is not associated with infections. Acute lymphadenitis is characterized by swelling, congestion, mobility and oedema of the lymph node which is painful and softer than normal. Chronic lymphadenitis is characterized by development of fibrous connective tissues of the node, enlargement and induration, sometimes oedema and congestion. Chronic lymphadenitis is seen in Johne's disease in which the mesenteric lymph nodes are most commonly affected.

Lymphadenitis is a characteristic lesion in many diseases such as tuberculosis, glanders, actinobacillosis, Johne's disease, epizootic lymphangitis, bovine lymphangitis, strangles in horses, caseous lymphadenitis in sheep, brucellosis in guinea pigs, tularemia in rodents, salmon poisoning in dogs *etc.*

Lymphoedema

Lymphoedema occurs due to lymphatic obstructions, injury, inflammation, tumours or congenital lymphatic defect (absence of lymphatics).

Caseation

The caseation in the lymph node is the necrosis with conversion of damaged tissue into a soft cheesy substance, which occurs in *Corynebacterium pseudotuberculosis* infection, glanders, actinobacillosis, and in the chronic infection like tuberculosis and Johne's disease.

Calcification

The calcification of lymph node is the formation of lime salts from pus, necrosed tissues *etc.* Caseation of lymph node is followed by calcification. The calcification in the lymph nodes most often result from prior granulomatous infections, especially tuberculosis, Johne's disease, actinobacillosis, glanders and histoplasmosis.

Hyperplasia

Hyperplasia of lymph nodes is a pathological condition that occurs due to an unusual reaction or chronic type of irritants. Hyperplasia of lymph nodes is characterized by enlargement due to the formation and growth of new cells, whitish-gray colouration and firm in consistency but not fibrosis or calcification.

Atrophy

Atrophy of lymph nodes is seen in old animals, especially those in poor condition. Atrophy of lymph nodes is characterized by very small size and dark brown pigmentation.

Neoplasm

Primary benign neoplasm of lymph nodes is uncommon. Secondary tumours in lymph nodes include carcinoma, malignant melanoma and occasionally sarcoma, which invade the lymph vessels.

Emphysema

In pigs, the emphysema of the mesenteric lymph nodes is seen in association with intestinal emphysema. In cattle, the emphysema of the bronchial lymph nodes is commonly seen in pulmonary interstitial emphysema. The emphysema of lymph nodes is characterized by enlargement, softness and puffy in nature.

Pigmentation

Pigmentation of lymph nodes may be exogenous or endogenous. Exogenous pigmentation of lymph nodes is most commonly found in the pulmonary and mesenteric lymph nodes. Anthracosis (deposition of coal dust) in the bronchial lymph nodes is seen in cattle and sheep grazing in urban areas and collieries and in dogs live in industrial areas. Endogenous pigmentation is most commonly associated with hemosiderin which is found in lymph nodes draining areas where haemorrhage has occurred. Xanthosis occurs in bovine lymph nodes and melanosis in sheep lymph nodes.

18

Pathology of the Food Animals

Developmental Abnormalities

The developmental abnormalities may include aplasia, anaplasia, dysplasia, hypoplasia, hyperplasia, metaplasia, atrophy, hypertrophy *etc.*

Aplasia

Aplasia is the absence of an organ in the body. Aplasia may be due to genetic factors, intrauterine infections and intake of teratogenic drugs.

Anaplasia

Anaplasia is the reversion of cells to a more primitive type and is a precursor of neoplasia. Anaplasia is characterized by occurrence of pleomorphism in the cells with abnormal constituents.

Dysplasia

Dysplasia is a retrogressive change in mature tissues. The etiology is same as aplasia, including chronic irritation. It mainly affects epithelial cells.

Hypoplasia

Hypoplasia is the failure of an organ to develop to its normal size due to decrease in size and number of cell. The etiology is same as aplasia.

Hyperplasia

Hyperplasia is an increase in the overall number of cells in an organ or tissues.

Metaplasia

Metaplasia is the change of one type of cell into another type. It is occasionally encountered in meat inspection.

Atrophy

Atrophy is the shrinking and reduction in size of an already fully developed organ. It may occur due to decrease in size and number of its cells as a result of necrosis and autolysis. It may be caused due to functional failure of thymus in old age, regression of uterus following parturition and regression of mammary gland after lactation, malnutrition, inadequate blood and nerve supply to the tissues, aging, and pressure from tumour. Pressure atrophy may occur as a result of inadequate blood and nutrient supplies to the tissues/organs, *e.g.*, hydronephrosis of kidney and atrophic rhinitis in pigs.

Hypertrophy

Hypertrophy is an increase in size of the cells and consequently size of an organ. Hypertrophy and hyperplasia usually coexist in the same organ of the body and occur due to increased functional demand by the organ.

Judgement of Developmental Abnormalities

Affected part/organ is rejected in case of localized conditions, whereas, total condemnation is recommended in generalised conditions like hypertrophy and hyperplasia.

Pathological Conditions/Lesions

During ante-mortem and post-mortem inspections, the following pathological conditions/lesions can be encountered:

Emaciation

Emaciation is a profound and marked state of constitutional disorder; general ill health and malnutrition. Emaciation may be due to chronic diseases like tuberculosis and Johne's disease and parasitic infestations. Advanced and generalised atrophy of skeletal muscle and certain organs with associated serous atrophy of fat deposits often associated with oedema. The animal may also be anaemic. The kidney fat diminishes. The heart may lose all of its fatty tissue and the ventricles also tend to become thin. The condemnation of the whole carcase is recommended in case of emaciation.

Immaturity

Immaturity in itself is not a pathological condition but the meat obtained from immature animal is not desirable. Meat of very young animals is less valuable due to high water content, very little amount of fat and more bone than meat. Therefore, the calf, lamb, kid, pig or any other animal below 21 days old should not be slaughtered. The sign of immaturity includes soft and watery meat with greyish-pink colour and weak muscle development (jelly between muscles).

Fever

Fever is an abnormal and high central body temperature. It is a cardinal sign of acute inflammation caused by infectious agent. Carcase from animal suffering from fever shows congestion, high microbial count, more prone to its spoilage and also has risk of spreading infection in the consumers. Therefore, total condemnation of the carcase is recommended.

Imperfect Bleeding

Imperfect bleeding may be due to fever, stress, moribund states, ineffective stunning or an ineffective bleeding cut, electrocution, lightning stroke, suffocation *etc.* The poorly bled carcase appears dark red colour due to blood contents inside the blood vessels. The liver and other organs may be dark purple-red in colour. The efficiency of carcase bleeding can be determined by a test known as malachite green test. The meat from such a carcase are more prone to spoilage, therefore poorly bled carcases are condemned.

Icterus

Icterus is the yellow discolouration of tissues especially in white tissue like mucous membranes, serous surfaces, cartilage, and fat as well as the endothelial lining of blood vessels due to an excess of bilirubin in the blood. Carcases are condemned due to aesthetic reasons. Icterus and carotene pigmentation must be differentiated using Rimmington and Fowrie test or other tests.

Neoplasm (Tumour)

The term neoplasm is derived from a Latin word meaning "new growth" or "new formation". A neoplasm is an abnormal mass of tissues, the growth of which exceeds and is uncoordinated with that of normal tissues. It persists in the same excessive manner after the cessation of the stimuli which evoked the cause. Tumours may be either malignant or benign. Malignant tumours grow quickly and expansively and infiltrate the surrounding tissues, usually causing severe damage and death of affected individual. Malignant tumours may undergo metastasis. Benign tumours like warts usually stay in one area and may disappear after time. There is local condemnation of tumour by trimming of affected body part or removal of affected organ. There is total condemnation of carcase and organs if the tumours are malignant and wide spread that is metastatic.

Papule

Papule is a small, circumscribed, raised solid elevated skin lesion.

Acne

Acne is also known as folliculitis or pyoderma. It is a pustular infection of hair follicles and sebaceous glands with a surrounding acute inflammation. It is usually caused by staphylococci.

Blister

Blister is also known as bulla or bleb, usually more than 5 mm in diameter. On skin, it contains clear fluid, while a pulmonary bleb contains air.

Abscess

Abscess is a localised, encapsulated collection of pus in a cavity formed by disintegrating tissues. The size of abscess may vary from microscopic to a large size. A single abscess may be removed if there is no chance of further spread of infection or contamination with pus to the rest of the carcase. In case of multiple abscesses in various organs, the total condemnation of carcase is recommended.

Injection Abscess

Subcutaneous, intramuscular or intravenous injection of antibiotic, hormones and other drugs may lead to abscess formation, localized oedema and necrosis. The most common regions of injection abscess are neck, shoulder and gluteal muscles.

Cyst

Cyst is a close cavity containing clear fluid, pus or semisolid material, which is usually 1-5 cm in diameter. Cysts may be of several types. Congenital cysts are frequently encountered in the kidneys of pigs and sheep. Dermoid cysts are of neoplastic in origin which contain sebaceous and sweat gland, hair follicle, bone, teeth *etc.* Epidermal inclusion cysts are small round nodules in the skin. Haematocycts are localized collection of extravagated blood in the organ or tissues. Hydatid cysts are larval stage of the parasite like *Echinococcus* spp.

Furuncle

Furuncle refers to the large size of abscess in the dermis.

Carbuncle

Carbuncle is a deep-seated skin abscess, caused commonly by staphylococci.

Vesicle

Vesicle is a small sac (usually less than 5 mm in diameter) on the skin, containing liquid. It represents an inflammatory condition which may be caused by injury or infection, usually by viruses, *e.g.*, vesicular stomatitis, foot and mouth disease and swine vesicular disease.

Ulcer

Ulcer is the localized destruction of an epithelial surface with exposure of underlying tissues. It may occur on the skin and organs due to bacterial or viral infection.

Bruise

Bruise is a condition which involves release of blood from ruptured blood vessels into surrounding tissues. Bruise occurs most commonly on hip area in cattle, ham area in pigs and hindleg in sheep. Bruising causes economic losses which may incur in the form of part rejection through trimming as well as rejection of the whole carcase. In addition to this, bruising may cause damage of hides and skins. Bruising may also cause spoilage of the meat. Bruising in the carcase also indicates the animal welfare situation that had been adopted while the animal was alive.

Bruise is one of the most common conditions seen during meat inspection and is a serious disadvantage in the meat trade. Bruising is the discolouration and actual haemorrhage at the site of injury. In the first 12 hours after injury the bruise is bright red, at 24 hours it is dark red, at 24–36 hours it loses its firmness and becomes watery, and at 3 or more days it is an orange-red colour and has a soapy feel. The degree of bruising may vary. Slight bruising requires little or no trimming, while extensive bruising requires condemnation of the whole carcase.

Necrosed Tissues

Necrosis is the death of tissues due to the action of lytic enzymes from lysosomes. Necrosis of tissue may occur due to chemical poisons, plant poison, toxins, bacteria (*Fusobacterium necrophorum*), viruses, physical damage, heat, burns, radiation *etc.* Necrosed tissues are paler than the normal and easily torn. There may be putrefactive odour in the necrosed tissues due to infection with *Clostridium* spp. Caseous necrosis is caused by *Mycobacterium tuberculosis* and *Corynebacterium pseudotuberculosis,* and characterized by softening of tissue structure, whitish, yellowish or greyish in colour and cheesy mass like appearance. Coagulative necrosis is caused by *Fusobacterium necrophorum* and *Corynebacterium pyogenes,* and characterized by denaturation of cellular protein and enzymes and whitish, yellowish or greyish in colour and swollen and soft tissues. Liquefactive necrosis is caused by *Fusobacterium necrophorum,* and characterized by liquefaction of tissues.

Gangrene

Gangrene is the death of body tissue (necrosis), generally in considerable mass, usually associated with loss of vascular supply and followed by bacterial invasion and putrefaction. Gangrene is the digested necrotic tissues by the bacterial actions. Primary or moist gangrene is caused by *Clostridium perfringens, Clostridium oedematiens* and *Clostridium septicum.* Secondary gangrene is caused by ischemia followed by digestion by saprophytic bacteria. Affected parts are rejected if gangrene is much localized and has no evidence of toxaemia. Carcase and offal is rejected if gangrene is extensive.

Intracellular Accumulations

Intracellular accumulations may result in calcification; pigmentation *etc.* which is discussed as follows:

Calcification

Calcification is the deposition of calcium salts, small amount of iron, magnesium and other minerals in the body tissues. It is associated with hypercalcaemia, hypervitaminosis D, secondary tumour formation and renal failure. Affected carcase may be partially or totally condemned, depending on situation.

Pigmentation

Abnormal colouration in the skin and other organs may occur due to deposition of certain type of pigments. Abnormal deposition of melanin is encountered mainly in the calf and ox. It is also seen in sheep and pig. In the ox, melanosis is commonly found in the lungs, liver and on the meninges and spinal cord. Black discolouration is also seen in the kidneys of adult cattle due to deposition of melanin and lipofuscin (an endogenous pigment closely related to melanin). Light yellowish-brown colouration in the kidneys is also encountered in healthy cattle due to depigmentation. Blackish colouration in the kidneys is seen in very young calves due to heavy deposition of bile pigment. Discolouration in the liver of sheep occurs due to deposition of lipofuscin.

In the pigs, the common melanotic deposits is seen in the belly fat or in the udder. Other common sites of deposition of melanin are pleura, peritoneum, fascia, bone and cartilage. Organs and other parts of the body affected with abnormal colouration are condemned due aesthetic point of view. Xanthosis is another type of pigmentation, commonly encountered in the animals. It is caused due to deposition of greyish-brown or bronze colour iron-free lipid granules in the heart of cattle. The muscles of masseter, tongue, diaphragm and limb are also affected.

Exogenous pigmentation is associated with the grazing of animals in proximity to certain industries. In, pneumoconiosis, the carbon or coal dusts are deposited in the lungs and associated lymph nodes. Endogenous pigments in animals include melanin, haematin, haemosiderin, bile pigments *etc.* Abnormal deposition of melanin is known as melanosis. This is mainly encountered in bovines and less commonly in sheep and pigs. In the ox, melanosis is commonly found in the lung, liver and meninges. There is partial condemnation in case of organs affected with melanosis, while total condemnation in generalized melanosis. Melanomas are malignant tumours which arise from melanocytes. Metastatic melanomas are totally condemned.

Icterus or jaundice is a pathological condition in which bile pigments are accumulated in the tissues, which are tinged yellow. This is caused due to reabsorption of bile pigments into the circulatory system. Icterus suspected carcase should be detained for 24 hours and then re-examined since the normal muscle enzymes may remove the yellow colour. If the pigmentation is slightly evident after 24 hours detention, the carcase may be safely released for consumption. This condition occurs in mild cases of obstructive jaundice, icterus associated with torsion of spleen, fracture *etc.* If the marked degree of icterus present even after a detention period of 24 hours, the total carcase is condemned.

Generalized Systemic Infections

Microorganisms present in the environment may gain entrance to the animal body by contact, ingestion, inhalation *etc.* Further spread of microorganisms depends on invasiveness, toxigenicity, ability to adhere mucosa, type of tissue invaded and host's body defense mechanism. Microorganisms are specific in establishing themselves and attacking a particular part of the body as per their nutritional demand. Some microorganisms localize at the original site of infection, while some are extremely selective in their tissue preference and predilection sites. Streptococci are less demanding and can invade and produce infection in any tissues. Pyogenic bacteria like *Streptococcus pyogenes* and *Staphylococcus aureus* produce local inflammation at the point of entry, while spirochaetes spread throughout the animal body. *Salmonella* spp. penetrates intestinal mucosa and reach to mesenteric lymph nodes before invading the bloodstream via the thoracic duct.

Septicaemia (Blood Poisoning)

Septicaemia is a condition where pathogenic organisms are present in the bloodstream. It may be caused by *Bacillus anthracis*, *Pasteurella* spp., *Actinomyces pyogenes*, *Streptococcus pyogenes* and *Staphylococcus aureus*, invasive strains of *Escherichia coli*, *Leptospira* spp., *Erysipelothrix* spp., *Corynebacterium pseudotuberculosis*, African swine fever virus, classic swine fever virus *etc.*

Septicaemia may be characterized by high fever, petechial haemorrhages and echymoses, lymph adenitis, fatty changes in heart, liver and kidneys, enlargement and softening of spleen, blood-stained serous exudate in thoracic and abdominal cavities *etc.* The meat from septicaemic animal may become dark, soft, alkaline and unpleasant odour due to acetone. Septicaemic carcase and organs are totally condemned.

Toxaemia

Toxaemia is the spreading of toxins produced by bacteria in the bloodstream. Lesions or changes depend on the type of toxins and its affinity for organs or systems in the body. Toxaemic carcase and organs are totally condemned.

Pyaemia

Pyaemia is the spreading of pus forming bacteria through the blood stream resulting in metastatic abscessation in other parts of the body. Pyaemia may develop in cattle due to penetration of sharp objects like nails, wire *etc.* in the myocardium. Pyaemia in the live animal is characterized by high fever, local area suppuration, swollen, hot and painful joints. The post-mortem findings reveal local suppuration, and distant haemorrhagic infarcts in the organs. Pyaemic carcase and organs are totally condemned.

Pathology of Blood and Haemopoetic Organs

Anaemia

The mucous membrane and body tissues are pallid and the blood is pale and watery in the animals suffering from anaemia. Depending on severity or cause of anaemia the carcase is condemned in case of an infectious disease or extreme anaemia. In less severe cases, conditional approval is provided after chemical and microbiological examinations are negative.

Oedema or Dropsy

Oedema is an excessive accumulation of fluids in the intercellular spaces and body cavities. There may be partial or total condemnation of carcase depending on the cause and degree of oedema. The carcases are passed in case of localized, minor forms of oedema and ascites and hydrothorax because these conditions are of less serous import than anasarca (oedema). The total condemnation is recommended in case of oedema involving the subcutaneous and connective tissues or any form of oedema accompanied by emaciation.

Wet Carcase Syndrome

A pathological condition in lambs in south Africa which is characterized by lesions in the form of layers of slimy translucent jelly-like encapsulated material on the buttocks, briskets and flanks. This condition is generally observed in animals of non-wool type, which come from regions where supplementary feeding with concentrates is necessary. Total condemnation is indicated in case of wet carcase syndrome.

Pathology of Gastrointestinal System

The organs of the gastrointestinal system may be affected due infection with bacteria, viruses, fungi and parasites. In addition to this, the system may be affected with chemicals and plant toxins; neoplasm *etc.* There may be injury to the gastrointestinal tract by items such as pieces of wires, nails, cloth, rope, rubber, plastic, coins *etc.* ingested along with foods. The ingested sharp objects may appear as pointed mainly in the ventral sac of the rumen, sometimes in the reticulum and occasionally in a tract through the diaphragm to the heart which causes traumatic reticulopericarditis.

Peritonitis

Peritonitis is an inflammatory condition of peritoneum which may be acute or chronic. This is characterized by haemorrhagic, granulomatous, purulent or non-purulent and local or widespread. Acute peritonitis is characterized by a straw-coloured exudate with fibrinous deposits in the peritoneal cavity with feeble adhesions. Acute peritonitis may result in toxaemia and death. Chronic peritonitis is characterized by yellowish cheese-like covering of the abdominal organs and a foul smell. In case of localized peritonitis, the affected part is rejected, while in case of extensive peritonitis with systemic involvement, the total condemnation is indicated.

Pancreatitis

Pancreatitis is the inflammation of pancreas. It rarely accompanies with abscess formation which may be due extension of infection from peritonitis. Interstitial pancreatitis is usually caused by infection with parasites such as *Eurytrema pancreaticum, Dicrocoelium dendriticum, Opisthorchis sinensis, Strongylus equinus, Stephanurus dentatus etc.*

Hepatitis

Hepatitis is inflammation of the liver, characterized by swollen with rounded borders. Condemnation of the liver with careful scrutinising of the rest of the carcase for signs of associated pathology is required.

Cholecystitis

Cholecystitis is the inflammation of gall bladder. The condition may occur in generalized disease or local infection. It may be caused due to infection by *Salmonella* spp., *E. coli* and streptococci.

Cholangitis

Cholangitis is the inflammation of bile ducts. It may be caused by many bacteria and parasites.

Cholelithiasis

Cholelithiasis is the formation of stones in the gall bladder (gallstones) and bile ducts (biliary calculi). The condition is commonly found in cattle but less frequently in sheep and swine. These stones vary in size and composed of bile salts, cholesterol, protein and calcium salts.

Pathology of Respiratory System

Pneumonia (Pneumonitis)

Pneumonia is an inflammatory condition of the lungs in which exudate is produced in the alveolar spaces that causes consolidation of the lung tissues. Pneumonia may be caused by bacteria, viruses, parasites, mineral dusts, irritant gases, allergens, plant toxins and aspiration of fluids following vomiting. The total condemnation is indicated in case of acute, septic, necrotic or gangrenous pneumonia.

Pulmonary Abscesses

Pulmonary abscesses may develop due to local septic pneumonia or infected emboli transferred by the blood from other septic organs and penetration of lung tissues by foreign bodies. The encapsulated abscess may be found on pleura that may contain pale green pus, caused by *Corynebacterium pyogenes*. The total condemnation is indicated in case of multiple pulmonary abscesses.

Pleurisy (Pleuritis)

Pleuritis is the inflammation of pleura. It is usually secondary to pneumonia or pulmonary abscess. The condition can develop due to infection with *Haemophilus* spp. in pigs and chlamydial infection in cattle. The disease can also develop due to penetration of foreign body from reticulum or oesophagus. The total condemnation is indicated in case of acute and septicaemic, diffuse fibrinous or serofibrinous pleurisy.

Tumours Formation in Respiratory System

Mesothelioma is most commonly encountered in cattle, horse and goat, while both primary and secondary tumours of the pleura are uncommon. The total condemnation is indicated in case of widespread tumour formation in respiratory system.

Pathology of Cardiovascular System

Pericarditis

Pericarditis is the inflammation of pericardium which may occur due to many infectious diseases like black disease, blackleg, clostridial haemoglobinuria, contagious bovine pleuropneumonia, pasteurellosis, salmonellosis, staphylococcosis, and tuberculosis or exist as extension from adjacent thoracic or anterior abdominal organs, *e.g.*, reticulum in traumatic reticulopericarditis in cattle. The early stage of pericarditis is characterized by hyperaemia with a thin and clear exudate. The total condemnation is indicated in case of acute septic and bovine traumatic pericarditis with fever and circulatory disturbance (systemic disease).

Epicarditis

Epicarditis is the inflammation of epicardium. It is characterized by petechial haemorrhages and ecchymosis in all animal species. It may develop in septicaemia and toxaemia, mulberry heart disease in swine and in improper stunning, especially electrical stunning.

Endocarditis

Endocarditis is the inflammation of endocardium. Subendocardial haemorrhages, especially in the left ventricle, are found in septicaemia and toxaemia, but are less common than the epicardial haemorrhages. Endocarditis is caused by bacteria, fungi and parasites. The heart valves are most commonly affected. Ulcerative valvular endocarditis is occasionally caused by streptococci or *Corynebacterium pyogenes*, which is characterized by formation of rough valvular plaques. The lesions like valve cysts and haematoma are frequently noticed in the heart of young calves. The total condemnation is indicated in case of acute verrucose endocarditis with associated lung and liver lesions (systemic disease).

Myocarditis

Myocarditis is the inflammation of myocardium (heart wall). It is caused by many bacteria like *Actinobacillus equi, Clostridium chauvoei, Haemophilus somnus, Listeria, Salmonella* Pullorum and rarely *Mycobacterium tuberculosis* and protozoa like *Neospora, Sarcocystis* and *Toxoplasma* spp.

Myocardial Necrosis

Myocardial necrosis is seen in the febrile diseases, especially foot and mouth disease, necrobacillosis and *Haemophilus somnus* infection. It also occurs in non-infectious diseases/conditions like porcine stress syndrome, and diseases due to deficiency of Vitamin E, selenium, copper, and certain plants like bracken fern and plants containing gossypol and glycosides.

Myocardial Abscess

Myocardial abscess may be present in the myocardium of cattle as result of foreign body penetration or resulting from metastasis in cases of pyaemia, septic metritis, clostridial infections, listeriosis, toxoplasmosis and Chagas' disease.

Myocardial Tumours

Myocardial tumours occur as a result of metastasis from other organs, especially in enzootic bovine leucosis. It is characterized by smooth, pale yellowish areas which can be nodular or diffuse form. The total condemnation is indicated in case of widespread lymphosarcoma.

Pathology of Lymphoreticular System

Lymphoreticular system is composed of the thymus gland and bone marrow and lymph nodes, spleen and lymphoid tissues. Total condemnation is indicated in lymphoreticular lesions involving systemic disease and marked circulatory disturbance. Pathology of lymphoreticular system is discussed as follows:

Lymphadenitis

Lymphadenitis is the inflammation of lymph nodes. It may be acute or chronic. It is a sequel in various infectious diseases. It must be distinguished from hyperplasia of the lymph nodes which is not associated with infections. Acute lymphadenitis occurs in *Streptococcus pyogenes* infection, anthrax, pasteurellosis, salmonellosis, strangle, tularaemia, rinderpest and acute swine fever. Acute lymphadenitis is characterized by swelling, congestion, mobility and oedema of the lymph node which is painful and softer than normal. Chronic lymphadenitis is characterized by development of fibrous connective tissues of the node, enlargement and induration, sometimes oedema and congestion.

Lymphangitis

Lymphangitis is the inflammation of lymphatic vessels. It occurs in diseases affecting lymph nodes and epizootic lymphangitis of equidae (*Histoplasma farciminosus*), ulcerative lymphangitis of horses (*Corynebacterium pseudotuberculosis*), glanders, Johne's disease, sporadic lymphangitis *etc.*

Lymphoedema

Lymphoedema is an abnormal collection of lymph. It occurs due to lymphatic obstructions, injury, inflammation, tumours or congenital lymphatic defect (absence of lymphatics).

Splenomegaly

Splenomegaly is the enlargement of spleen. Splenomegaly with hyperaemia and other inflammatory changes are seen especially in acute specific diseases, *e.g.*, anthrax, salmonellosis *etc.* Chronic enlargement of spleen is seen in liver stasis resulting from swine erysipelas, traumatic pericarditis, leukaemia, lymphoma *etc.*

Splenic Tuberculosis

Tuberculosis of spleen occurs due to haematogenous spread of tuberculosis. It is characterized by enlargement of spleen and development of tubercles in the spleen. Tuberculosis of spleen is common in pigs but less commonly found in cattle.

Splenic Abscess

Abscess formation in the spleen may be associated with melioidosis in all the food animals especially pigs. The size of abscess is less than 1 cm in diameter. It contains thick, creamy or yellowish-green pus which later becomes inspissated.

Thymus Tumours

Thymus tumours are most often seen in cattle and goats, include benign thymoma, lymphoma, lymphoid thymoma and lymphosarcoma.

Pathology of Urinary System

Nephritis

Nephritis is the inflammation of kidney. Depending on the part involved in the kidney, nephritis may be broadly divided into glomerulonephritis (inflammation of glomeruli), interstitial nephritis (inflammation of interstitial tissues), acute tubular necrosis and pyelonephritis (inflammation of renal pelvis and calyces). Nephritis without systemic involvement is approved as fit for human consumption, while total condemnation is indicated in case of nephritis with uraemia.

Nephrosis

Nephrosis is a pathological condition of kidney, characterized by acute tubular necrosis, and tubular degenerative and inflammatory changes.

Uroliths (Urinary Calculi)

Urolith is a condition in which the stones are formed in the urinary system. Uroliths are formed due to concretions of carbonate, oxalate, urate, silica and xanthenes. Urinary calculi are formed in any part of the urinary system. Uroliths are more common in males due to their anatomical structures. In males, the urinary calculi are most commonly found in the sigmoid flexure of the urethra of cattle and sheep. The carcase having renal calculi without systemic involvement is approved as fit for human consumption.

Cystitis

Cystitis is the inflammation of urinary bladder. It is caused by some bacterial and viral diseases, *e.g.*, salmonellosis in pigs, African swine fever and perpura haemorrhagica. Acute cystitis is characterized by inflammation, congestion, oedema and haemorrhages in the wall of the urinary bladder. Chronic cystitis is characterized by irregular thickening and folding of mucosa and formation of polyps (tumours). Chronic cystitis and cystitis without systemic involvement is approved as fit for human consumption, while total condemnation is indicated in case of cystitis with systemic involvement.

Pathology of Reproductive System

Metritis

Metritis is the inflammation of the uterus. It is caused by bacteria and usually occurs as a result of retained placenta, and injury of uterus during parturition. It is characterized by large size than the normal and containing large amount of chocolate or greyish-coloured fluid composed of blood and foetal membrane. Contagious equine metritis is a highly contagious disease caused by *Taylorella equigenitalis* (coccobacillus). It is characterized by a thick, mucopurulent discharge accompanies the endometritis. The carcase is condemned in case of acute metritis associated with septicaemia or toxaemia. In case of chronic metritis with no toxaemic signs, the carcase may be passed after thorough inspection.

Pyometra

Pyometra is an accumulation of pus in the uterus which may result in adhesion between ovary and the oviduct.

Ovarian Tumours

Ovarian tumours may be of primary or secondary origin. Primary tumour like granulosa-theca cell tumour is benign in nature, while secondary tumour like lymphosarcoma is metastatic in nature.

Uterine and Vaginal Tumours

There may be formation of different types of tumours such as squamous cell carcinoma in the uterus of cows and mares; adenocarcinoma in the endometrium and cervix of cows, mare and rabbits; fibropapilloma in the vulva of cows; and diffuse and nodular forms of lymphosarcoma in uterus and vagina of all species, especially in cows.

Mastitis

Mastitis is the inflammation of mammary tissues which is more often seen in dairy cows. The udder is swollen, hot and painful to touch. The milk usually contains small lumps, which can be seen when the milk is drained through a sieve. Mastitis may occur in acute as well as chronic forms which may be gangrenous and involve systemic changes. In chronic cases, the udder is removed and condemned. Acute or gangrenous mastitis with systemic involvement and inflamed and congested lymph nodes warrants the condemnation of the whole carcase.

Orchitis

Orchitis is the inflammation of the testes. It may occur in the animals suffering from brucellosis, trypanosomiasis, anaplasmosis, babesiosis; ingestion of toxic plants and metals; and ionizing radiation and hormonal disturbances. It is characterized initially by enlargement and oedema, and later on decrease in size, fibrosis and calcinosis due to degenerative changes.

Epididymitis

Epididymitis is the inflammation of epididymis. It is caused by infections with organisms *viz., Brucella, Actinobacillus seminis, Actinomyces pyogenes, Corynebacterium pseudotuberculosis, Mycoplasma* spp. and *E. coli.*

Tumours of the Penis and Prepuce

In the penis and prepuce, there may be formation of different types of tumours such as fibropapilloma in bull; fibropapilloma of transmissible form in prepuce of boars; squamous cell carcinoma in bull and horse; and squamous papilloma, lymphosarcoma and haemangioma in all species.

Judgement of Conditions of the Reproductive System

All chronic and localized conditions without systemic effects and satisfactory carcase condition are approved as fit for food with disposal of affected part. Total condemnation is indicated in the following conditions:

1. Acute septic, necrotic or diphtheritic metritis with circulatory disturbance.
2. Dystocia with acute metritis and vaginitis.
3. Prolapsed, torsion and rupture of uterus with peritonitis and fever.
4. Presence of putrefied foetus with systemic effects.
5. Retained placenta with systemic changes.
6. Septic, necrotic and gangrenous mastitis with systemic effects.
7. Septic and gangrenous mastitis.
8. Malignant neoplasm with extensive metastasis.

Oestrus

Oestrus is a normal physiological process but it may lead to imperfect bleeding in slaughtered animal especially heifers. Moreover, sows and gilts in heat produce poor keeping quality of bacon. Female animals in heat are under stress; therefore, it is important to provide adequate rest to them before their slaughter.

Advanced Pregnancy/Recent Parturition/Abortion

Animals especially bovines in advanced pregnancy are not recommended for slaughter due to the reasons of loss of valuable fetus and stress in animals subjected to slaughter especially during advanced pregnancy. The hind quarters are moister in animal that recently given birth. Moreover, the genital tract shows hyperaemia and the liver shows fatty changes in such animals. Abortion in animals may occur due to brucellosis, listeriosis, toxoplasmosis and campylobacteriosis. Animals in advanced pregnancy and recent parturition should not be sent for slaughter until 10-14 days after parturition or abortion. Poor condition and evidence of general disease require total condemnation of carcase.

Pathology of Skeletal System

Rickets

Rickets is a disease of young animals which occurs due to defective mineralization of growing bones. It is characterized by softening, swelling and curvature of the bone and widening of the epiphyseal plate at joints. The condition occurs mainly due to deficiency of vitamin D as a result of impaired metabolism or malabsorption of vitamin D and phosphorus.

Osteomalacia

Osteomalacia is the softening of bones in adult animals due to deficiency of phosphorus and vitamin D, which develops in grazing animals in areas of phosphorus deficient soil. Long bones, ribs and pelvis become liable to fracture and distortions of the skeleton.

Osteomyelitis (Osteitis)

Osteomyelitis is the inflammation of medullary cavity. It is caused by *Fusobacterium necrophorum*, streptococci, staphylococci, *Actinomyces pyogenes, Brucella suis, E. coli and Salmonella.* Total condemnation is indicated in case of gangrenous/ suppurative osteomyelitis.

Tumours of Bones and Cartilage

Tumours of bones may be primary or secondary, arising as malignant metastasis from other sites such as lung, skin and mammary gland. Benign tumours include osteoma, chondrioma and fibroma. Malignant tumours include osteosarcoma, chondrosarcoma and fibrosarcoma. Total condemnation is indicated in case of generalized tumour.

Arthritis

Arthritis is the inflammation of joint, synovial membrane and articular surfaces. Arthritis may be due to infectious or non-infectious origin. Arthritis may be acute or chronic, septic or non-septic and fibrous. Infectious arthritis is caused by *Actinobacillus suis* in pigs, *Actinomyces pyogenes* in cattle and pigs, *Brucella abortus* in cattle, *Corynebacterium pseudotuberculosis* in sheep, *Erysipelothrix rhusiopathiae* in pigs and sheep, *Haemophilus* spp. in sheep and pigs, *Mycobacterium* spp. in cattle, and sheep and *Salmonella* spp. in pigs. Non-infectious arthritis is caused due to deposition of sodium urate crystals in the synovial fluid and tendon sheaths, which occurs in articular gout in poultry. Total condemnation is indicated in case of acute infectious, purulent and fibrinous arthritis.

Joint Tumours

Synovioma and sarcoma of the synovial membrane are rare.

Fracture

Fracture is a break in the continuity of a bone and is usually due to trauma. Total condemnation is indicated in case of fracture with infection and systemic changes.

Pathology of Muscular System

Myositis

Myositis is the inflammation of muscle which may be due to infectious diseases and some other conditions. Myositis may occur in foot and mouth disease, bluetongue, toxoplasmosis and clostridial infections (blackleg, braxy, malignant oedema). Myositis may also occur in localized abscess formation due to *Haemophilus* spp., *Streptococcus* spp., *Staphylococcus* spp., *Actinobacillus* spp., *Actinomyces pyogenes*, and *Corynebacterium pseudotuberculosis*.

Parasitic Diseases of Muscles

Eosinophilic myositis occurs manly in cattle but it is relatively rare condition. Sheep are less commonly affected. It occurs due to accumulation of eosinophils as a result of degenerating *Sarcocystis* spp. between the muscle fibres that cause necrosis of muscles. It is characterized by greenish, brown-green or grey-green focal strips or patches which become whitish on exposure to air. The condition may be present in the myocardium. Parasitic infestations in the muscles also occur due to other parasitic infestations such as cysticercosis, trichinellosis *etc*.

Muscular Degeneration

Muscular degeneration is usually seen in haemoglobinaemia in horses and milk fever in cows. It may also affect the diaphragm, abdomen and intercostal muscles in pigs. Muscular degeneration may occur due nutritional deficiency especially in young Friesian cattle. Deficiency of selenium leads to muscular degeneration which is seen in many diseases such as white muscle disease, nutritional myopathy, stiff lamb disease and enzootic muscular dystrophy of calves and lambs.

Muscular Haemorrhages or Splashing

Muscular haemorrhage may occur in carcase at various locations such as muscular portion of diaphragm, inner aspect of the thoracic and abdominal walls, large intestine of sheep, endocardium of calves, beneath the splenic capsule of cattle, neck and longisimus dorsi muscles. Muscular haemorrhages is characterized by dark-coloured streaks resembling brush marks or dark spots of varying size which may be up to 3 cm. It is caused mainly due to delay between stunning and slaughter, and increased blood pressure rising from violent muscle contraction during stunning. The condition occurs more commonly when animals are stunned by high voltage electrical stunning than the stunning by captive bolt pistol. Local condemnation is indicated in case of lesser degree of splashing involving restricted area, while total condemnation in extensive splashing.

Pathology of Nervous System

Encephalitis

Encephalitis is the inflammation of brain. Encephalitis may occur in some infectious diseases like listeriosis, rabies, bovine spongiform encephalopathy, classical swine fever, African swine fever, scrapie *etc*. The carcase is approved as fit in case of chronic encephalitis, while total condemnation is indicated in case of acute encephalitis.

Meningitis

Meningitis is the inflammation of meninges. Leptomeningitis is purulent in nature and hematogenous in origin. Tuberculous meningitis in bovines occurs due to hematogenous infection of congenital origin. It generally occurs in young cattle. The carcase is approved as fit in case of chronic meningitis without fever, while total condemnation is indicated in case of acute meningitis.

Abscess in the Brain and Spinal Cord

Abscess formation may occur due to spread of infection from blood, peripheral nerves, and from adjacent infected structures in the skull, sinuses, or vertebral column or directly by trauma. It is caused by pyogenic organisms, *viz., Actinomyces pyogenes, Erysipelothrix rhusiopathiae, Pseudomonas aeruginosa, Listeria monocytogenes, Haemophilus somnus, Pasteurella haemolytica* and *Pasteurella multocida.* It occurs mostly in pigs, usually as a result of tail biting, and in sheep due to careless and unhygienic docking. The carcase is approved as fit in case of localized brain or spinal cord abscess, while total condemnation is indicated in case of brain abscess associated with pyaemia.

Tumours in the Nervous System

Tumours in the nervous system of food animals generally develop secondarily. Tumours in various types of nervous tissues may arise as meningioma, sarcoma, glioma, astrocytoma, neurofibroma, meduloblastoma, swan cells tumour *etc.* Total condemnation is indicated in case of malignant tumours and multiple tumours.

Parasitic Diseases in the Nervous System

Parasites such as *Cysticercus bovis* in cattle, *Cysticercus cellulosae* in pig, and *Cysticercus tenuicollis* in sheep may be found occasionally in the brain. *Coenurus cerebalis* is commonly found in the brain of sheep, and less commonly in the cattle. *Setaria* spp. may be found in the central nervous system of horses, sheep, and goats. *Stephanurus dentatus* commonly encysts in the meninges of pigs. The first stage larvae of *Hypoderma bovis* is commonly found in the spinal canal and epidural fat. The larvae of *Oestrus ovis* are frequently found in the central nervous system of sheep.

Pathology of Sensory Organs

The pathological conditions of sensory organs like eye, ear *etc.* are discussed as follows:

Eye

The condition of eye may indicate disease or abnormal condition. The eye ball is usually retracted in its socket in emaciated carcase. Yellow colouration of the conjunctivae may be indicative of obstructive jaundice, haemolytic jaundice, or hepatocellular icterus. Keratitis and conjunctivitis may occur in local infection or part of systemic disease. Squamous cell carcinoma of eye may occur in bovine which requires rejection of the carcase. Partial condemnation is indicated in localized eye condition, while total condemnation is indicated in case of eye lesions as part of systemic disease with fever, and malignant tumours.

Ear

The ear may be infected with parasites, ringworm or generalized disease. Ear pinna may be necrosed due to infection with *Borrelia* spp., sarcoptic mites, cannibalism and ergot poisoning. Squamous cell carcinoma, adenoma and chondrioma are the most common tumours in cattle. Partial condemnation is indicated in localized ear condition, while total condemnation is indicated in case of ear lesions as part of systemic disease with fever and malignant tumours.

Pathology of Skin

Skin is an important organ of the body which grossly reflects the health status of the animal. In disease and abnormal conditions, it shows alterations in its appearance. The pathological conditions of skin are discussed as follows:

Skin Infections

Skin infections may be caused by bacteria, fungi and parasites. Streptococci and staphylococci cause pyoderma. *Actinomyces pyogenes* causes hygroma on knee and hock. *Mycoplasma kansasii* causes skin tuberculosis. *Dermatophilus congolensis* causes lumpy wool. *Dermatophilus pedis* causes strawberry foot rot. Fungi cause ringworm. Parasites such as ticks, mites, lice, flies *etc.* cause infestation and damage the skin.

Skin Lesions

Skin lesions may be due primary or secondary in origin. The primary skin lesions are directly associated with skin disease which include abscess, wound, bites, cuts, abrasions, cysts, haematoma, nodule, papule, pustules, tumours *etc.* Secondary skin lesions are associated with systemic diseases. Secondary skin lesions may include alopecia, crust, fissure, keratosis, ulcer *etc.*

Neoplasm

Different types of cutaneous neoplasm like cutaneous lymphosarcoma, fibroma, and squamous cell carcinoma may occur.

Genetic Defects

Genetic defects may lead to a condition known as epitheliogenesis imperfect which is characterized by absence/deformity of skin of ear, muzzle and mouth. Genetic defects may cause development of epidermal dysplasia in calves (baldy calves) which is characterized by alopecia and scaly, thickened, folded skin.

Judgement of Skin Lesions

Skin involving local lesions is partly rejected, while, carcase with generalized skin lesions or evidence of systemic disease showing septicaemia, fever, jaundice *etc.* are totally condemned.

Handling of Dead Animals at Abattoir

Dead or dying animal should not be brought into the abattoir premises. All "dead on arrival" and "dead in pen" animals must be disposed of as condemned materials. Prior to flaying or cutting up for disposal or inspection of dead animals, a blood smear examination is required to rule out the possibility of anthrax. Dead animals must be properly and hygienically disposed to avoid contamination of environment and spread of diseases in the animals and humans.

19

Post-Mortem Inspection

Post-mortem inspection means inspection of slaughtered animal that is carcase, visceral organs and other parts to detect abnormalities/pathological conditions, and also constant supervision of all procedures from the stunning of the animal prior to slaughter until the carcase dressing and its washing is completed, including making judgement for acceptance/rejection of carcase, and edible offal/other parts to ensure the suitability and safety of the consumers. All slaughtered animals should be subjected to post-mortem inspection, which should be carried out as soon as practicable after slaughter. Post-mortem inspection procedures and judgements should be determined by the competent authority. A perfectly healthy carcase or meat from it can be rendered unfit for human consumption by contact with contaminated equipment, intestinal content or a variety of other causes.

The main purpose of post-mortem inspection is to detect and eliminate abnormalities including contamination to ensure the wholesomeness of meat for human consumption. Many of the diseases and conditions rendering carcase unfit at post-mortem, are not detectable at the time of ante-mortem inspection. The important subsidiary aspect of post-mortem inspection is checking the efficiency of slaughter and carcase dressing technique; and diagnosis of disease conditions for disease control purposes.

Principles of Post-Mortem Inspection

Routine post-mortem inspection of red meat carcases is based on the examination of heads and their lymph nodes, thoracic and abdominal viscera and their lymph nodes, and the exposed parts of the carcase. The principles of post-mortem inspection are as follows:

1. There should be sufficient time and adequate light facility for inspection of slaughter.

2. A systematic procedure of inspection must be followed for all carcases.
3. Animal should be slaughtered in the presence of meat inspector.
4. Lymph nodes associated with the organs must be examined carefully.
5. A diligent post-mortem inspection of partially dressed carcase presented under an approved procedure is performed to ensure that those organs and exposed carcase parts that are presented have no abnormalities.
6. It is still necessary to conduct the full routine post-mortem inspection even it is obvious that a portion is to be condemned, *e.g.*, incision of lymph nodes and masseter and internal pterygoid muscles is still required on a contaminated beef head.
7. The dressed carcase and all its corresponding detached parts having significant deviations from the normal is held and referred to a veterinarian for final inspection and disposition.

Objectives of Post-Mortem Inspection

1. To protect the consumers against meat-borne infections, intoxication and hazard associated with drugs, pesticides and hormones residues in meat.
2. To protect the consumers and meat industries against economic losses.
3. To protect the meat handlers against occupational zoonotic diseases.
4. To protect the livestock and consumers from spread of zoonotic diseases.

Prerequisites for Post-Mortem Inspection

There are some basic requirements for conducting efficient post-mortem inspection of slaughtered animal. The following are the prerequisites for post-mortem inspection:

1. Adequate light
2. Protective clothing
3. Disinfection of equipment
4. Identification of carcase
5. Carcases that are dressed without flaying (*e.g.*, pigs) must be thoroughly washed before making any incision for the purpose of evisceration.
6. Serous membranes should not be removed until the inspection is completed.
7. Slaughterhouse personnel should wash their hands frequently and thoroughly with soap or detergent under running warm and potable water.

Facilities Required for Post-Mortem Inspection

In addition to structural and mechanical facilities, the facilities of sufficient light are also required for proper post-mortem inspection. There should be one or two hand washing units with a supply of hot and cold running water, a mixing faucet (a fixture for drawing liquid from a pipe), soap and towels. At each inspection location, there should be one or more units consisting of a lavatory and a sanitizer with provisions for liquid soap and hot water at 82.2°C for sanitizing contaminated equipment. Tags, brands or seals are required for the identification and control of diseased and condemned carcase and parts. Vegetable dye such as purple GL-31 is used under USDA and FDA rules as it is deemed safe and harmless.

Procedure of Post-Mortem Inspection

1. Post-mortem inspection can be carried out by visual examination, palpation of tissues and organs and incision where necessary.
2. Inspector's sense of smell for judging the carcase and laboratory tests (if indicated) can be used during post-mortem inspection.
3. Post-mortem inspection must always be carried out in a systematic and hygienic manner and care must be taken to avoid contamination especially septic nature.
4. Incision must be made in such a way as to stand a clean and undistorted surface.
5. Careless incision of abscess must be avoided otherwise it may lead to widespread contamination of carcase and visceral organs which will require unnecessary trimming and local condemnation.
6. The head and visceral organs must remain with the carcase to facilitate identification until the inspection is completed. All organs and carcase portions should be kept together and correlated for inspection before their removal.
7. Care must be taken to avoid contamination of carcase during dressing from the hide or fleece or as a consequence of insufficient/improper evisceration technique.
8. The carcase should be kept separate both during dressing and before adequate cooling in order to avoid contamination.
9. Contaminated knife must be discarded for sterilization.
10. A systematic inspection of carcase and their organs should proceed in an order like head, lungs, heart, liver, oesophagus, stomach, intestine, kidney, spleen, uterus, udder, testes and carcase.
11. The carcase and organs of each animal are examined to eliminate the whole or any part of it, if found diseased or unfit.
12. There are three routine post-mortem inspection locations such as the place where the head and cervical lymph nodes are inspected; the place where visceral organs are inspected and the place where the eviscerated carcase is examined before leaving the slaughterhall.
13. If disease or abnormal condition is detected in slaughtered animal, the meat inspector should affix the "Retained" tags to various parts of the carcase and its visceral organs.

14. If a condition is found which requires condemnation of carcase or a part thereof, the word "condemned" is prominently displayed.
15. The words retained on the tags and condemned on the brands are accompanied with an identification of the agency having jurisdiction, *e.g.*, US retrained and US condemned.
16. In small number of slaughter, the head is removed from carcase and cleaned thoroughly and then placed individually on loops, which present the ventral and posterior surface of the head for the inspection.
17. Visceral organs of cattle are usually presented for inspection in hand truck specially designed for the purpose or a moving table.
18. The detailed post-mortem inspection of the head is not warranted in case of veal carcase, but a visual examination of the mouth and tongue should be made for foot and mouth disease and calf diphtheria. Examine for dysentery and white scour in small intestine. Examine the portal lymph node and posterior mediastinal lymph nodes for congenital tuberculosis. Examine for septic omphalophlebitis in the umbilicus and joints.
19. The final judgement at post-mortem inspection of a carcase or its part is based on the information collected by observation, palpation, incision, smell, ante-mortem reports and results of any laboratory test.
20. Carcase and their visceral organs which are found diseased/abnormal on routine post-mortem inspection are subjected to final inspection with a more thorough inspection to determine whether they are suitable for human consumption. In doubtful cases, the carcase is held pending till laboratory finding.

Post-Mortem Inspection of Cattle/Buffaloes

A systematic post-mortem inspection of carcase and their organs should proceed in the following order:

Head

1. In post-mortem inspection, the head is inspected first thereafter the visceral organs, carcase *etc.* Do not commence the inspection until the head is clean, properly prepared (free of hair, pieces of skin, contamination *etc.*) and presented in a satisfactory manner.
2. The inspector must perform a visual examination of the head, including the eyes and the tongue to detect any abnormality. Palpate the tongue to detect abscesses, actinobacillosis, foot and mouth disease and other abnormal conditions.
3. Examine the outer surface of head and eyes followed by gum, lips and tongue for foot and mouth disease, stomatitis, actinobacillosis and actinomycosis.
4. Incise the internal and external masseter muscles parallel with the lower jaw for detection of *Cysticercus bovis*.

5. Incise the retropharyngeal, submaxillary and parotid lymph nodes for detection of tuberculous lesions, abscess and actinobacillosis.

Lungs

1. Examine the lungs visually followed by palpation for evidence of pneumonia, pleurisy, tuberculosis, abscess, tumor, hydatid cysts *etc.*
2. Incise and examine the bronchial and mediastinal lymph nodes.

Heart

1. Visually examine the exterior and interior of the heart (that is the valves and the endocardium).
2. Examine the pericardium for evidence of pericarditis, hemorrhage *etc.*
3. Examine the epicardium for evidence of petechial hemorrhages.
4. Incise the ventricles and examine for evidence of cysticerci, hydatid cysts and linguatulae in the myocardium.
5. Examine the endocardium for evidence of petechial hemorrhages.

Liver

1. Examine the liver visually and palpate thoroughly.
2. Observe and palpate the liver for fatty change, abscess, telangiectasis, actinobacillosis, *Cysticercus bovis*, linguatulae, fascioliasis and hydatid disease.
3. Incise the gastric surface and base of caudate lobe of liver and examine the bile ducts.
4. Incise and examine portal and hepatic lymph nodes.

Oesophagus, Stomach and Intestine

1. Visually examine the rumen, reticulum, omasum and abomasum. Visually examine the rumino-reticular junction to detect any abnormalities such as existing inflammatory conditions, abscesses, presence of protruding foreign bodies as a result of reticular puncture *etc.*
2. Observe and palpate these organs for evidence of tuberculosis and actinobacillosis.
3. Examine the gastric and mesenteric lymph nodes for evidence of tuberculosis. Incise the enlarged mesenteric lymph nodes and other lymph nodes having suspicious lesions. Submit the lymph nodes showing granulomatous lesions to the laboratory for diagnosis of bovine tuberculosis.

Kidneys

1. Kidneys may be examined, either in the carcase or separately, (*e.g.*, with the other visceral organs).

2. Kidneys should be fully exposed by the operator prior to inspection and visually examined by the meat inspector.
3. Examine the kidneys for evidence of petechial haemorrhages.
4. Incise the kidneys and examine for pyelonephritis, pyaemic abscess and infarcts.
5. Examine the renal lymph nodes for tuberculosis and other conditions.

Spleen

1. Examine the spleen through observation, palpation or incision for evidence of tuberculosis, haematoma, *Echinococcus* cyst and infarcts.
2. Examine the splenic lymph node for tuberculosis and other conditions.

Uterus

1. Observe, palpate and incise (if necessary) the uterus for evidence of metritis, recent parturition or pregnancy.
2. Examine iliac lymph nodes for any abnormality.
3. Do not incise or handle the uterus with bare hands in brucellosis reactors.

Udder

1. Incise the udder and examine for mastitis and abscess.
2. Incise and examine the supramammary lymph node for tuberculosis and abscess.
3. Remove the intact udder without incision in brucellosis reactors.

Testes

1. Examine testes for orchitis.
2. Examine superficial inguinal lymph nodes for any abnormality.

Carcase

1. Post-mortem inspection of the carcase is carried out after removal of visceral organs, but before carcase washing.
2. The inspection consists of a careful examination of the external surfaces of the carcase, the internal cavity, including a visual inspection of the iliac lymph nodes and of the cut surface of the vertebrae, as applicable. Kidneys should be examined if these have been left in the carcase.
3. Examine carcases including musculature, exposed bones, joints, tendon sheaths *etc.* to determine any signs of disease or abnormality.
4. Examine cut surfaces of bone and muscle, carcase exterior, pleura, peritoneum and diaphragm for several conditions such as efficiency of bleeding, colour, cleanliness, odours and evidence of bruising and other abnormalities.

5. If necessary, palpate and incise the parts such as triceps brachii muscle for *Cysticercus bovis*.
6. Examine the thoracic and abdominal cavity for inflammation, abscess, actinobacillosis, mesothelioma and tuberculosis.
7. Examine the superficial inguinal, external and internal iliac, prepectoral and renal lymph nodes.
8. If the above routine examination reveals no evidence of abnormality, the carcase may be passed for human consumption.

Post-Mortem Inspection of Calves

1. The routine post-mortem inspection of calves is similar to those for adult bovines with special attention to particular sites.
2. Examine the mouth and tongue for foot and mouth disease and calf diphtheria.
3. Examine the abomasum for evidence of peptic ulcer and small intestine for white scour or dysentery.
4. Examine the portal lymph nodes and posterior mediastinal lymph nodes for congenital tuberculosis.
5. Examine the umbilicus for omphalophlebitis.

Post-Mortem Inspection of Sheep and Goats

1. The routine post-mortem inspection procedures for adult sheep and goat consist of a visual examination of head, visceral organs and the dressed carcase.
2. Examine the carcase for satisfactory bleeding.
3. Examine the lungs for hydatid disease and nematodes
4. Examine the liver for hydatid cysts and fascioliasis.
5. Examine the lymph nodes visually and incise when necessary for evidence of abscesses indicating caseous lymphadenitis, granulomas *etc.* Granulomas consistent with a tuberculosis lesion should be submitted to the laboratory for tests.

Post-Mortem Inspection of Horses

1. Post-mortem inspection procedures for horses are similar to those for cattle, except that the examination for *Cysticercus bovis* is not required.
2. In post-mortem inspection of head including the guttural pouch is important.
3. Examine the neck region for fistulous conditions near the first two cervical vertebrae.
4. Carefully examine the abdominal walls for evidence of encysted parasites.

5. Examine the axillary and subscapular spaces of white and gray horses for evidence of melanosis.
6. Visually examine the mesenteric lymph nodes, palpate and incise if necessary.

Post-Mortem Inspection of Pigs

1. Post-mortem inspection procedures for pigs are same as for cattle with some special attention.
2. Examine the skin for swine erysipelas, swine fever and urticaria and shotty eruptions.
3. Examine the tail for necrosis, the feet for abscess formation and the udder for actinomycosis.
4. Examine the visceral organs in the manner detailed for cattle with particular attention to the lungs for pneumonia and the secondary complication such as pleurisy, pericarditis and peritonitis.
5. Examine the submaxillary lymph nodes for tuberculosis.

Post-Mortem Inspection of Poultry

1. Examine whole-carcase after defeathering and washing.
2. Examine the carcase for fracture, injuries or any other disease
3. Examine the partially eviscerated carcase and relate carcase and visceral organs.

Judgement

The differentiation between a localized and a generalized condition is important in the judgement of an animal carcase. On post-mortem inspection of slaughtered animal the judgement varies according to existing conditions as follows:

Localized Condition

In a localized condition, a lesion is restricted by the animal defense mechanisms to a certain area or organ. Systemic changes associated with a localized condition may also occur, *e.g.*, jaundice caused by liver infection or toxaemia following pyometra.

Generalized Conditions

In a generalized condition, the animal's defense mechanisms are unable to stop the spread of pathogen and thereby the disease process, by way of the circulatory or lymphatic systems. The signs of a generalized disease may include generalized inflammation of lymph nodes, inflammation of joints, lesions in liver, spleen, kidneys, heart and other organs, the presence of multiple abscesses in different portions of the carcase including the spine of ruminants.

Generalized lesions usually require more careful judgement than localized lesions. As a general rule the acute and generalized condition requires the total condemnation of the carcase and the visceral organs, while the chronic and localized conditions may require only local condemnation or in some cases no condemnation at all. Meat is to be declared unfit for human consumption if it derives from animals that have not undergone ante-mortem inspection, except for hunted wild game. Meat from animals that undergo slaughter following an accident in a slaughterhouse may be used for human consumption in the following conditions:

1. If the animal had a veterinary ante-mortem inspection either before or after the accident.
2. If the post-mortem inspection is performed personally by the veterinarian.
3. If at post-mortem inspection no serious lesions other than those due to the accident are found.
4. If a bovine is over 4 years old, transmissible spongiform encephalopathy (TSE) testing will be required.

Acute Condition

The acute stage of disease or condition is manifested by inflammation of different organs or tissues, enlarged haemorrhagic lymph nodes and often by petechial haemorrhage of the mucosal and serous membranes and different organs such as heart, kidney and liver. An acute stage parallels with the generalized disease complex, when an acute infection tends to overcome the animal's immune system and becomes generalized.

Chronic Conditions

In a chronic condition, inflammation associated with congestion is replaced by adhesions, necrotic and fibrotic tissue or abscesses. The judgement in the chronic stage is less severe and frequently the removal of affected portions is required without the condemnation of the carcase. However, judgement on the animal or carcase judgement tends to be more complicated in sub-chronic and sometimes in peracute stages. If generalized necrotic tissue is associated with previous infection, carcase must be condemned.

Post-Mortem Judgement Categories

1. Safe and suitable for human consumption.
2. Safe and suitable for human consumption but subject to application of a prescribed process, *e.g.*, cooking, freezing *etc.*
3. Held on suspicion of being unsafe or unsuitable, pending the outcome of further procedures and/or tests.
4. Unsafe for human consumption but able to be used for some other purpose, *e.g.*, pet food, feed and feed ingredients, and industrial non-food use under adequate hygiene controls to prevent contamination of human food chain.
5. Unsafe for human consumption and requiring condemnation and destruction.
6. Unsuitable for human consumption and requiring condemnation and destruction.

7. Unsafe for animal health reasons, as specified in national legislation, and disposed of accordingly.

Precautions to be Taken While Conducting Post-Mortem Inspection

1. Post-mortem inspection should be carried out in adequate light.
2. There should be sufficient time for conducting efficient post-mortem inspection.
3. Lymph nodes associated with organs must be examined carefully.
4. Serous membranes should not be removed until the inspection is completed.
5. Careless incision of abscess must be avoided to prevent contamination of carcase, visceral organs, knife and other equipment.
6. Contaminated knife should not be used for post-mortem inspection purpose and it must be discarded for sterilization.

20

Differentiation of Meat of Different Food Animal Species

Consumption of meat products containing mislabeled flesh not only cause allergic reactions in the sensitized individuals but also hurt the religious statements of various communities. Adulteration of meat may be defined as "the fraudulent practice which involves substitution or mixing of flesh of cheaper variety which is objectionable for the reasons of health, religion and economy". This act of cheating is punishable under the Prevention of Food Adulteration (PFA) Act, 1954. Meat of various species of animal can be differentiated by the following methods:

Anatomical Method

Meat of various animal species can be differentiated on the basis of anatomical peculiarities. Bones and cartilage when present along with the meat can be used for species differentiation. This method has their own limitations as they cannot be employed for the species detection when meat is minced form. Absence of characteristic osseous structures also restricts its application for identification. By anatomical method, the meat/carcase of various species of animal can be differentiated to some extent as mentioned in the Tables 20.1 and 20.2.

Physical Method

Physical method has some limitations and cannot be employed in minced meat, lean meat, boiled meat, cooked or processed meat. By this method, the meat of various species of animal can be differentiated to some extent as mentioned in the Table 20.3.

Table 20.1. Anatomical Differentiation of Carcases of Horse and Ox

Sl. No.	Feature	Horse	Ox
1.	Forequarters	The ulna extends only half the length of the radius.	The ulna extended and articulates with the carpus.
2.	Thoracic cavity	Longer (18 pairs of ribs).	Smaller (13 pairs of ribs).
3.	Ribs	Narrower and more markedly curved.	Wider and less curved.
4.	Vertebrae	☆ The superior spinous processes of the first six dorsal vertebrae are more markedly developed and less inclined posteriorly. ☆ The last three lumbar transverse processes articulate with each other, the sixth articulating in a similar manner with the sacrum.	☆ Less developed and more inclined. ☆ Do not articulate.
5.	Hindquarters	☆ Great muscular development. ☆ The fibula extends the two thirds the length of the tibia.	☆ Less muscular development. ☆ The fibula is only a small pointed projection.
6.	Flesh	Well defined muscle fibres.	Fine in texture.
7.	Fat	Soft yellow fat beneath the peritoneum in the gelding and mare. Fat is lighter colour and almost white in stallion.	The kidney fat is always firmer, whiter and more abundant than the horse.

Table 20.2. Anatomical Differentiation of Carcases of Sheep and Goat

Sl. No.	Feature	Sheep	Goat
1.	Back and wither	Round and well fleshed.	Sharp and little flesh.
2.	Thorax	Barrel- shaped.	Flattened laterally.
3.	Tail	Fairly broad.	Thin.
4.	Radius	One and half times the length of metacarpus.	Twice as long as metacarpus.
5.	Scapula	Short and broad. Superior spine bends back and thickened.	Possesses distinct neck. Spine straight and narrow.
6.	Sacrum	Lateral borders thickened in the form of rolls.	Lateral borders thin and sharp.
7.	Flesh	Fine in texture.	Coarse.

Table 20.3. Physical Differentiation of Meat of Different Animal Species

Sl. No.	Animal Species	Characteristics of Meat	Characteristics of Fat and Its Distribution in Meat/Carcase
1.	Cattle (Zebu)	Red.	Yellow, firm and uniformly distributed with muscle tissues (marbling).
2.	Cattle (European)	Red.	White, firm and uniformly distributed with muscle tissues (marbling).
3.	Buffalo	Darkened.	Pure white, firm and uniformly distributed with muscle tissues (marbling).

Sl. No.	Animal Species	Characteristics of Meat	Characteristics of Fat and Its Distribution in Meat/Carcase
4.	Sheep	Pale red and fine in texture.	White, first odourless and not intermixed with meat.
5.	Goat	Dark red and coarse.	Fat is practically absent between the muscle and at subcutaneous tissues. However kidneys are covered with abundant of fat.
6.	Pig	Pale red (nearly white on boiling).	Grayish, very soft and present subcutaneously.
7.	Horse	Dark bluish- red.	White, sometimes yellow and soft.

Chemical Method

The meat of different animal species can be differentiated on the following chemical basis:

1. **Glycogen Content in Muscle and Liver:** The meat of different animal species can be differentiated on the basis of glycogen content in muscle and liver. The horse flesh is rich in glycogen than the flesh of other food animals. Glycogen however, starts to disappear from meat from the time of slaughter and unless the examination is made soon after the carcase is cut up, very little may be found. The liver of pig contains more quantity of glycogen in comparison to other food animals.
2. **Linoleic Acid in Fat:** Horse fat contains 1-2 per cent of linoleic acid. The fat of other food animals contain 0.1 per cent or below 0.1 per cent linoleic acid.
3. **Iodine Value of Fat:** This test is based on the amount of iodine absorbed by the unsaturated fatty acids present in the fat. The iodine value of fat varies in the different animal species as mentioned in the following table:

Table 20.4. Iodine Value of Fat of Various Animal Species

Sl.No.	Animal Species	Iodine Value of Fat
1.	Sheep	35-46
2.	Ox	38-46
3.	Pig	50-70
4.	Horse	71-86

4. **Refractive Index of Fat:** The refractive index of fat varies in different animal species as mentioned in the following table:

Table 20.5. Refractive Index of Fat in Different Animal Species

Sl.No.	Animal Species	Refractive Iindex of Fat
1.	Ox	Less than 40
2.	Pig	Not more than 51.9
3.	Horse	53.5

Biochemical Method

Meat of different animal species can be identified by biochemical method by using various electrophoretic techniques such as polyacrylamide gel electrophoresis (PAGE), polyacrylamide gel isoelectric focusing (PAGIF), and sodium dodecyl sulphate polyacrylamide gel electrophoresis (SDS-PAGE). The basic principle of applying of these electrophoretic methods is based on the assumption that the protein composition of meat is specific and uniform within a species but differs in different species of animals.

In all these methods muscle/offal proteins are separated in gel under electric field and the isolated analyte is recognized or estimated by comparing with the bands of standard molecular weight proteins and species specific protein. While determining the protein pattern specific to the species, it should be considered that these differences are species specific or an expression of different conditions such as stress, nutritional status, age of animal *etc.* In addition post-mortem changes like rigor mortis and changes due to other treatments on protein pattern also need to be taken into consideration *e.g.*, in pale, soft exudative (PSE) meat/carcase, there is a partial denaturation of sarcoplasmic muscle proteins shortly after slaughter due to simultaneous high body temperature and low pH of muscle resulting into their low solubility. Consequently this change influences the electrophoretic pattern of sarcoplasmic muscle protein.

Immunological Method

The increased regulation of foodstuffs particularly meat and meat products in modern society requires analytical methods that should be easy to perform, sensitive, specific and relatively inexpensive. This is important from economic as well as religious point of view. The basic antigen-antibody reaction provides means for very specific analytical procedures. Immunoassays are powerful analytical tools that permit the specific and rapid detection or measurement of antigens and haptens to which antibodies can be produced. There are several immunological methods such as precipitating tests in the tubes, precipitating tests in agar gels on slide (AGPT/DID), enzyme linked immunosorbent assay (ELISA), immunoelectrophoresis (IE) and counter immunoelectrophoresis (CIE) can be used to identify meat proteins as follows:

Precipitating Tests in the Tubes

This test is also known as single immunodiffusion test. In this test the antigen is layered over the antiserum in capillary tube. The reaction occurs at the interface of the layers in the form of immunoprecipitate.

Agar Gel Precipitation Test/Double Immunodiffusion Test (AGPT/DID)

This technique is more sensitive than the single diffusion technique. This technique is widely accepted and used for identification of species origin of meat samples as well as for the detection of adulteration of one species of meat with other species in the field.

Principle

The antigen and antibody both are allowed to diffuse towards each other in an inert medium such as agar gel.

Preparation of Field Antigens

Homogenize 100 g meat / offal with 40 ml of normal saline solution. Centrifuge the slurry at 5000 rpm for 15 minutes. Filter through Whatman filter paper No. 41. This filtrate serves as antigen.

Preparation of Standard Antigen

This can be prepared as above from respective species of animal.

Preparation of Diagnostic Antibody

Diagnostic antibody or species specific antibody can be prepared in rabbits by immunization with respective species antigen (serum, tissue) or by cross immunization (cattle injected with buffalo antigen and vice versa).

Procedure of AGPT/DID

1. Prepare 0.6 per cent agarose gel in normal saline solution.
2. Pipette out 4 ml of molten agarose gel on a precoated glass slide with 0.1 per cent agarose.
3. Place on leveled platform.
4. Allow the agarose gel to set and punch wells.
5. Place one or two drops of antiserum in central well and unknown antigen or test antigen in the peripheral wells.
6. Place the homologous antigen in one of the peripheral wells.
7. Incubate the slide at 37°C for 4 hrs.
8. Transfer the slide to a refrigerator for 48 hrs.
9. Observe the development of precipitation line with homologous antigen as well as with the unknown antigen.
10. Stain precipitation lines with 0.1 per cent aqueous amido black solution for 5 minutes.
11. Destain with several changes of destaining solution till excess stain is removed.
12. Interpret the result on the basis of observations.

Enzyme Linked Immunosorbent Assay (ELISA)

This is a powerful qualitative as well as quantitative immunological technique of which different versions such as solid phase-indirect and capture ELISA are used. Indirect ELISA uses meat extracts to coat the solid phase (Polyvinyl chloride, PVC disposal well trays). The species specific antibody added to the wells recognizes species antigen adsorbed to the PVC and the enzyme labeled antispecies IgG detects species specific antibody. The capture assay employs an antibody adsorbed to the PVC wells to bind meat antigen and enzyme labeled species specific antibody to detect the presence of bound antigen.

Requirements

1. **Antigen:** Meat sample/meat extract.
2. **Antiserum:** Raised in rabbit against antigen or meat extract.
3. **Phosphate Buffered Saline (PBS):** Containing 0.05 per cent tween 20.
4. **PBS and Pig Serum:** 300 ml PBS + 20 ml pig serum.
5. **Horse Radish Peroxidase (HRP):** Conjugate HRP anti-bovine IgG (diluted to 1:5000 in PBS) and pig serum diluted to 1:500 in 20 ml PBS) and put 50 µl to each well.
6. **Hydrogen Peroxide:** 3 per cent solution.
7. **2 N H_2SO_4:**10 ml conc. H_2SO_4 + 93.8 ml distilled water.
8. **Tetramethyl Benzidine (TMB) Substrate:** (0.05 g TMB in 5ml dimethyl sulphoxide 20 µl in 20 ml sodium acetate buffer for 2 plates + 15 µl H_2O_2, and put 100 µl to each well).
9. **0.1 M TMB Substrate Buffer (8.203 g/lit.):** 0.1 M sodium acetate adjusted to pH 6.0 with 0.1 M citric acid.
10. **Coating Buffer (Reggriardo's Buffer):** (18.5 g Glycine + 10.45 g NaF + 29.2 g NaCl + 0.186 g EDTA + 500 ml distilled water).

Procedure

1. Inoculate 50 µl of antigen in buffer to rows 1 to 10 of micro-test plate (96 well).
2. Incubate at 37°C for 3 hrs. and then at 4°C for overnight.
3. Add 200 µl of PBS containing pig serum to each well and incubate at 37°C for 1 hour.
4. Wash the plate in BPS thrice and add 50 µl of PBS containing pig serum to wells 2-10.
5. Add 50 µl 25 times diluted test sample (diluted in PBS containing pig serum) to well 1, 2, 11 and 12. Dilute from row 2 to 9.
6. Leave row 10 as antigen control, while rows 11 and 12 as sample control and keep at room temperature for 1 hour.
7. Wash the plate 3 times and add 50 µl of HRP to each well and keep at room temperature for 1 hour.
8. Wash thrice in PBS and then add 100 µl of TMB substrate and observe the development of colour in 10 minutes.
9. Add 25 µl of H_2SO_4 per well.
10. Read the optical density of ELISA plate reader. Take the average of wells 11 and 12 and any well above that value is positive.

Immunoelectrophoresis (IE)

This technique is based on the separation of proteins by their differential migration through a supporting medium under the influence of electric field. The protein bands thus resolved are visualized by general, enzymological, chemical or immunological means. The sample can be prepared by homogenization with 10 volume of 0.03 N phosphate buffer (pH 7.4). This is followed by extraction of protein with 0.1 N phosphate buffer and myofibrillar protein with 1.1 M KI (pH 7.4) by centrifugation at 10000 rpm for 20 min. The supernatant is filtered and subjected to gel electrophoresis.

Counter Immunoelectrophoresis (CIE)

CIE is an immunodiffusion test in which protein is subjected to accelerated movement under electric filed. In this method an alkaline gel is used which causes electro-osmosis. In counter CIE, the antibodies and proteins from meat extract moves towards each other and in case of homologous condition they form a precipitation band at the point of its joining. This technique is rapid and more sensitive.

Molecular Methods

Polymerase chain reaction (PCR) is a rapid method because in this technique we can obtain multiple copies of specific piece of DNA sequence in vitro and it has high degree of selectivity and sensitivity. PCR is a qualitative test for meat species specification. PCR amplifies a target DNA sequence in an exponential phase, being capable of detecting even a single copy sequence from a single cell sample. There are two main techniques for amplification of genetic marker, *i.e.*, mono-locus-specific primers for amplification of a concrete DNA fragment and multi-locus amplification of non-targeted DNA. By this technique the meat of closely related animal species can be identified. The technique is also useful in differentiating the raw meat of male and female of the same animal species.

21

Chemical and Antimicrobial Residues in Meat

Residues, as defined by the European Union (EU) and the Center for Veterinary Medicine, an agency under the Food and Drug Administration (FDA/CVM) in the USA are "pharmacologically active substances (whether active principles, recipients or degradation products) and their metabolites which remain in foodstuffs obtained from animals to which the veterinary medicinal products (VMPs) has been administered". Under the normal physiological conditions, following administration of a drug to an animal, most drugs are metabolized in order to facilitate elimination, and to a large extent detoxification as well. In general, most of the parent product and its metabolites are excreted in urine and a lesser extent via faeces. However, these substances may also be found in the milk, meat and eggs.

Sources of Chemical Residues in Meat

Different types of chemical substances are used for various purposes. Veterinary drugs, hormones, anthelmintics, heavy metals, radioactive substances, pesticides, natural toxins *etc.* are the main chemicals that may enter the animal body and cause their residual effects in their products. Veterinary drugs, hormones and anthelmintics are used in the livestock for preventive, therapeutic and other purposes. Heavy metals and radioactive substances may enter the animal body through feed, fodder, water and air contaminated with them. Pesticides may enter the milk or meat through dipping animals to prevent the tick infestations and tick-borne diseases, control of insects in stables, feeding stuffs and contamination from the environment.

The chemical substances to which animals may become exposed during their production cycle which have been identified to date could come from drugs and growth promoters aimed at treating diseases and improving production parameters, biologically-derived toxins (mycotoxins, phycotoxins, phytotoxins) and/or environmental contaminants linked to atmospheric pollution, from the soil and/or water. Ivermectin is widely used for the treatment and prevention of both nematode and arthropod parasites in food-producing animals. Ivermectin is lipophilic and residues are found in the tissues of the treated animal, particularly in horse with a high fat content. The meat withdrawal time of ivermectin in livestock is long.

Veterinary Drugs

Veterinary drugs residues include the parent compounds and/or their metabolites in any edible portion of the animal product and include associated impurities of the veterinary drug concerned. **Veterinary drugs** residues usually accumulate in the liver and kidney rather than other tissues. Drug residue level varies in different tissues of the body. The most likely reason for drug residues are improper usage of drugs (dosages, frequency of drug administration *etc.*) and failure to keep to the withdrawal period including using overdose and long acting drugs. Improper sanitary care during animal or product transportation, including the cross contamination of animal feeding stuffs with inadvertently applied drugs, environmental and animal to animal transfer of drugs may also cause residues.

Hormones

Hormones are used for treatment of animals and as growth promoters. Hormones are also used for augmentation of milk and meat production in the animals in various parts of the world. Hormones used for these purposes may reach in the consumers through animal products and may adversely affect their health. Hormones have been found association with neoplasia. Diethylstilbestrol is the most commonly used hormone. In females, it has been associated with cervical adenocarcinoma. Blood and faeces are the most conventional samples for testing of hormones. Screening test for growth promoting hormone residues are based on immunoassays and radioimmunoassay. ELISA is more advance technique including rapid, sensitive and cost-effective. The synthetic hormones can be confirmed by liquid chromatography.

Anthelmintics

Anthelmintics are widely used to prevent and control parasitic infections in the animals. Some common anthelmintics and their maximum residue levels are given in the Table 21.1.

Pesticides

Pesticide residue is any specified substance in food, agricultural commodities or animal feed resulting from the use of a pesticide. The term includes any derivatives of a pesticide, such as conversion products, metabolites, reaction products and impurities considered to be of toxicological significance. Pesticides may be of different types depending upon the use against different types of pests, *e.g.*, insecticides, fungicides *etc.* Majority of the pesticides used for agricultural and domestic pest control purpose degrade rapidly in the environment, a few are trans-located from soil to plant tissues and then to animals where they have specific affinity for adipose tissue and ultimately lead to contamination of livestock products. Some of the older, cheaper pesticides can remain for years in soil and water. These chemicals have been banned from agricultural use in developed countries, but they are still used in many developing countries. Some pesticides such as dichlorodiphenyltrichloroethane (DDT) and lindane can remain for years in soil and water. These pesticides are banned in many countries.

Table 21.1. Maximum Residue Levels for Common Anthelmintics

Anthelmintic	*Animal Species*	*Target Tissue*	*Maximum Residue Level (µg/kg)*
Levamisole	Bovine, ovine, porcine, poultry	Muscle, kidney, fat	10
		Liver	100
Ivermectin	Bovine	Liver	100
	Ovine	Fat	40
	Porcine	Liver	15
		Fat	30
Closantel	Bovine	Muscle, liver	1000
	Ovine	Kidney, fat	3000
		Muscle, liver	1500
		Kidney	5000
		Fat	2000
Febantel, Fenbendazole	All food-producing species	Liver	1000
		Muscle, kidney, fat	10
Oxfendazole	Bovine, ovine	Milk	10
Triclabendazole	Bovine, ovine	Muscle, kidney, liver	150
		Fat	50
Thiabendazole	Bovine, ovine, caprine	Muscle	1000
Netobimin	Bovine, ovine, caprine	Liver	1000
		Kidney	500
		Muscle, fat	100
		Milk	100

There are various sources of contamination of pesticides in meat, milk or other products. Pesticide may enter the milk or meat by using it for dipping cattle to prevent the tick infestations as well as prevention of tick-borne diseases, control of insects in stables, feeding stuffs and contamination from the environment. Skin contact with organochlorine pesticides and intake of pesticides contained fodder and soil give rise to residues in the body fat of animal and in the milk. Milk and milk products may be contaminated with residues of the organophosphates and carbamates from animal feed treated with insecticides, animals feed manufactured plant material which has been treated during the growing season and use of insecticide for animal dwellings.

The persistence nature of some pesticides lead to their accumulation in animal tissues and subsequently causes human dietary exposure to pesticides through consumption of animal products *viz.*, meat, milk, eggs and seafoods. Scientific evidence suggests that even such low dose but long term exposure can cause serious health hazards to human health and environment as well. Pesticides are potentially toxic to humans and can have both acute and chronic health effects, depending on the quantity and ways in which a person is exposed. The rate of elimination of pesticides and duration of any toxic effects are major factors causing health hazards in the consumers. Some products of pesticide metabolism are more toxic than the parent compound and are responsible for many of the undesirable toxic effects, both acute and chronic. The pesticides can be detected by chemical methods, *e.g.*, spectrographic method, chromatographic method *etc.*

Heavy Metals

Toxic metals present in soil or emission are expected in feed through different sources. Lead and cadmium are important in this concern. Unclarified industrial wastes may be particularly responsible for increased cadmium concentration in soil. This leads to increased availability and utilization by the lactating dairy animals. Organic mercury compounds are particularly toxic as they are lipophilic and easy to absorb. Heavy metals may cause toxicities, teratogenicity and carcinogenesis including damage of particular organs in the consumers.

Intake of excess amount of heavy metals such as lead, arsenic, mercury, cadmium *etc.* in foods causes intoxication. Lead is accumulated in the tissue of animals grazing close to smelting plants or in animals ingesting paints or substances with high lead contents. Cattle and other animals graze freely on environment and drink water from ponds, streams, rivers and other possible contaminated water sources. Animals in this process may be exposed to high levels of these metals in the environment. Ruminants are more commonly affected than other farm animal species. Acute lead poisoning should be detected during ante-mortem inspection. At post-mortem inspection the muscle of acutely-poisoned bovine is usually pale.

Arsenic

Arsenic is probably the second most important poisoning hazard for the farm animals. The animals may be exposed to arsenic through contamination of feed, fodder and water from arsenical herbicides, rodenticides or insecticides. Chronic toxicity in animals may occur when arsenical compounds are fed to animals in small amounts because arsenic is accumulated in liver, kidneys and bones. A withholding period of 40 days is needed in case of chronic toxicity with arsenic. Following normal therapeutic dosing, a withholding period of 5 days is sufficient. The concentration of arsenic in shellfish may be high if taken from water contaminated with high amount of arsenic. Generally, the concentration of arsenic in bottom feeders is higher than the surface feeders.

Lead

Lead is found in the environment in sources such as dust and soil. It can also be found in water. Industrial waste water containing large amount of lead may reach the animal body through its consumption. The residues of lead in meat may cause neurological disorders, reproductive problems and diminished intelligence in the consumers.

Mercury

Mercury occurs naturally in rocks and soils and can be found in lakes, streams and oceans. Combustion of fossil fuels, mining, pulp and paper industries and burning garbage can also release mercury into the environment. Mercury compounds are used in agriculture and veterinary medicines. Mercury poisoning in animals may occur due to feeding of grains treated with mercury-containing dressings to prevent fungal growth. The poisoning in animals may also occur due to grazing of pastures contaminated with industrial sewage containing mercury or feeding of plants/crops grown in contaminated sewage as fertilizer.

Mercury is found in inorganic as well as organic forms. Methyl mercury (organic mercury) is the most common form of mercury found in the aquatic environment and most fish have trace amounts in their body. It has been found that larger and older fish tend to have the highest levels of mercury, due to bioaccumulation. Mercury causes damage to the nervous system, kidneys, and the developing foetus. It also causes brain damage, irritability, tremors, memory problems, changes in vision and hearing. Children are more sensitive to mercury than adults. Methyl mercury is suspected to be human carcinogen. Inorganic mercury is stored mainly in liver and kidneys but organic mercury (*e.g.*, methyl mercury) is widely distributed in the body tissues. Now, mercury containing compounds have been replaced by less toxic compounds, therefore, there is little risk of mercury to consumers from the meat of farm animals.

Cadmium

Cadmium is used in the manufacture of batteries, pigments, coatings, plating, stabilizers for plastics, ore processing and smelting, thus it finds its way into the environment through waste, waste water and soil uptake. The main source of cadmium is industrial sewage dealing with cadmium compounds or organocadmium fungicides. In meat-producing animals, the cadmium enters the body through consumption of plants grown in soil contaminated with cadmium and its compounds through industrial waste water. Cadmium and its compounds are highly toxic and carcinogenic. Cadmium is accumulated in higher concentrations mainly in the kidneys and liver of the animals. Higher concentrations of cadmium in humans cause kidney failure.

Radioactive Substances

Radioactive contamination may occur through natural or man-made sources. Natural sources of radioactive material include the presence of uranium, radium, thorium, polonium, K_{40}, C_{14} *etc.* in soil, rocks, and water, and cosmic rays. Radioactive contamination may also occur through man-made sources, especially during nuclear leakage or explosion. Radioactive contamination may heavily occur through man-made sources that may cause heavy contamination of soil, water and air. This causes accumulation in the animal body through these sources and pose a serious threat to the meat consumers.

Mycotoxins

Mycotoxins are toxic secondary metabolites released from moulds growing in food and foodstuffs. Mycotoxins are produced mainly from *Aspergillus* spp., *Fusarium* spp. and *Penicillium* spp. Mycotoxins cause many problems in animals and the potential for residues in meat, poultry or dairy products. The toxins also cause severe health hazards in public on consuming animal products containing such toxins. Aflatoxins are potent mycotoxins produced by *Aspergillus flavus* and *Aspergillus parasiticus*. There are four major toxins *viz.*, AGB1, AFB2, AFG1 and AFG2. AFG1 is the most commonly produced and the most toxic aflatoxin. The toxin is retained for long periods in liver and kidneys than the other tissues of the body. Excretion of the toxin is almost completed within 96 hours after ingestion.

Ochratoxins are produced by some *Penicillium* spp. and some *Aspergillus* spp. Ochratoxin A is the most common and the most toxic to mammals, birds and fish. It is mainly toxic to kidneys and some extent to liver. It is accumulated mainly in the kidneys of the consumers and cause potential risk. It is accumulated in less concentration in liver, fat and muscles. The toxins can be detected by commercially available immunoassay based kits. The animals found positive for toxins, should be retained on toxin-free diet for 4 weeks prior to slaughter to ensure that the residue levels in kidneys have decreased. A withholding period of 2 days is sufficient to clear muscle. The accumulation of ochratoxin A in ruminants is highly unlikely because of detoxification of the toxin in the rumen.

Shellfish Toxin

Shellfish tissues can contain three different groups of natural toxins according to clinical symptoms in the consumers. These toxins include paralytic shellfish toxin, diarrhoetic shellfish toxin and anamnestic shellfish toxin. These toxins are produced by algae in the environment which are consumed by shellfish. These toxins occur in most shellfish harvesting areas and concentration can vary considerably depending on seasonal conditions. There are no overt signs by which toxin-containing shellfish can be detected.

Other Residues Substances in Meat

Fluorine is accumulated in cattle that graze pastures contaminated with industrial effluents. Chronic disease associated with fluorine is staining of teeth and degenerative changes in skeletal system and internal organs. Non-steroidal anti-inflammatory drugs (NSAIDs) can be categorized into two main groups such as carboxylic acids (*e.g.*, salicylate, aspirin, ibuprofen, ketoprofen *etc.*) and enolic acids (*e.g.*, phenylbutazone, dipyrone, isopyrin *etc.*). NSAIDs are toxic substances, therefore their use in farm animals for meat production is not recommended.

Maximum Residue Levels (MRLs)

MRLs for pesticide and veterinary drugs are the maximum concentrations of residues to be legally permitted in or on a food. MRLs for pesticides may also be applicable to animal feeds. MRLs are set by the Codex Alimentarius Commission (CAC), acting as the risk manager. The amount of pesticide residues in soil may influence the nature and level of residues in crops, particularly for soil or seed treatments. Crop rotational studies are designed to define the nature and level of pesticide residues that might occur in a crop sowed or planted subsequent to the original crop that received the pesticide treatment.

Method of Residues Detection in Meat

Detection of drug residues from tissues and other animal products could be quite an expensive, time consuming and laborious venture. Typically, muscle, liver, kidney and fat are analyzed because they are the tissues that are typically eaten in large amounts, tissues that function as storage points for fat-soluble residues, or tissues that metabolize the major portion of the drug in the process of body elimination (microbiological methods are quite suitable for the detection of antimicrobial residues especially as they are less expensive than immunochemical and chromatographic methods) and are able to screen a large number of samples at minimal cost.

Initially there were microbial growth inhibition tests, followed by more sensitive and specific methods based on receptor binding, immunochemical, and chromatographic principles. Chromatography is a technique used for the separation of constituents from the mixture. In chromatography, two phases are involved in separation. The extract from the sample contains mixture of pesticide and co-extractives; various chromatographic techniques can be employed for separating them or removal of the co-extractives from the pesticides. High performance liquid chromatography (HPLC), gas chromatography *etc.* are important methods for analysis of residues in the samples.

Health Hazards due to Chemical Contamination of Meat

Contaminant is any substance not intentionally added to food, which is present in such food as a result of the production (including operations carried out in crop husbandry, animal husbandry and veterinary medicine), manufacture, processing, preparation, treatment, packaging, transport or holding of such food or as a result of environmental contamination. The term does not include insect fragments, rodent hairs and other extraneous matter.

The residue of a substance has a pharmacological action of their metabolites and of other substances transmitted to animal products adversely affects human health. High level of residues of chemical substances such as pesticides, metal, hormones *etc.* in meat have their toxic effects in the meat consumers. Chemical contamination of meat may lead to the development of resistant pathogens, hypersensitivity reaction, effects on intestinal microflora, carcinogenic effect, teratogenic effect *etc.*

Prevention and Control of Pesticide Residues in Food

Insecticide Act and Prevention of Food Adulteration (PFA) Act are enforced in India for judicious use of pesticides and safety of consumers of animal products. The Government of India has taken steps to ensure the safe use of pesticides. The Insecticide Act, promulgated in 1968 and enforced on 1st August, 1971envisages to regulate the import, manufacture, sale, transport, distribution, and use of insecticides, with a view to prevent risks to humans and animals, and for matters connected therewith. ICMR's National Institute of Occupational Health (NIOH), Ahmadabad, and several other national laboratories, universities and other research and development organizations have been actively engaged in toxicological evaluation of pesticides, synthesis of safer molecules and evaluation of environmental contamination due to pesticides.

WHO and FAO have jointly developed an "International Code of Conduct on Pesticide Management". The most recent edition of the voluntary framework was published in 2014. It guides government regulators, the private sector, civil society, and other stakeholders on best practices in managing pesticides throughout their lifecycle from production to disposal.

Antimicrobial Residues in Meat

Antimicrobial residues in food have become a serious issue in recent years due to a widespread uncertainty among consumers about food safety. There are many factors influencing the occurrence of residues in animal products such as drug's properties and their pharmacokinetic characteristics, physico-chemical or biological processes of animals and their products. The most likely reason for drug residues might be due to improper drug usage and failure to keep the withdrawal period.

Sources of Antimicrobial Residues in Meat

Antimicrobial agents are used for prophylactic and therapeutic purposes, growth promotion *etc.* Antimicrobial agents are used in the treatment of several systemic as well as localized infectious diseases. Antimicrobial agents persist in the meat obtained from recently treated slaughtered animals.

Commonly Used Antimicrobials

Commonly used antimicrobials include penicillins, aminoglycosides, tetracyclines, cephalosporins and sulphonamides. Nystatin and griseofulvin are antifungal agents. Bacitracin, Polymyxin B and other sulphonamides have been used as additives in feed. Therefore, it is essential that the inspection team should be constantly vigilant for the potential problems of antibiotics throughout the inspection procedure. At the time of ante-mortem and post-mortem inspection the presence of disease conditions such as pleuropneumonia or arthritis in pigs and sheep or chronic mastitis in cows, may suggest that an antibiotic has been administered recently. Recent injection sites may be seen as area of discolouration or bruising.

Table 21.2. The Maximum Residue Levels of Some Antimicrobials for Bovines

Compound	*Target tissue*	*Concentration (μg/g)*
Benzylpenicillin	Muscle, liver, kidney, fat	50
	Milk	4
Ampicillin	Muscle, liver, kidney, fat	50
	Milk	4
Cloxacillin	Muscle, liver, kidney, fat	300
	Milk	30
Tetracycline	Kidney	600
	Liver	300
	Muscle	100
	Milk	100
Streptomycin	Kidney	1000
	Muscle, liver, fat	500
	Milk	200
Erythromycin	Muscle, liver, kidney, fat	400
	Milk	40
Tylosin	Muscle, liver, kidney, fat	100
	Milk	50
Cefquinone	Kidney	200
	Liver	100
	Muscle	50
	fat	50
Sulphonamides	Muscle, liver, kidney, fat	100
Trimethoprim	Muscle, liver, kidney, fat, milk	50

The residues of antimicrobials beyond their maximum permissible limit of concentrations have several adverse effects on the heath of consumers and economy of the product obtained from them. Higher levels of antimicrobial residues in the meat may cause damage of liver, kidneys and other organs in the consumers, emergence of antimicrobial resistant microorganisms, failure of manufacturing of desired fermented products, unsighty lesions and discolouration at the site of injection, formation of hard and fibrous nodules at the site of oily base injection, yellowish colouration at the site of tetracycline injection *etc.*

Potential Effects of Antimicrobial Residues on Public Health

Low-level contamination of antimicrobial agents generally may not generate a violation problem on public health. However, extensive use of these substances may increase the risk of an adverse effect of residues including the occurrence of antimicrobial resistance and hypersensitivity reaction. The major public health significances of drug residue are development of antimicrobial resistance, hypersensitivity reactions, carcinogenicity, mutagenicity, teratogenicity and disruption of intestinal normal microflora.

Antimicrobial agents are used as therapeutic agents as well as feed additives. The presence of residues of banned substances/substances permitted but exceeding their permissible limits in veterinary drugs, pharmaceutical products and pharmaceutically active substances in products of animal origin like muscle, liver, kidney, fish-flesh, milk, egg *etc.* is a matter of concern for public health. Antimicrobial residues are considered undesirable for the following reasons:

1. Antimicrobial residues may cause allergic reactions in highly sensitized consumers.
2. Antimicrobial residues also create resistant bacteria in farm animals which may then pass to the consumer.
3. Some antimicrobials have carcinogenic property.
4. Antimicrobial agents alter the ruminal microflora which may reduce efficiency of digestion, growth and weight gain.
5. Antimicrobial agents also reduce the symptoms of disease and may result in unhealthy animals being accepted at ante-mortem inspection.
6. Antimicrobial residues produce unsighty lesions when administered by injection. The site of injection is discoloured and may be haemorrhagic if administered shortly before slaughter. Long standing injection sites, particularly those that incorporate an oil base, may be hard, fibrous nodules within a muscle. Long acting tetracycline may leave a yellowish stained area with a distinctive odour on injection site.
7. Incorporation of antimicrobial agents in feed poses particular problems for the meat inspection services for several reasons. First there are no injection sites to be detected by visual observation. Second the residue levels are likely to be low probably close to or below the maximum residue level.

Therefore, it is essential that the inspection team should be constantly vigilant for the potential abuse of antimicrobial agents throughout the inspection procedure. Particular attention must be paid to suspect casualty animal, which should be sampled as a matter of routine. Recent injection sites may appear as an area of discolouration or bruising but deep intramuscular injections can only be detected as a very slight swelling or lack of symmetry in the muscle. Suspected carcases must be retained until samples prove negative to laboratory investigation.

Methods of Detection of Antimicrobial Residues in Meat

Chemical detection of antimicrobials is difficult owing to their diverse and great variation in the chemical structure and molecular weights. The most commonly used test for antimicrobial agents are based on the detection of residual antimicrobial activity. Antimicrobial residues in food can be detected in several ways:

1. **ELISA Test:** An ELISA test allows a precise quantification of certain antimicrobial agent.

2. **HPLC Test:** By performing a sample preparation using immuno-affinity columns and a subsequent HPLC test, even complex matrices can be analyzed.
3. **Growth of Bacterial Test:** Another test is a microbial test in which a population of bacteria is added to the sample and observed for their growth. Antimicrobial residues can be detected on the basis of growth of bacteria and decided whether the sample contains antimicrobial agent or not. This method allows an easy and cost-effective screening for a wide range of antimicrobial agents. In addition, this test can be performed directly on-site at the breeder, at the slaughterhouse or at meat processing plants. Antimicrobial residues in meat can be detected by the following procedure:
   i. Inoculate young culture of *Sarcina lutea* in nutrient broth.
   ii. Dilute the test organism (*Sarcina lutea*) by 1 ml of culture in 5 ml normal saline.
   iii. Pour the diluted culture over the peptone-free nutrient agar plate (pH 7.4-7.6).
   iv. Discard the excess fluid after few minutes and allow to dry the plate.
   v. Incubate at 37°C for 30 minutes.
   vi. Sterilize the surface of the kidney with a hot spatula.
   vii. Incise kidney and place a small piece of renal cortex and medulla over the incubated culture plate.
   viii. Allow the plate for pre-diffusion at 4°C for 3 hrs.
   ix. Keep moistened strip paper inside the plate to prevent drying.
   x. Incubate the plate at 30°C for 48 hrs.
   xi. If kidney tissues have antimicrobial residue, zone of inhibition around the tissue is evident.
4. **Four-Plate Test (FPT):** The most frequently used tests for antimicrobial agents are based on the detection of residual antimicrobial activity. The basic microbial method is the four-plate test (FPT). This is an agar diffusion test. This test has the advantage of requiring simple apparatus and of having rapid turn around and broad spectrum detection and well-suited to routine laboratory work. The disadvantage of this system is the lack of specificity and its reliability. False positive results arise most frequently from the use of deep frozen tissues, particularly kidney. To avoid these problems, each tissue can be tested in duplicate, one piece being placed directly on the agar surface and the other being placed on a dialysis membrane placed on the surface. The membrane prevents the diffusion of inhibitory cell components into the medium. The procedure of the test is given as follows:
   i. Take 3 agar plates inoculated with *Bacillus subtilis* spores at pH 6, 7.2 and 8.0.

ii. Add trimethoprim into the pH 7.2 medium to enhance the sensitivity of the test for sulphonamides residues.
iii. Take another agar plate (fourth agar plate) inoculated with *Micrococcus luteus* at pH 8.0.
iv. Apply the meat sample to four plates of agar medium.
v. Incubate the plates at 37°C for overnight.
vi. Observe diffusion of the active antimicrobial agent by the formation clear zone of inhibition on one or more plates.

5. **High Voltage Electrophoresis (HVE) Test:** The procedure of the test is given as follows:
   i. Prepare 2 gels (agar and agarose).
   ii. Place a piece of meat on each medium.
   iii. Allow the antimicrobials to diffuse in the medium.
   iv. Pass high voltage through the medium for a period of 2 ½ hrs.
   v. Overlay the plates with media containing sensitive species of bacteria (*Bacillus subtilis* or *Micrococcus leutus*).
   vi. Incubate at 37°C for overnight.
   vii. Observe the inhibition of growth of bacteria by antimicrobials.
   viii. Interpret the result on the basis of zone of inhibition.

Method of Reducing the Antimicrobial Residues in Animal Products

The methods of reducing the antimicrobial residues in animal products are as follows:

1. **Herd Health Management:** All food animals should be maintained in a clean and healthy environment whenever possible. Antimicrobial residues are best avoided by implementing management practices (good nutrition to meet growth, maintenance and lactation needs) and herd health programme that keep animals healthy and producing efficiently.
2. **Use of Approved Drugs:** Dairy and beef producers should not use or store un-approved drugs, special mixes, or products within adequate labels as unapproved drugs have no data regarding efficacy, safety, or withholding time. There should be correct and proper use of antimicrobials in animals in order to avoid their residues in the animal products. Unnecessary use of antimicrobial chemoprophylaxis and chemotherapy must be avoided.
3. **Proper Drug Administration and Identification of Treated Animals:** Before administering drugs, it is important to know that the drugs approved for all classes of cattle on the farm and be familiar with approved dosage, route of administration, and withholding time; when animals are treated with antimicrobials, a waiting period is required before their slaughter.

4. **Withdrawal Period:** The withdrawal period is also known as the depletion or clearance period. It is the interval necessary between the last administration of the drug to the animals under normal condition of used and the time when treated animal can be slaughtered for the production of safe foodstuffs. Depending on the drug product, dosage form, and route of administration, the withdrawal period may vary from a few hours to several days or even weeks.
5. **Proper Maintenance of Treatment Records and Identification of Treated Animals:** Institution of a workable health record for each animal to record all health related events including administration of medication is important. Record the identification of all animals in the permanent health record book.
6. **Proper Drug Residue Testing:** There should be proper drug residue testing capabilities on and off the farm. This control point addresses the conditions under which residue testing should be considered. There should be proper selection and interpretation of tests, the inherent limitation and potential misuse of residue testing *etc.*
7. **Awareness:** Creating awareness of proper drug use, and methods to avoid marketing adulterated products principally educational, total residue avoidance programme based upon the objective of improving the livestock producer's management and quality control of marketing animals with emphasis on avoidance of drug residues.

22
Meat Microbiology

It is important to provide adequate rest and good facilities in the lairage to the animals in order to keep them free from stress and injury. This helps to prevent the unnecessarily depletion of muscle glycogen reserves in them. Cattle and buffaloes are rested preferably for a period of 24 hours in the lairage before their slaughter. Adequate muscle glycogen reserve in the animals serves as a source of lactic acid in the meat that help in maintaining its quality. Lactic acid in the muscle retards the growth of bacteria that have contaminated the carcase during slaughter and dressing. These bacteria cause spoilage of the meat during storage, particularly at higher temperatures and the meat changes such as off-smells, colour changes, rancidity and slime may occur. Spoilage of meat decreases its shelflife which results into wastage of meat.

Contamination of meat with food poisoning organisms may lead to infections and intoxications in the meat consumers resulting into severe health hazards and economic losses in treatment, hospitalization, loss of manpower and wages. Thus, the meat from animals, which have suffered from stress or injuries during handling, transport and slaughter, is likely to have a shorter shelflife due to spoilage. This is perhaps the biggest cause for meat wastage during the production processes.

Spoilage of Meat

Pre-slaughter handling of livestock and post-slaughter handling of meat play an important part in deterioration of meat quality. Some of the pre-slaughter handling that influence the spoilage of meat includes nutrition, transportation, marketing, lairaging and stunning. The main causes of spoilage of meat and meat products after slaughtering and during processing and storage are microorganisms, lipid oxidation and autolytic enzymatic spoilage. Spoilage in meat and meat products is caused mainly by bacteria and fungi.

Bacterial Spoilage

1. **Bone Taint:** This is associated with the growth of putrefactive bacteria (mostly anaerobic spore-forming bacteria). The rate of cooling is slow in deep seated musculature especially in heavy carcase that is carcase with greater thickness of musculature or excess amount of fat on the carcase. This may result in growth of anaerobic spore-forming bacteria and cause development of taint in both the musculature around the bone (*e.g.*, femur) and in the bone marrow. The development of bone taint can be prevented by preventing the contamination of carcase and its rapid and efficient cooling. The affected part of the carcase requires local condemnation.
2. **Phosphorescence:** This is a type of microbial spoilage which is caused by *Pseudomonas phosphorescens*. The organism is widely distributed in nature especially in sea water and may contaminate chilling room and the surface of the meat when seen in a dark room, shows luminous areas scattered on the surface of the meat.

Fungal Spoilage

1. **Black Spot:** This is type of fungal spoilage of meat which occurs on the surface of the meat. Black spot is caused mainly by *Cladosporium herbarum*. The condition is also caused by *Pullularia* spp. and *Rhizopus* spp. This defect is mostly found in imported meat. In beef, black spot is commonly found on the neck, diaphragm and pleura, while in frozen mutton, it is found on the legs, inside the neck or in the thoracic or abdominal cavity. The condition arises when quarters of the chilled beef are taken from ships and placed in the cold store at temperature above -8°C. Black spots are about 6-13 mm in diameter; occur on the surface of the meat. Black spot may be accompanied by bacterial spoilage, associated with *Achromobacter* spp. Less extensive black spots on the surface of the carcase without bacterial spoilage can be removed by trimming. Too extensive black spots on the surface of the carcase accompanied by bacterial spoilage require condemnation of the whole quarter.
2. **White Spot:** This is a type of fungal spoilage that occurs on the surface of the meat which is characterized by development of small, flat, wooly spots frequently accompanying black spots of similar size. This defect is most commonly found in imported meat. This is caused by *Sporotrichum* spp. and *Chrysosporium* spp. The spores of these organisms can develop at -8°C and become profuse at temperature above 0°C.
3. **Whiskers:** This is a type of fungal spoilage of meat which is caused by *Thamnidium* spp. and *Mucour* spp. This is characterized by hair-like growth of fungi. Hyphae grow well at 0°C and may project more than 2.5 cm on the surface of the meat. The growth of these fungi ceases at temperature below -7.5°C.

Microbial Toxins in Foods and their Health Hazards

Microorganisms in food may cause its spoilage as well as health hazards in consumers resulting from infections and intoxications caused by them. Food poisoning includes bacterial and viral infections; chemical contamination of food, plant or animal toxins; and food allergies. Pre-slaughter conditions of the animals may affect the quality of meat in terms of the numbers and types of microorganisms. Pathogenic microorganisms contaminating meat from the gastrointestinal tract are *Salmonella* spp., *Yersinia enterocolitica, Campylobacter jejuni, Aeromonas hydrophila, Listeria monocytogenes* and *Escherichia coli*. Foods are rapidly colonized by bacteria, moulds or yeasts, which are the main causes of spoilage, a consequence of decomposition by factors such as enzyme action and oxidation. Microorganisms in meat causing health hazards are discussed as follows:

Clostridium botulinum

Clostridium botulinum is a Gram-positive, anaerobic, sporulating, non-motile, catalase negative rod-shaped organism. It produces neurotoxins, *viz.*, A, B, C, D, E, F and G. The organism produces heat-stable toxins. Human botulism is caused by *Clostridium botulinum* type A, B, E and rarely F. The classical food-borne botulism occurs due to ingestion of preformed toxin in food.

Clostridium perfringens

Clostridium perfringens is a Gram-positive, anaerobic, sporulating, non-motile, catalase negative rod-shaped organism. It may be of different types such as type A, B, C, D, E, and F. α-toxin is produced by all strains of *Clostridium perfringens*. This toxin causes haemolysis and necrosis. β-toxin is produced by type B, C and F. This toxin causes inflammation of intestine and damage of mucosal cells (enteritis). ε-toxin is produced by type B and D. This toxin is highly lethal and causes necrosis. ß-toxin is produced by type E. *Clostridium perfringens* type A causes alteration in the permeability of intestinal mucosa. This increases intestinal outputting of sodium and chloride ions and water and cause diarrhoea.

Escherichia coli

Escherichia coli is a Gram-negative, straight, rod-shaped bacterium which belongs to the family Enterobacteriaceae. *E. coli* produces different types of toxins. Enterotoxic *E. coli* (ETEC) produces enterotoxins, *viz.*, heat labile toxins (LT) and heat stable toxins (ST). LT activates adenylate cyclase enzyme and causes loss of fluid and electrolytes. ST is of two types, *viz.*, STa and STb. STa toxin affects calves, and STb toxin affects piglets. Enterohaemorrhagic *E. coli* (EHEC) produces STx toxin which causes haemorrhagic colitis (HC), hemolytic uremic syndrome (HUS) and thrombotic thrombocytopenic purpura (TTP) in humans. Enteroaggressive *E. coli* (EaggEC) produces EAST-1 toxin which causes chronic enteritis.

Salmonella spp.

Salmonella is a Gram-negative, straight coccobacillus, filamentous, rod-shaped, motile bacterium which belongs to the family Enterobacteriaceae. The genus

Salmonella consists of two species, *viz.*, *Salmonella enterica* and *Salmonella bongori*. *Salmonella enterica* is further classified into six subspecies, *viz.*, *Salmonella enterica enterica, Salmonella enterica salamae, Salmonella enterica arizonae, Salmonella enterica diarizonae, Salmonella enterica houtenae* and *Salmonella enterica indica*. Each subspecies of *Salmonella enterica* has several serovars with a total number more than 2579 serovars. *Salmonella bongori* has more than 20 serovars. *Salmonella* produces 3 types of toxins. First toxin is heat-stable trypsin sensitive cytotoxin. Second toxin is cytotoxin and third toxin is haemolysin. These toxins cause diarrhoea and gastroenteritis.

Shigella dysenteriae

Shigella dysenteriae is a Gram-negative, rod-shaped organism, which belongs to the family Enterobacteriaceae. *Shigella dysenteriae* produces toxins like Stx1 and Stx2. These toxins increase the activity of adenylate cyclase enzyme which causes diarrhoea.

Staphylococcus aureus

Staphylococcus aureus is an aerobic, spherical or ovoid, Gram-positive organism which produces enterotoxin B (SEB). It stimulates T-lymphocytes which lead to release of massive amount of cytokines in the lamina propria of intestine, which causes nausea, vomiting and diarrhoea.

Vibrio parahaemolyticus

Vibrio parahaemolyticus is a straight or curved, Gram-negative bacterium which belongs to the family Vibrionaceae. *Vibrio parahaemolyticus* produces Kanagawa haemolysin. It also produces TDH aka enterotoxin.

Yersinia enterocolitica

Yersinia enterocolitica is a Gram-negative, straight rod, facultative anaerobic and motile organism below30 °C but non-motile at 37 °C. It produces YST toxin which is a heat-stable enterotoxin. It increases the activity of guanylate cyclase enzyme and causes diarrhoea.

Fungi

Fungi such as *Aspergillus* spp., *Fusarium* spp. and *Penicillium* spp. produce mycotoxins which are their metabolic products. *Aspergillus* spp. produces aflatoxin M. The main foodstuffs involved are milk and milk products. This toxin is carcinogenic. Some other fungi produce ochratoxin A. The main foodstuffs involved are blood and kidney of pigs; blood and liver of poultry; and occasionally milk and milk products. This toxin is carcinogenic, teratogenic, immunosuppressive and nephrotoxic.

Isolation and Identification of Microorganisms from Meat and Meat Products

The typical protocol for isolation of most of the microorganisms includes selective enrichment followed by selective plating. Isolation and identification of important microorganisms from meat and meat products are discussed as follows:

Aeromonas spp.

The enrichment media for isolation of *Aeromonas* spp. include alkaline peptone water (APW), alkaline peptone water-cephalothin (APW-C) and *Aeromonas* enrichment broth. The selective medium for isolation of *Aeromonas* includes Ampicillin dextrin agar (ADA).

Isolation Procedure

1. Inoculate food sample in APW-C and AEB and inoculate at 37 °C for 18 hours.
2. Streak a loopful culture from enrichment broth onto ADA and incubate at 37 °C for 24 hours. Avoid over incubation, as yellow colour colonies of *Aeromonas* spp. is changed to blue colour.
3. Pick up characteristic colonies and subculture onto nutrient agar for further characterization.

Identification

On ADA, it forms yellow-coloured, small, smooth and honey-drop like colonies. It is a Gram-negative short rod. It is arginine decarboxylase, ornithine decarboxylase, catalase, citrate and oxidase positive but urease negative. It can grow in KCN and 0 per cent NaCl. It hydrolyses gelatin. It gives positive test for nitrate reduction but does not produce H_2S. It does not produce gas from glucose but ferments arabinose, sucrose, maltose and mannitol.

Bacillus cereus

Trypticase-soy-polymyxin (TSP) broth is used as enrichment medium for isolation of *Bacillus cereus*. Mannitol egg-yolk polymyxin (MEYP) agar and polymyxin egg-yolk mannitol bromothymol blue agar (PEMBA) are used for isolation of *Bacillus cereus*. Heat shock at 70°C for 15 min before incubation enhances germination of spores.

Isolation Procedure

1. Inoculate food sample in nutrient broth or trypticase-soy-polymyxin (TSP) broth and inoculate at 37 °C for 18 hours. A heat shock of 70 °C for 15 minutes before incubation is recommended.
2. Streak a loopful culture from enrichment broth/homogenate onto MEYPagar or PEMBA and incubate at 37 °C for 24 hours.
3. Pick up characteristic colonies and subculture onto nutrient agar for further characterization.

Identification

On nutrient agar, it forms large, grey-white and granular colonies. On blood agar, it is haemolytic and on egg-yolk agar, it gives a strong lecithinase reaction which is seen as a halo of opalescence and opacity around the colonies. *Bacillus cereus* is a Gram-positive, rod-shaped, non-capsulated and motile organism. It causes liquefaction of gelatin stabs. It haemolyses sheep RBCs. It gives positive test with catalase and nitrate reduction.

Campylobacter spp.

Bolton broth and Preston broth along with selective supplements can be used as enrichment media for isolation of *Campylobacter* spp. Selective supplements such as ferrous sulphate, sodium metabisulphite and sodium pyruvate are used when blood is not available. The enriched sample is incubated at 42-43 °C for 24 h under microaerophilic atmosphere (5 per cent O_2, 10 per cent CO_2 and 85 per cent N_2). Skirrow's selective medium, Butzler's selective medium, Blaser-Wang selective medium and Preston medium are selective media for isolation of *Campylobacter* spp.

Isolation Procedure

1. The sample should be processed as soon as possible after its collection. The samples that cannot be processed within 4 hours or those likely to contain small number of organisms should be placed in a transport medium, *e.g.*, Cary-Blair medium.
2. Mix a total of 25 g meat sample in 100 ml enrichment broth and incubate at 42-43°C for 24 hours under micro-aerobic atmosphere (5 per cent O_2, 10 per cent CO_2 and 85 per cent N_2).
3. Streak a loopful of culture on selective medium from growth in enrichment broth and incubate at 42-43°C for 24-48 hours under micro-aerobic conditions.
4. Examine for *Campylobacter* spp. colonies after 24-48 hours.

Identification

The colonies of *Campylobacter* spp. are small, flat and translucent. They appear as water droplet and tend to show confluent growth on a wet surface. The colonies on blood agar are non-haemolytic, flat or slightly raised having irregular edges. *Campylobacter* spp. is a Gram-negative, curved rod appearing in comma, S-shaped or spiral forms. *Campylobacter* can be recognized by its peculiar darting motility, which has a corkscrew-like movement. Most thermophilic campylobacters are oxidase, nitrate and catalase positive, and urease negative (exceptionally *Campylobacter lari* is urease positive). It generally does not produce H_2S on triple-sugar iron (TSI) medium.

Clostridium botulinum

The classical food-borne botulism occurs due to ingestion of pre-formed toxin in food. Hence detection of toxin is of more significance than the isolation of the organism. Isolation of *Clostridium botulinum* invariably requires enrichment and subsequent plating. Robertson cooked meat (RCM) medium, trypticase-peptone glucose yeast extract (TPGY) broth and thioglycolate broths are enrich media for isolation of *Clostridium botulinum.* Sample inoculated in RCM medium is incubated at 35 °C, while TPGY at 20°C for 5 days. Horse blood agar and anaerobic egg yolk agar can be used as selective media for isolation of *Clostridium botulinum.* Streaked plate is incubated at 35 °C for 48 hours. It is a strict anaerobe and grows optimally at 30-37 °C, depending on strain.

Isolation Procedure

1. Remove dissolved oxygen from media by steaming for 15 minutes and cool.
2. Inoculate 2 g sample in two tubes, each of RCM and TPGY media.
3. Incubate at 35 °C for RCM medium and 20°C for TPGY medium.
4. Examine after 5 days for turbidity and gas production. If there is no growth, extend the incubation for next 10 days.
5. Add equal amount of filter-sterilized ethanol to 2 ml of culture. Mix and incubate for 1 hour at room temperature.
6. Streak on selective medium with treated culture and incubate at 35 °C for 48 hours under anaerobic condition.
7. Observe for the growth of typical colonies.
8. Transfer 10 colonies in each of TPGY broth for *Clostridium botulinum* Type E and RCM medium for other toxin types.
9. Incubate at 35 °C for RCM medium and 20 °C for TPGY medium.
10. Streak culture in duplicate on egg yolk agar from step 8 and incubate one plate anaerobically and other plate aerobically at 35 °C. Growth of *Clostridium botulinum* on anaerobic plate may be pure if there is absence of growth on aerobic plate.

Identification

Clostridium botulinum colonies on solid medium are irregularly circular, large, smooth, greyish and translucent. Most strains are haemolytic on horse blood agar. *Clostridium botulinum* is a Gram-positive, anaerobic, spore-forming, rod-shaped organism. It is straight or slightly curved rod, arranged singly or less commonly in pairs or short chains. Spores are oval, sub-terminal and may slightly distend the cell. The organism is motile.

There are four metabolic groups of *Clostridium botulinum, viz.,* Group 1- Type A and proteolytic strains of type B and F, Group 2- Type E and sachharolytic strains of types B and F, Group 3- Type C and D, and Group 4- Type G. All the four groups of *Clostridium botulinum* hydrolyze gelatin, but give nitrate reduction negative test. Group 1 hydrolyzes aesculin, while other groups do not hydrolyze. Group 2 ferments sucrose and trehalose, while other groups do not ferment. Group 4 is lipase negative, while other groups are lipase positive.

Clostridium perfringens

On the basis of toxin production, *Clostridium perfringens* is grouped into 5 types, *viz.,* A, B, C, D and E. Among these, type A is the most important enterotoxin producing and food poisoning causing organism. Selective enrichment followed by selective plating can be approached for isolation of *Clostridium perfringens.* Robertson cooked meat (RCM) medium and thioglycolate broth are used as enrichment media for isolation of *Clostridium perfringens*. Tryptose sulphate cycloserine (TSC) agar, sulphate polymyxin sulphadiazine (SPS) agar, Shahid-Ferguson perfringens (SFP) agar and neomycin blood agar are used as selective media for isolation of *Clostridium perfringens*. Sample inoculated in enrichment broths are incubated at 44 °C for 24-48 h, while solid media at 37 °C for 24 h.

Isolation Procedure

1. Homogenize the sample and inoculate in thioglycollate enrichment broth in 1:10 ratio.
2. Incubate at 44 °C for 48 hours.
3. Take 0.1 ml enrichment broth grown culture in an empty sterile petri plate.
4. Pour sterile molten selective plating medium over the inoculum and mix properly.
5. Incubate at 37 °C for 24 hours in anaerobic condition (in McIntas anaerobic jar with hydrogen gas filled in commercially available anaerobic gas pack). Alternatively, take inoculum from enrichment broth in a sterile test tube and add molten selective plating medium (SPS agar). Mix well and allow to solidify. Again add a layer of molten selective plating medium in the tube to cover it, allow to solidify and incubate at 37 °C for 48 hours.
6. Observe for growth of characteristic colonies of the organism.

Identification

The characteristic colony of the organism is picked up and identified on the basis of morphological characters and biochemical tests. *Clostridium perfringens* is anaerobic but can grow microaerophilically. Optimum growth temperature range is 37-45°C. On horse blood agar, colonies are usually surrounded by a variable zone of complete haemolysis and a wider darker zone of incomplete haemolysis. In TSC agar, sodium metabisulphite and ferric ammonium citrate act as indicators of sulphite reduction, indicated by black coloured colonies. Kanamycin and polymyxin are added to inhibit accompanying flora. On SFP agar, it forms black colonies surrounded by opaque zone. In SFP tube, it forms black cotton wool like growth. On SPS agar, it forms black colonies because it reduces the sulphite to sulphide which reacts with iron and forms black iron sulphide precipitate.

Clostridium perfringens is relatively a large Gram-positive rod with stubby ends, occur single or in pairs and often capsulate in tissues. It is a non-motile and sporulating organism. It is actively saccharolytic which ferments glucose, lactose, sucrose, maltose, mannose, inositol, and starch and produces gas. It produces acid and gas in litmus milk medium and gives "stormy clot" reaction. It causes gelatin liquefaction but usually does not cause liquefaction of coagulated serum. It actively reduces sulphite and nitrates.

Escherichia coli

Escherichia coli can be isolated from meat sample by enrichment in selective broth followed by selective plating. Enterobacteriaceae enrichment (EE) broth and

MacConkey broth are commonly used as enrichment media for isolation of *E. coli.* MacConkey agar and eosin methylene blue (EMB) ager are used for isolation of *E. coli.* MUG (4-methyl-umbellyferyl glucuronide) sorbitol agar is a selective plating medium for isolation of *E. coli* O157:H7.

Isolation Procedure

1. Homogenize 10g meat sample with 90 ml enrichment broth and incubate at 37°C for 24 hours.
2. Streak a loopful of culture from enrichment broth on MacConkey agar and incubate at 37°C for 24 hours.
3. Streak a very small amount of the colony exhibiting small, round and pink colour from MacConkey agar onto eosin-methylene blue agar and incubate at 37°C for 24 hours. Use 4-methyle-umbliferyl-beta-D-glucuronide (MUG) sorbitol agar for isolation of *E. coli* O157:H7.
4. Observe the growth of characteristic colonies of *E. coli.*
5. Identify the organism on the basis of cultural, morphological, biochemical, serological and molecular characterization.

Identification

E. coli ferments lactose and produces smooth, glossy, pink colonies on MacConkey agar. EMB is used for the isolation and differentiation of Gram-negative enterobacteria from clinical and non-clinical specimens. The colonies of *E. coli* show metallic sheen on EMB agar. It forms yellow colonies on cystine lactose electrolyte-deficient (CLED) agar or xylose lysine deoxycholate (XLD) agar. Microorganisms utilizing sorbitol exhibit yellow colonies whereas sorbitol-negative strains (such as *E. coli* O157:H7) grow as colourless colonies on MUG sorbitol agar. MUG is cleaved by beta-glucuronidase forming pathogens and can be detected by fluorescence under UV light. The plates are exposed to ammonia fumes to increase fluorescence. All commensal *E. coli* produce β-glucuronidase and therefore cleave MUG and appear fluorescent when observed under long wave UV light (366 nm). *E. coli* O157:H7 is not capable of forming β-glucuronidase, thus when exposed to long-wave UV light, no fluorescence is observed.

E. coli is a Gram-negative rod-shaped organism. It gives indole and methyl red positive tests. It gives negative tests with Voges-Proskauer, citrate, phenylalanine deaminase and urease. It does not produce H_2S in triple sugar iron (TSI) agar and does not grow in potassium cyanide (KCN) medium. Most of its strains form gas from glucose.

Listeria monocytogenes

Listeria monocytogenes is aerobic and facultative anaerobic organism. Optimum temperature for growth is 35-37 °C, but capable to grow at 2-43 °C. University of Vermont (UVM) broth I and II and Fraser broth I and II are two step enrichment broths for isolation of *Listeria monocytogenes*. One step enrichment broth such as polymyxin acriflavin lithium chloride ceftazidime aesculin mannitol (PALCAM) broth, *Listeria* enrichment broth and modified UVM broth are also used for isolation of *Listeria monocytogenes*. Drominguiz-Rodringuiz agar (DRA), PALCAM agar and lithium chloride phenyl ethanol moxalactum (LPM) agar are used as selective plating media for isolation of *Listeria monocytogenes*.

Isolation Procedure

1. Take 10 g meat sample and homogenize with 90 ml of enrichment broth (UVM-I or Fraser-I).
2. Incubate at 30°C for 24 hours.
3. Transfer 0.1 ml inoculum from UVM-I or Fraser-I broth to UVM-II or Fraser-II broth and incubate at 30°C for 24 hours.
4. Streak a loopful inoculum from enrichment broth onto selective agar and incubate the plate at 30°C for 48 hours.
5. Pick up characteristic colonies of *Listeria* and process for identification on the basis of cultural, morphological and biochemical characters and other tests.

Identification

On blood agar, *Listeria monocytogenes* forms zone of slightly hazy β- haemolysis. On DRA, it produces greenish-yellow glistening colonies. On LPM agar, colonies are observed under oblique trans-illumination and colonies appear bluish with ground glass appearance. On PALCAM agar, it forms black halo around colonies (*i.e.* grey-green in colour with black sunken center and black halo against a cherry-red medium).

Listeria monocytogenes is a Gram-positive rod, occurs singly or in pairs which are often angled at the point of contact and may resemble diphtheroids or diplococci. It is non-sporing, non-capsulating and non-acid-fast organism. *Listeria* spp. gives positive tests for catalase, Voges-Proskauer and methyl red. It gives negative tests for oxidase, urease, indole and H_2S production. It ferments glucose without gas production. Christie Atkins Munch Peterson (CAMP) test is used to differentiate *Listeria monocytogenes* (CAMP positive) from other *Listeria* spp. which are CAMP negative.

Salmonella spp.

Salmonella occurs in small numbers in the food, therefore, three sequential steps, *viz.*, pre-enrichment, selective enrichment and selective plating are used for its isolation. However, pre-enrichment step can be omitted for isolation of *Salmonella* spp. from faecal sample. Buffered peptone water is used as pre-enrichment medium in which inoculated sample is incubated at 37°C for 20-24 hours. Rappaport-Vassiliadis broth, selenite F broths and tetrathionate broth are used as selective enrichment media in which pre-enriched culture is inoculated. Inoculated selective enrichment media like tetrathionate broth and selenite F broth are incubated for 18-24 h at 37°C, while Rappaport-Vassiliadis broth is incubated at 43°C. Brilliant green agar (BGA), bismuth sulphite agar (BSA), xylose lysine deoxycholate (XLD) agar and Hektoen enteric agar (HEA) are used as selective media for isolation of *Salmonella* spp. Brilliant green agar is an excellent selective as well as differential medium for *Salmonella* except *S.* Typhi. Brilliant green dye present in the medium inhibits the growth of *E. coli, Proteus and* other commensal enterobacteria.

Isolation Procedure

1. Homogenize 50 g meat sample in 250 ml 0.1 per cent buffered peptone water.
2. Add 50 ml volumes of the homogenate to two 50 ml volumes of selenite broth; incubate the cultures for 24 hrs., the one at 37°C and the other at 43°C.
3. Streak a loop full of selectively enriched broth culture onto selective plating media *viz.*, brilliant green agar (BGA), Hektoen-enteric agar (HEA), xylose lysine deoxycholate (XLD) agar, bismuth sulphite agar (BSA) and incubate the plates at 37°C for 24 hours.
4. Examine for *Salmonella* colonies after 24 hrs. and 48 hrs.

Identification

On BGA, *Salmonella* produces pink colour colonies surrounded by red medium. On BSA, XLD and HEA, *Salmonella* produces black colour colonies. *Salmonella* forms pale yellow or nearly colourless colonies on MacConkey agar. *Salmonella* spp. is a Gram-negative, non-capsulated and non-sporing rod-shaped organism. Most strains of the organism are motile. It gives positive tests with methyl red, citrate and H_2S production. It gives negative tests with indole, Voges-Proskauer and urease. It ferments glucose, mannitol, maltose, dulcitol and sorbitol, while does not ferment lactose, sucrose and salicin. It does not liquefy gelatin and does not grow in potassium cyanide (KCN) medium.

Staphylococcus aureus

Staphylococcal enrichment broth and trypticase soy broth with 10 per cent NaCl and 1 per cent pyruvate are used as enrichment media for isolation of *Staphylococcus aureus.* It can be isolated from meat by direct plating. However, enrichment step can be included for better recovery. Baired-Parker agar (egg yolk glycine tellurite pyruvate agar), mannitol salt agar (MSA) and *Staphylococcus* medium No. 110 (Oxoid) are selective plating media for isolation of *Staphylococcus aureus.*

Isolation Procedure

1. Take 10 g meat sample and homogenize with 90 ml staphylococcal enrichment broth. Alternatively, trypticase soy broth with 10 per cent sodium chloride and 1 per cent sodium pyruvate can be used.
2. Inoculate meat sample in enrichment broth and incubate at 37°C for 24 hours.
3. Streak a loopful of culture from enrichment broth onto Baird-Parker agar and incubate at 37°C for 24 hours.

4. Observe the growth of characteristic colonies of *Staphylococcus aureus.*
5. Identify the organism on the basis of cultural, morphological, biochemical, serological and molecular characterization.

Identification

On Baird-Parker agar, it forms shiny jet-black colonies. On MSA, it forms small (1 mm diameter) yellow colonies surrounded yellow medium due to acid formation. Sodium chloride present in MSA inhibits the growth of most other bacteria but *Staphylococcus aureus* is tolerant of sodium chloride. On *Staphylococcus* medium No. 110 (Oxoid), it forms orange colour colonies. It causes haemolysis on blood agar. The organism is Gram-positive cocci and shows clusters like arrangement. The organism is catalase and coagulase positive and oxidase negative. It forms acid from glucose, maltose, mannitol, lactose, sucrose and salicin.

Vibrio spp.

The genus *Vibrio* is a member of the family Vibrionaceae. The organism is Gram-negative, short rod, often curved, and motile by means of a single polar, sheathed flagellum. Motility is optimum in liquid-grown cultures. Alkaline peptone water with 3 per cent NaCl (APWS) is used as enrichment medium for isolation of *Vibrio* spp. A secondary enrichment in glucose salt teepol broth (GSTB) is recommended for isolation of *Vibrio parahaemolyticus*. Thiosulphate citrate bile salt sucrose (TCBS) agar is used as selective plating medium for isolation of *Vibrio* spp. Sample inoculated in enrichment medium is incubated at 37 °C for 6-8 h, while streaked plate is incubated at 37 °C for 24 h.

Isolation Procedure

1. Inoculate meat sample in alkaline peptone water with 3 per cent sodium chloride (APWS) or trypticase soya broth with 3 per cent sodium chloride (TSBS) and incubate at 37°C for 6-8 hours.
2. A secondary enrichment broth "glucose salt teepol broth (GSTB)" may be inoculated by transferring 1 ml from the first enrichment broth and incubate at 37°C for 24 hours.
3. Streak a loopful of culture from enrichment broth onto TCBS agar and incubate at 37°C for 24 hours.
4. Pick up characteristic colonies of the organism and streak onto nutrient agar for further characterization.
5. Test the culture for oxidase reaction and other biochemical tests.
6. Send the identified *Vibrio cholerae* for serotyping at National Institute of Cholera and Enteric Diseases (NICED), Kolkata.

Identification

Strains of *V. cholerae* produce yellow colonies on TCBS agar because of fermentation of sucrose. *V. alginolyticus* also produces yellow colonies. A few strains of *V. cholerae* may appear green or colourless on TCBS agar due to delayed sucrose fermentation. *V. parahaemolyticus* is a sucrose non-fermenting organism, and therefore, produces blue-green colonies, as does *V. vulnificus*. TCBS agar is highly selective for *Vibrio* spp. However, occasional isolates of *Pseudomonas* and *Aeromonas* may also form blue green colonies on TCBS agar.

Vibrios are facultative anaerobes, producing acid from fermentation of a range of carbohydrates. They are oxidase positive (except *V. metschnikovii*) and reduce nitrate to nitrite. Most of vibrios are arginine dihydrolase positive except *V. fluvialis*. *V. cholerae* and *V. parahaemolyticus* are lysine decarboxylase and ornithine decarboxylase positive and produce indole. *V. cholerae* and *V. parahaemolyticus* grow in peptone water with 3 per cent NaCl and do not grow in 10 per cent NaCl. *V. cholerae* grows in peptone water with 0 per cent NaCl, while *V. parahaemolyticus* does not grow.

Yersinia enterocolitica

Yersinia enterocolitica is a Gram-negative straight rod or small cocci and non-motile at 37°C but motile at temperature below 30°C. It is a facultative anaerobic organism. Non-selective enrichment broth (1/15M phosphate buffered saline) can be used for isolation of *Yersinia enterocolitica.* Selective enrichment broths such as bile oxalate sorbose (BOS) broth, Irgasn ticarcillin chocolate (ITC) broth and peptone sorbitol bile salt (PSB) broth are used for isolation of *Yersinia enterocolitica.* Sample inoculated in non-selective enrichment broth is incubated at 4°C for 21 days, while selective enrichment broth at 22-25°C for 2-5 days. As many media used for Enterobacteriaceae can be used for isolation of *Yersinia enterocolitica.* Cefsulodin Irgasan novobiocin (CIN) agar, Virulent *Y. enterocolitica* (VYE) agar, *Salmonella-Shigella* deoxycholate calcium chloride (SSDC) agar and MacConkey agar can be used for isolation of *Yersinia enterocolitica.*

Isolation Procedure

1. Enrich the meat sample in non-selective enrichment broth and incubate at 4 °C for 21 days or enrich the sample in selective enrichment broth and incubate at 22-25 °C for 2-5 days.
2. For direct plating, mix 0.5 ml sample homogenate with 0.5 ml dilute aqueous alkaline solution (0.72 per cent KOH: 0.5 per cent NaCl) for 30 seconds and plate immediately.
3. For post-enrichment, mix 0.5 ml enrichment broth with 4.5 ml dilute aqueous alkaline solution (0.5 per cent KOH: 0.5 per cent NaCl) for 15 seconds and plate immediately.
4. Streak onto selective medium from enriched culture (step 1).
5. Incubate at 25 °C for 48 hours.
6. Pick up characteristic colonies of the organism and streak onto nutrient agar for further characterization.

Identification

Yersinia enterocolitica forms small colony (less than 1mm in diameter). On CIN agar, it forms red bull's eye colonies. On VYE agar, it forms red colonies. On SSDC agar and MacConkey agar, it forms small, round colourless colonies. Yersiniae are Gram-negative coccobacilli showing pleomorphism in older cultures. They are motile by means of peritrichous flagella when grown at 22 °C, while non-motile when grown at 37 °C. The organism is halophilic at 5 per cent but not at 7 per cent NaCl, urease positive and ferments mannitol. It does not produce H_2S, lysine decarboxylase and phenyl alanine deaminase.

23

Physico-Chemical Methods for Detecting Microbial Spoilage in Meat

There are certain changes in the meat due to contamination and other factors which lead to undesirable changes and spoilage in the meat. Such meat may be unsuitable for further processing and unfit for human consumption. Therefore, the quality evaluation of meat is important from health and economic point of view. Microbial spoilage in meat can be detected by various physico-chemical methods as follows:

Meat Swelling Capacity (MSC)

The freshness of meat can be determined on the basis of meat swelling capacity. Due to breakdown (spoilage) of protein the spaces between peptide chains is increased, resulting into penetration of more amount of water in protein matrix, which swells and increases meat swelling capacity. A test is applied to assess the quality of meat by determining the meat swelling capacity. Meat swelling capacity increases during spoilage due to breakdown of protein and penetration of more amounts of water in protein matrix. The procedure of determining the meat swelling capacity is given as follows:

1. Homogenize 25 g of meat sample with 100 ml distilled water for 2 minutes.
2. Centrifuge 35 ml of meat homogenate at 2000 rpm for 15 minutes.
3. Measure the amount of supernatant.

4. Calculate the percentage of meat swelling by using the following formula:

$$\text{Per cent of meat swelling} = \frac{(35\text{-S-P})}{\text{P}} \times 100$$

where, S= supernatant, P = pellet. The spoilt meat swells than the fresh meat.

Extract Release Volume (ERV)

The technique of determining the ERV of meat (beef) is useful for assessing the incipient spoilage in meat as well as predicting the refrigerated shelflife of meat. The ERV method of determining meat spoilage reveals two aspects of the spoilage mechanisms. First, low-temperature meat spoilage occurs in the absence of any significant breakdown of primary proteins. The second aspect of meat spoilage revealed by ERV is the increase in hydration capacity of meat proteins.

The technique is based on the volume of aqueous extract released by a homogenate of meat when allowed to pass through a filter paper for a given period of time. Meat of good organoleptic and microbial quality releases large volume of extract while meat of poor microbial quality releases smaller volume or none. The procedure of determining the ERV is given as follows:

1. Take 25 g of meat sample and homogenize with 100 ml distilled water for 2 minutes.
2. Filter the homogenate by funnel lined with Whatman's filter paper No. 1.
3. Collect the extract in a measuring cylinder.
4. Record the extract release volume and interpret the result on the basis of standards given in the following table:

Table 23.1. Determination of Quality of Meat on the Basis of ERV of Meat

ERV (ml)	*Meat Quality*
> 25	Good
> 20	Incipient spoilage
< 20	Spoilt

The fresh meat releases more extract volume because bacteria utilize glucose during the initial stages then attack protein. So protein of fresh meat absorbs less water. Spoilt meat releases less volume than fresh meat because bacteria breakdown proteins and protein matrix absorb more water.

Boiling Test

This is a type of sensory test for determining the quality of meat. The procedure of boiling test is given as follows:

1. Take a piece of fat rich meat (about 100 g), 24 hrs. after slaughter.

2. Place the piece of meat in a pan containing cold water.
3. Boil and simmer with a lid on the pan.
4. Remove the piece of meat from water after adequate cooking.
5. Incise the meat piece and observe the smell of cut surface of meat.
6. If abnormal odour is present, the meat also has abnormal taste.

Nitrazine Yellow Test

Nitrazine yellow test is used for determining the pH of meat. Most of the microorganisms grow best at pH around 7.0 (6.6-7.5), while few grow below pH 4.0. Growth of microorganisms in meat and meat products cause breakdown of their various constituents and lead to spoilage. Bacteria tend to be more fastidious in their relationships to pH than moulds and yeasts. Pathogenic bacteria are even most fastidious. Therefore, the pH measurement is useful for evaluation of meat quality for further processing, control of ripening of raw fermented products and control of acidity of ingredients such as brines, marinades *etc*. The pH of meat can be measured by using digital pH meter and chemical indicator method using nitrazine yellow. The procedure of nitrazine yellow test is given as follows:

1. Take a piece of meat (free of blood, fat and connective tissue) in a petri dish.
2. Add sufficient amount of nitrazine yellow indicator (0.01 per cent) to cover the meat piece.
3. Mix with stirring rod and note the colour change.
4. Interpret the result on the basis of standards given in the following table:

Table 23.2. Determination of Quality of Meat on the Basis of pH of Meat

pH	*Colour*	*Inference*
6.0	Yellow	Good keeping quality
6.4	Olive Green	Not having same good keeping quality
6.8	Bluish violet	Suspect on signs of incipient spoilage

Rimmington and Fowrie Test

Rimmington and Fowrie test is used for differentiation of jaundice and carotene pigmentation in the carcase. The bile salts are soluble in water, so the bottom layer is yellow in jaundiced carcase. Carotene is soluble in ether with the result that the top layer is yellow. The procedure of Rimmington and Fowrie test is given as follows:

1. Place 2 g finely chopped fat from suspected carcase in a test tube.
2. Add 5ml sodium hydroxide (5 per cent) to the tube.
3. Boil for 1 minute and shake frequently until fat is dissolved.
4. Cool the tube under running tap water.

5. Add 5 ml volume of ether, mix gently and allow to settle.
6. Interpret the result on the basis of standards given in the following table:

Table 23.3. Determination of Quality of Meat on the Basis of Colour Reaction

Colour Reaction	*Inference*
Yellow colour in top layer	Carotene pigmentation
Yellowish green colour in bottom layer	Jaundice (bile salts)
Yellow colour in top layer and yellowish green colour in bottom layer	Jaundice and carotene pigmentation (that is both conditions are existing)

Rothera's Test

Rothera's test is also known as sodium nitroprusside test. The test is useful for detection of acetone in carcase affected with acetonaemia/ketosis. Acetonaemia/ ketosis is a metabolic disturbance. Whenever, the glucose levels in blood plasma declines (*e.g.*, starvation) the concentration of free fatty acids in the plasma rises. This rise is paralleled by increase in the concentration of ketone bodies (acetone) which provide a normal physiological third source of energy. The carcase possesses a sweet odour, so the liver should always be incised and smelt in suspected cases.

The test reagent can be prepared by adding 100 g ammonium sulphate, 50 g anhydrous sodium carbonate and 3 g sodium nitroprusside in a flask. Add 100 ml distilled water and dissolve these ingredients. The procedure of Rothera's test is given as follows:

1. Place 10 g minced meat in a large test tube.
2. Add 15 ml distilled water and 5 ml test reagent.
3. Shake and allow to stand for 2-3 minutes.
4. Development of purple colour indicates the presence of peptones (acetone) in the carcase/meat.

Peroxide Test

Peroxide test is used for detection of freshness of meat. The procedure of peroxide test is given as follows:

1. Take 10 g of minced meat sample in a flask and add 30 ml distilled water.
2. Keep it for 30 minutes.
3. Filter the solution through filter paper.
4. Take 2 ml of filtrate in a test tube.
5. Add 5 drops of benzidine solution and 2 drops of hydrogen peroxide.
6. Interpret the result on the basis of standards given in the following table:

Table 23.4. Determination of Quality of Meat on the Basis of Colour Reaction

Colour Reaction	Inference
Change from bluish colour to brown turbidity	Fresh meat
Appearance of bluish colour after considerable time	Doubtful
No development of any colour	Unfit for consumption

Copper Sulphate Test

Copper sulphate test is used for detection of staleness of meat. The procedure of copper sulphate test for determining the quality of meat is given as follows:

1. Take 10 g of minced meat sample in a flask and add 30 ml distilled water.
2. Boil and filter it through filter paper.
3. Cool it to room temperature.
4. Take 5 ml of filtrate in a test tube and add 4 drops of 1 per cent copper sulphate solution.
5. Interpret the result on the basis of standards given in the following table:

Table 23.5. Determination of Quality of Meat on the Basis of Development of Turbidity

Development of Turbidity	Inference
Clear broth	Fresh meat
Turbid broth	Doubtful
Deep turbid broth	Unfit for consumption

Picric Acid Turbidity (PAT) Test

Picric acid turbidity test is used for detection of spoilage in meat. Proteins of meat are broken down into free amino acids due to spoilage. Addition of picric acid to meat homogenate forms picrates from different amines causing increased turbidity. The procedure of PAT test for determining the quality of meat is given as follows:

1. Homogenize 5 g of meat sample with 20 ml of ethyl alcohol (70 per cent) for 30 seconds.
2. Add 5 ml of saturated aqueous solution of picric acid to the homogenate and homogenize it again for 30 seconds.
3. Filter the slurry through Whatman's filter paper No.41.
4. Collect 10 ml of filtrate and record the optical density of the filtrate at 540 nm wavelength using spectrophotometer.

The picric acid turbidity (PAT) of fresh mutton is approximately 0.086 and of mutton stored at ambient temperature (28 ± 2°C) for 18 hrs. is approximately 0.126.

Malachite Green Test

Malachite green test is used for determining the efficiency of bleeding of a carcase. It is important to distinguish between the congestion of a carcase resulting from active hyperemia and that of carcase which is imperfectly bled due to mechanical causes. Congestion of a carcase is associated with fever, and is therefore likely to be accompanied by systemic changes in the organs. Imperfect bleeding due to mechanical causes accompanied by systemic changes may occur due to injury, heart failure, suffocation and lightening stroke.

Imperfect bleeding of a carcase is characterized by blood in left ventricle, injected subcutaneous blood vessels, dark coloured flesh, and congestion in the organs. The high blood content of the visceral organs is most apparent in the lungs, while the lymph nodes, particularly the pre-scapular are suffused with blood though they are not enlarged. Poorly bled carcases undergo rapid decomposition and have low keeping quality; therefore, these are not fit for human consumption. Therefore, the determination of the efficiency of bleeding of a carcase is important which can be done by malachite green test.

Principle

The hemoglobin of blood when exposed to oxygen is converted into oxyhaemoglobin. The presence of oxyhaemoglobin forms green colour complex with the malachite green dye.

Preparation of Malachite Green Solution

1. Add 100 mg malachite green in a beaker containing 25 ml of 30 per cent glacial acetic acid.
2. Dissolve the dye by heating at 60°C for 10 minutes.
3. Add 100 ml of distilled water and mix it properly.
4. Store this solution in a dark coloured stopper bottle.

Preparation of Meat Extract

1. Weigh 15 g of meat sample and triturate with mortar and pestle.
2. Transfer the triturated meat sample in a flask containing 35 ml distilled water.
3. Allow to stand for 15 minutes and filter through a fitter paper.

Procedure

1. Transfer 0.7 ml of the meat extract to an agglutination tube
2. Add 1 drop of malachite green solution and mix.
3. Add 1 drop of 3 per cent H_2O_2 and mix.
4. Allow to stand for 20 minutes and observe the colour changes.
5. Interpret the result on the basis of standards given in the following table:

Table 23.6. Determination of Quality of Meat on the Basis of Colour Reaction

Colour Reaction	Inference
Clear fluid with blue colour	Satisfactory bleeding
Cloudy fluid with green colour	Moderate degree of imperfect bleeding
Cloudy fluid with olive-green colour	Marked degree of imperfect bleeding

Resazurin Dye Reduction Test

Resazurin dye reduction test is used for determining the dye reducing capacity of meat. The main objective of the test is to estimate the bacterial load in a meat sample indirectly. The procedure of resazurin dye reduction test for determining the quality of meat is given as follows:

1. Dip the filter paper strips in 0.05 per cent resazurin solution (violet colour).
2. Dry the filter paper strips and use for the test.
3. Moist the filter paper strips before use.
4. Place the meat sample on this moist filter paper strip.
5. Record the dye reduction time.
6. Interpret the results on the basis of standards given in the following table:

Table 23.7. Determination of Quality of Meat on the Basis of Time Required for Changing the Colour from Violet to Colourless

Dye (Colour) Reduction Time	Inference
10 minutes	Unfit
10-30 minutes	May be accepted
More than 60 minutes	Very good quality

24

Microbiological Methods for Detecting Microbial Spoilage in Meat

A successful commercial production of meat requires control of microbial contamination, activity to achieve maximum shelflife and safety of the meat product. The public concerns over the wholesomeness of the meat supply have increased greatly in recent years. These concerns have resulted in increased efforts by the meat industry to improve the microbiological status of meat, and by regulatory authorities to implement advanced technologies.

Objectives of Microbiological Testing of Meat

The objectives of microbiological testing of meat are as follows:

1. To validate and verify the process control measures associated with implemented HACCP system.
2. To determine the types and numbers of microorganisms present in meat.
3. To determine the presence/absence of pathogenic organism.
4. To assess the standard of slaughter.
5. To assess the carcase dressing methods.
6. To assess the standard of animal hygiene.
7. To assess the standard of personal hygiene.
8. To assess the standard of environmental hygiene.
9. To assess the quality of meat for further processing.
10. To find out the evidence of spoilage due to bacteria/moulds, this may occur in improperly stored meat.

Sampling of Meat for Testing

A sample is a small portion of a large amount taken for different types of examinations. Sampling is a complex process that requires careful planning, specialized equipment and trained sampling staff. The basic principle of sampling is the sample should be true representative of the bulk or lot.

Sampling Plans

Each sampling plan has several characteristics as follows:

n = Number of samples to be taken from the lot.

c = Number of samples permitted to fail.

m = Microbial count below which the sample is considered to be satisfactory.

M = Microbial count above which the sample is considered unsatisfactory.

Table 24.1. ICMSF's Sampling Plan and Recommended Microbiological Limits for Raw Meat

Product	*Test*	*Case*	*Plan Class*	*n*	*c*	*m*	*M*
Carcase meat, before chilling	APC*	1	3	5	3	10/5	10/6
Carcase meat, chilled	APC	1	3	5	3	10/6	10/7
Edible offal, chilled	APC	1	3	5	3	10/6	10/7
Carcase meat, frozen	APC	1	3	5	3	5 x 10/5	10/7
Boneless meat, frozen (beef, pork, mutton)	APC	1	3	5	3	5 x 10/5	10/7
Comminuted meat, frozen	APC	1	3	5	3	10/6	10/7
Edible offal, frozen	APC	1	3	5	3	5 x 10/5	10/7
Raw chicken, fresh or frozen	APC	1	3	5	3	5 x 10/7	10/7

*APC = Aerobic plate count

Table 24.2. Provisions of Directive 88/657/EEC

Organism	*n*	*c*	*M*	*m*
Aerobic mesophile bacteria (total viable count)	5	2	5 x 10^6	5 x 10^5
E. coli	5	2	5 x 10^2	50
Clostridium perfringens (sulphite-reducing anaerobes)	5	1	10^2	10
Staphylococci	5	1	5 x 10^2	50
Salmonella	5	0	Absence in 25 g	

Bacteriological Examination of Carcases

In the healthy and physiologically normal animal, the organ which have no direct contact with the exterior may be considered as virtually sterile. But the bacteria may enter the blood, tissues and organs during slaughter and dressing operations.

Type of Bacteria in Meat

The most frequently isolated bacteria from meat and poultry includes *Aeromonas, Campylobacter, Corynebacterium, Enterococcus, Listeria, Micrococcus, Moraxella, Pseudomonas, Psychrobacter, Virgococcus* spp. *etc.* The invasion of bloodstream by the naturally occurring intestinal bacteria and the non-hemolytic staphylococci on the skin is of a secondary nature occasioned by some other pathological condition and does not indicate that the animal was affected with a generalized disease.

Obligation of Bacteriological Examination

Bacteriological examination may be considered obligatory in the following conditions:

1. Animal slaughtered in emergency.
2. Slaughter of animal suffering from systemic disease.
3. Carcase and offal exhibiting pathological changes on post-mortem inspection.
4. Animal excreting food-poisoning organisms prior to slaughter.
5. Delay in evisceration (more than 1hour) or absence of parts of slaughtered animal.
6. Animal slaughtered without prior ante-mortem inspection.

Submission of Material for Bacteriological Examination

The samples from muscles, prescapular, internal iliac, axillary and mesenteric lymph nodes, spleen, liver, and parts showing pathological changes should be submitted for bacteriological examination.

Microbiological Analysis

The meat or the rinse from swabs or sponges forms the basis for microbiological analysis. The solid samples are homogenized in a sufficient quantity of sterile diluent. The homogenate can be further diluted serially to avoid overgrowth of organisms. The diluted sample is mixed properly with the molten agar medium (total count agar). After solidification the plates are incubated aerobically at 20 °C and 35 °C for growth of mesophilic organisms. The plated samples can also be incubated anaerobically for anaerobic organisms. The growth on plating medium is counted and the result is presented as colony forming unit per gram (CFU/g). The colonies can be further subjected to cultural and morphological characterization and biochemical tests for identification of the organisms. Serological and molecular tests are also used for further identification/confirmation of the organisms.

Determination of Standard Plate Count (SPC) in Meat

Standard plate count is used to determine the microbial load in foods. The procedure of standard plate count is given as follows:

1. Triturate 10 g of meat sample in a sterile mortar with 10 ml sterile peptone water (0.1 per cent)

2. Add 80 ml peptone water to the triturate to obtain 1:10 dilution.
3. Prepare a series of 10 fold dilutions of the triturate, by successively transferring 1 ml volumes through a series of tubes holding 9 ml normal saline as a diluent.
4. Starting with the highest dilution of the sample and working towards the lower dilution. Aseptically pipette 1 ml of dilution into sterile petri dishes (2 petri dishes for each dilution).
5. Maintain molten plate count agar at 50°C in a water bath.
6. To each plate add 15 ml molten plate count agar and mix the sample properly by rotating the plate clockwise and anti-clockwise, moving the plate to and fro and also moving at right angle. Then leave the medium to solidity.
7. Incubate the plates at 37°C for 24 hrs.
8. Select the plates having colonies range from 30 to 300.
9. Find out the mean counts of colonies from selected plates of a particular dilution.
10. Multiply the mean count of colonies by the dilution factor to obtain plate count.
11. Express the result as number of organisms per g (CFU/g) of meat sample.
12. Interpret the result on the basis of standards for standard plate count.

Enumerations of Coliforms in Meat

Coliforms are indicator organisms that likely indicate the presence of pathogens in the food and water. Coliform count is used for determining the sanitary quality of food. The procedure of enumeration of coliforms in meat is given as follows:

1. Triturate 10 g of meat sample with 10 ml sterile peptone water (0.1 per cent).
2. Add 80 ml of diluent to this mixture to obtain 1:10 dilution.
3. Prepare a series of 10 fold dilution of the sample by transferring 1ml volumes through a series of tubes holding 9ml diluent.
4. Transfer 1ml of the diluted sample to the plates and incorporate with 12 ml of Violet Red Bile Lactose (VRBL) agar.
5. When the medium has set, each plate should be over-layered with 4 ml VRBL agar.
6. Incubate the plates at 30 $\pm$ 1°C for 24 hrs.
7. Select the plates having colonies range from 30 to 300.
8. Find out the coliform counts per g of meat by multiplying the mean count with dilution factor.

Enumeration of *Staphylococcus aureus* in Meat

Food commonly contaminated with *Staphylococcus aureus* includes synthetic cream, custards, trifles and high-salt food such as ham. The procedure of *Staphylococcus aureus* counts in meat is given as follows:

1. Homogenize 10g meat sample in 90 ml 0.1 per cent peptone water.
2. Prepare 10 fold serial dilutions of the homogenate.
3. Plate the homogenate or dilutions of it on mannitol salt agar.
4. Incubate at 37°C for 24 hrs.
5. Identify colonies on mannitol salt agar, presumptive *Staphylococcus aureus* colonies are surrounded by a yellow zone because of mannitol fermentation.
6. Find out the number of *Staphylococcus aureus* per g of meat sample.

Microbiological Examination of Surfaces

It is important to maintain the food contact surfaces in a hygienic state. The hygienic status of food contact surfaces can be assessed by various methods, but none of the methods can recover all microorganisms from the surfaces. The most commonly used methods for surface assessment in food operations are described as follows:

Swab/Swab-Rinse Method

Swabbing is the oldest, and most commonly and widely used method for microbiological examination of surfaces not only in the food industries but also in hospitals and restaurants. Either cotton or calcium alginate swabs can be used. When cotton swabs are used, the organisms must be dislodged from the cotton fibres. When calcium alginate swabs are used, the organisms are released into the diluent upon dissolution of the alginate by sodium hexametaphosphate. The procedure of the method is discussed as follows:

1. Prepare a template with openings corresponding to the size of the area to be swabbed (*e.g.*, 1 inch^2 or cm^2).
2. Place the sterile template over the surface and rub the exposed area thoroughly with a moistened swab.
3. Return the exposed swab to its holder (test tube) containing a suitable diluent and store at refrigeration temperature until plated.
4. Enumerate the organisms in the diluent by SPC or other methods.
5. Use differential and selective culture media for enumeration/isolation of specific organism.

Alternatively, swab-rinse method can also be performed by adding 1.5 ml of fluid to a flat surface and swab for 15 seconds over a 3 cm^2 area, and collet 0.1 and 0.5 ml volume in a microliter pipette. The fluid may be surface or pour plated using plate count agar or selective media.

Contact Plate Method

The replicate organisms direct agar contact (RODAC) method employs special petri dishes which are poured with 16 ml of an appropriate plating medium, resulting in a raised agar surface. When the plate is inverted, the hardened agar makes direct contact with the surface. The main drawback of this method is ineffectiveness for heavily contaminated surfaces that may lead to spreading of colonies on the agar surface. These problems can be minimized by using dried agar surfaces and by using selective media.

Agar Syringe/Agar Sausage Method

This method has also same disadvantages as RODAC method. The procedure of the method is discussed as follows:

1. Take a 100 ml syringe and modify it by removing the needle end to create a hollow cylinder that is filled with agar.
2. Push a layer of agar beyond the end of the barrel by means of plunger and press against the surface to be examined.
3. Cut off the exposed layer and place in a petri dish.
4. Incubate at optimum temperature for a specific time period.
5. Count the colonies after incubation.

25

Preservation of Meat

Preservation is a process of treatment of meat by various means to prolong the shelflife of meat. The preservation of food is being practiced since very long time. It goes back to the last Ice Age, about 15,000 BC, when Cro-Magnon humans discovered for the first time a way of preserving food by smoking it. But at present there are several methods of food preservation. The primary purpose of meat preservation is to prevent its spoilage. Meat can be preserved by several methods such as drying, curing, cold, heat, chemicals, irradiation and high pressure. Chemical preservation can be done by smoking of meat and fish usually as an adjuvant to commercial salting and pickling.

Meat is a highly perishable product and soon becomes unfit for consumption and possibly dangerous to health through microbial growth, chemical change and breakdown by endogenous enzymes. These processes can be curtailed by reducing the temperature sufficiently to slow down or inhibit the growth of microorganisms, by heating to destroy organisms and enzymes (cooking, canning), or by removal of water by drying or osmotic control (binding the water with salt or other substances so that it becomes unavailable to the organisms). It is also possible to use chemicals to inhibit the growth and, recently, ionizing radiation. However, ionization is not allowed in some countries. Therefore, preservation of meat is important for maintaining its keeping quality and safety to the consumers.

Objectives of Meat Preservation

The important objectives of meat preservation are as follows:

1. To prevent the spoilage of meat due to the action of microorganisms.
2. To prevent economic losses due to spoilage of meat.
3. To maintain the hygienic condition of meat and meat products.

4. To maintain the quality of meat and meat products.
5. To store the meat and meat products for long duration.
6. To prevent the meat-borne zoonotic diseases.

Principles of Meat Preservation

The creation of unfavourable conditions against the microorganisms to prevent or reduce their growth by implementing various techniques and measures such as deprivation of water and sometimes oxygen, excess saltiness, increased acidity of the medium in which the microorganism exist, extreme heat or cold *etc.*

Methods of Meat Preservation

There are various methods of meat preservation depending on its use, available facilities and duration of storage. Some important methods of meat preservation are discussed as follows:

Drying

Drying is a common method of meat preservation. Sun drying is the oldest and most widely used method of food preservation. Nowadays foodstuffs may be dried in air, superheated steam, in vacuum, in inert gas and by direct application of heat. Many types of dryers exist, the particular type being dependent on the nature of the material, the desired form of finished product *etc.* Dehydration is a process in which heat is transferred into the water in the food, which is vapourized. The water vapour is then removed. Preservation of meat by drying in wind and sun, salting and smoking has been traditionally used since thousands of years. Drying is not used as a commercial method of meat preservation. The method can be used to fulfill the requirements of protein diet.

Drying can be used for preservation of different types of meat. For preservation of jerked beef of South America, the meat is cut into strips and dried in the air. For preservation of pemmican of North America, the beef, venison or fish is smoked before drying and is then pulverized and an equal amount of fat is added. For preservation of biltong of South Africa, the beef and sometimes the flesh of game animals, zebra and ostrich are cut into strips and cured in salt and sometimes immerging in vinegar and then dried. There are two public health dangers associated with consumption of biltong. The first is the meat obtained from animals infected with *Salmonella* and second is the contamination of meat with this organism during careless and unhygienic handling, preparation and storage of meat.

Cysticerci in meat are rapidly destroyed by using higher salt concentration (the ratio of salt to moisture in the meat should not be less than 1 to 4. At this concentration the cysticerci are viable after 16 hours but 50 per cent are destroyed after 90 hours and 100 per cent after 6 days.

Freeze Drying (Lyophilization)

Freeze drying is the process of removal of water content from the frozen foods. In order to ensure the removal of water content, the meat should be sliced or diced and packaged in completely moisture proof packaging material. In freeze drying, the material is frozen and placed in a sealed chamber. The pressure of chamber is reduced and maintained at a value below 1 mm Hg. Heat is applied to the material, the surface ice heats up and the resultant water vapour is drawn off by the vacuum system. As the ice boundary recedes into the material, the ice sublimes in situ and the water percolates to the surface through the pore structure of the material.

Smoking

Meat is being treated with smoke since ancient times. Smoking is carried out traditionally over a wood fire. At present the smoking is carried out by producing smoke from wood sawdust in a generator and conducting the smoke over the meat. Light smoking is carried out by depositing preservatives on meat to enhance its flavour and appearance. It has limited preservative effect and the products need refrigeration storage. Intensive smoking is carried out both by heavier deposition of preservatives and by the drying effect of the hot air. Consequently preservation by smoking is regarded as an emergency measure when other methods cannot be used.

In modern method of smoking, an aqueous solution of the constituents of smoke is used which reduces the amount of strongly flavoured and other unwanted substances. The bacon is first dusted with pea meal and then smoked with smoke emitting from a smouldering. Formaldehyde present in wood smoke acts as bactericidal agent. The purpose of smoking of bacon is to preserve the colour and improve the flavour of meat.

Curing

Curing is a method of preservation of foods such as meat, fish and vegetables by adding the combination of salt, nitrates, nitrites, sugars *etc.* with the aim to draw the moisture from the foods by the process of osmosis. Curing can be traced back to antiquity, and was the primary way of preserving meat and fish until the late 19th century. Dehydration was the earliest form of food curing. Curing process may involve preservation and improvement in the flavour and colour of the foods. Many curing processes also involve smoking, spicing or cooking. The curing ingredients can be rubbed on to the food surface or mixed into foods is known as dry curing. When the curing ingredients are dissolved in water, it is known as brine, wet, or pickle curing. In the wet curing, the food is submerged in the brine until completely covered. With large cuts of meat, brine may also be injected into the muscle. In curing, the term pickle is used for any brine solution or a brine cure solution that has sugar added.

Originally, the term curing was applied for preservation in general but is now restricted to preservation with table salt (sodium chloride) and sodium or potassium nitrite or nitrate or a mixture of these salts. The nitrate serves as a source for nitrite which acts as an active compound. Since bacteria in the curing solution form nitrite from the nitrate. Curing is suitable for meat containing high content of fat, *e.g.*, pork or fine-fibred beef intermixed with fat. Therefore, high quality of pickled meat can be prepared from brisket and flank of beef; while lean beef, veal or mutton is not suitable for pickling due to its unpalatability and dryness.

Curing involves the conversion of the meat pigments, predominantly myoglobin to the nitroso or cured form. Myoglobin in freshly-cut uncured meat is in the reduced form (purple colour), which in contact with air is rapidly oxygenated to oxymyoglobin (bright red colour) which is responsible for the "bloom on meat". If oxidized, these pigments are converted to metmyoglobin (brown or grey colour). Under suitable conditions, these pigments can be converted to the nitrosoform (nitrosomyoglobin) by the addition of nitric oxide. During the curing process, nitric oxide is formed by reduction of nitrites formed by bacteria from nitrates. Nitrosomyoglobin gives freshly-cut cured meat its bright red colour, which is unstable and rapidly oxidises to brown and grey forms. However, on heating, nitrosomyoglobin is converted to nitrosohaemochrome (pink colour) which is distinct from the grey and brown of the cooked uncured meat.

Importance of Curing

Unpreserved/untreated meat decomposes rapidly and the rate of decomposition depends on several factors such as ambient humidity, temperature, and the presence of microorganisms. If the meat is kept under these conditions for a long time, it leads to decomposition, colour change and foul odour. Ingestion of such spoilt meat can cause serious food poisonings, like botulism. Curing is able to significantly extend the shelflife of meat before it spoils, by making it unfavourable to the growth of spoilage microorganisms. Because curing increases the solute concentration in the food and hence decreases its water potential, the food becomes unfavourable for the growth of microorganisms that cause food spoilage. Curing also contributes to the characteristic pink colour.

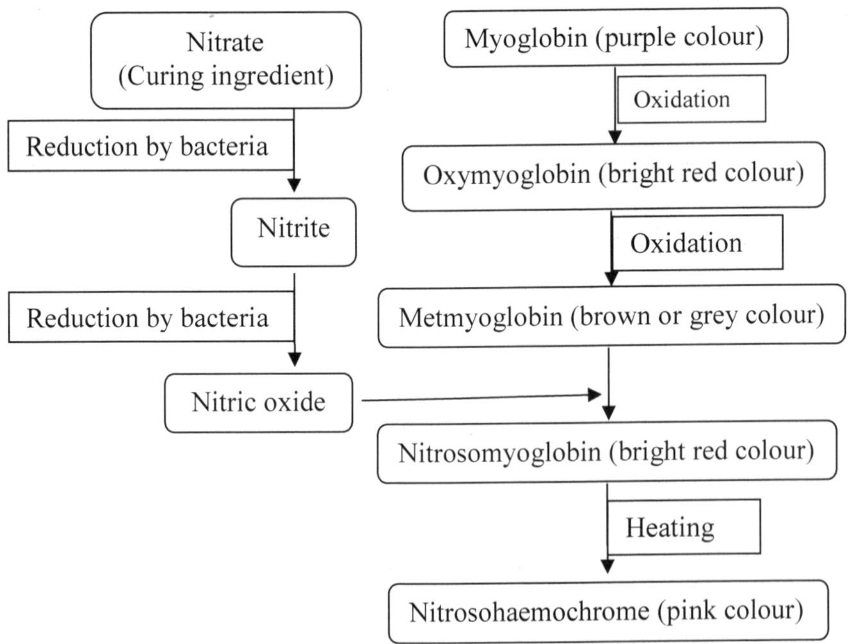

Figure 25.1. Mechanism of Curing.

Table 25.1. Curing Ingredients and their Functions

Ingredient	Level in Curing Brine	Function
Sodium chloride	15-30 per cent	Preservative
Sodium nitrate	0.15-1.5 per cent	A source of nitrite
Sodium nitrite	500-1000 ppm	Preservative
Sodium ascorbate	0.2-1.0 per cent	Reducing agent (reduction of nitrate to nitrite)
Polyphosphates	2-4 per cent	Reduce the cooking losses
Sugars	1-4 per cent	Alleviate the harsh flavour the salt
Liquid smoke	1 per cent	Flavouring agent

Curing Ingredients and their Role

The important curing ingredients and their role are discussed as follows:

Salt

The use of salt is one of the oldest methods of preserving meat since at concentrations greater than 4 per cent in the aqueous phase it inhibits the growth of most spoilage organisms. However, for a complete preservative action, a salt concentration of around 17 per cent is needed, but at this concentration the product becomes unpalatable. In most cured meat products the salt concentration is kept between 2.5 and 5 per cent.

Table salt (sodium chloride) is the primary ingredient used in meat curing. Removal of water and addition of salt to meat creates a solute-rich environment where osmotic pressure draws water out of microorganisms, slowing down their growth. In addition, salt also causes the soluble meat proteins to come to the surface of the meat particles within sausages. These proteins coagulate when the sausage is heated, helping to hold the sausage together.

Sugar

The sugar is available in different forms like honey, corn syrup, and maple syrup. Sugar is added to meat for the purpose of curing. It alleviates the harsh flavour of the salt. Sugar also enhances the growth of beneficial bacteria such as *Lactobacillus* spp. by feeding them.

Nitrates and Nitrites

The use of nitrite and nitrate salts for meat curing goes back to the Middle Ages, and in the US has been formally used since 1925. Because of the relatively high toxicity of nitrite (the lethal dose in humans is about 22 mg/kg of body weight) the maximum allowed nitrite concentration in meat products is 200 ppm. Nitrates (NO_3) and nitrites (NO_2) not only help to kill bacteria, but also produce a characteristic flavour and give meat a pink or red colour. Nitrite is generally supplied by sodium nitrite or (indirectly) by potassium nitrate. Nitrite salts are most often used in curing. Nitrate is specifically used only in a few curing conditions and products where nitrite (which may be generated from nitrate) must be generated in the product over long periods of time. Nitrate reduction step is not required if nitrite is used as the curing agent. The cure colour is more rapidly developed when nitrite is used.

Nitrite inhibits the growth of organisms and also reacts with proteins when heated to form some compounds, called "Perigotype factors". These factors prevent the formation of spores of *Clostridium botulinum.* The early curing procedures were lengthy in which the meat was first injected with the curing solution and the process completed in 1-2 weeks. But recent developments have led to a reduction in the time required in which thin slices of meat such as bacon can be cured in a few hours, and the processing time can be reduced to a few minutes if heated and if the cure is completed in the final package.

The use of nitrites in food preservation is controversial due to formation of nitrosamines in high concentrations of nitrite and cooking of product at high temperatures. This effect occurs in red or processed meat, but not in white meat or fish. Nitrates and nitrites may cause cancer and the production of carcinogenic nitrosamines. Nitrosamines formation during curing can be prevented by the use of the antioxidants, vitamin C and the alpha-tocopherol form of vitamin E. Moreover, the addition of ascorbate to cured meat reduces formation of nitrosamines, but increases the nitrosylation of iron. Still, nitrates and nitrites are considered irreplaceable in the prevention of botulism from consumption of cured dry sausages by preventing germination of spores.

Smoke

Meat can also be preserved by smoking in which the meat is exposed to smoke from burning or smouldering plant materials, usually wood. The meat becomes tender when exposed to hot smoke. Smoking can also be carried out in smokehouse using damp wood chips or sawdust. In North America, hardwoods such as hickory, mesquite, and maple are commonly used for smoking. Smoking helps in sealing of outer layer of the food being cured, making it more difficult for bacteria to enter. It can be done in combination with other curing methods such as salting.

Hot smoking, smoke roasting (pit barbecuing) and cold smoking are the common method of smoking. Smoke roasting and hot smoking cook the meat while cold smoking does not. If the meat is cold smoked, it should be dried quickly to reduce the bacterial growth during the critical period where the meat is not yet dry. This can be achieved, as with jerky, by slicing the meat thinly. The smoking of food directly with wood smoke may cause contamination of the food with carcinogenic polycyclic aromatic hydrocarbons.

Cure Accelerators

The curing process can be accelerated (development of cured colouring short time) by using cure accelerators, *e.g.*, ascorbic acid, erythorbic acid, or their derivatives. Cure accelerators rapidly convert nitric acid to nitric oxide. Sodium ascorbate has reducing capacity and allows smaller amounts of nitrite to be used, so there is less residual nitrite in the meat which reduces the possibilities of the formation of nitrosamines. Moreover, residual ascorbic acid has an antioxidant effect in stabilizing the colour and preventing rancidity.

Polyphosphates

Polyphosphate is added to improve the water-holding capacity of the meat by reaction between the salt and the structural proteins, aided by the polyphosphate.

Tumbling and Massaging

Tumbling is a technique of injecting the curing salt solution in the meat pieces or immersing the chopped meat in the curing salt solution followed by mechanical shaking (tumbling). Different concentrations of salt solutions (2-8 per cent) are used, sometimes with the addition of polyphosphate, when there is some extraction of water-soluble protein, mainly myosin. The extracted proteins set to a strong gel on heating and so bind together the pieces of meat, which can then be shaped or sliced. Massaging is a process which involves a relatively gentle mechanical treatment, while tumbling involves an accelerated and more vigorous action.

Common Defects in Cured Meat

1. **Browning:** Browning of meat occurs when the cured meat pigment nitrosomyoglobin is changed to brown metmyoglobin due to dehydration. The dehydration occurs due to high temperature, low humidity, and oxidation (caused by prolong exposure to air, excessive nitrate and poor packaging).
2. **Greening:** This may be caused by excessive nitrate and by bacterial contamination.
3. **Fiery red:** Fiery red areas are caused by lack of available nitrate-miscure. This defect may occur in deep meat cuts.
4. **Discolouration:** Areas of discolouration may be caused by blood splashing or bruising.
5. **Jelly pockets:** These are caused by injection of brine into connective tissue, which causes its denaturation.
6. **Rancidity:** Rancidity in frozen bacon may appear as pronounced yellowing of the fat due chemical reactions.

Refrigeration

Refrigeration is the most commonly and widely used method of meat preservation at commercial as well as domestic levels. Low-temperature processes involve storage in a cold store freezing and deep-freezing, which allows foodstuffs to be preserved in their naturally fresh state. The temperature of cold store is determined according to the nature of the products. The principle of meat preservation by refrigeration is based on cessation of metabolic activity and prevention of multiplication of organisms like bacteria, moulds and yeasts by removal of water as ice. The multiplication of all organisms is stopped at -8°C.

Mechanical refrigeration may be carried out by using carbon dioxide, sulphur dioxide, ammonia, chlorofluorocarbons (CFCs) and hydrofluorocarbons (HFCs). Carbon dioxide and sulphur dioxide were used in earlier times. Carbon dioxide is uneconomical while sulphur dioxide is corrosive and toxic. Ammonia is most widely used refrigerant which has lesser effect on the ozone layer. It is somewhat corrosive and has a penetration odour so that leakage of ammonia can affect the quality of stored products. CFCs have been replaced by HFCs because of absence of chlorine in HFCs which have lower potential to breakdown the ozone and deplete the ozone layer. Chilling and freezing are the two main methods of meat preservation by refrigeration. For obtaining optimum result in refrigeration and preventing the growth of the organisms, the following criteria should be adopted:

1. Ensure the flow of cold air around the carcases.
2. Ensure the minimum opening and closing of the door.
3. Ensure the maintenance of constant temperature.
4. Avoid the overloading of the carcases in the chilling hall and ensure that they should not touch each other.

Chilling

Chilling is useful for short-term preservation of meat. Mechanical refrigeration is a modern method of meat preservation. However, it was used by ancient Romans to keep the food cool with ice. Chilled meat is usually stored at temperatures around 1-4°C for keeping it well for several days. Edible offal is more perishable due to higher glycogen contents and lesser fat covering. Therefore, these should be kept at a refrigeration temperature not higher than 3°C. Meat can be kept very cool (1°C to 0°C) during slaughter and meat cutting under strict hygienic conditions, modern packaging techniques including storage under carbon dioxide or nitrogen or in vacuum, which can extend this period to about 2 months. Chilling should be carried out in the dark to avoid oxidation of fat and liberation of free fatty acids and the production of rancidity.

For preservation of meat by chilling, the atmosphere should be kept dry to avoid growth of moulds. Chilling scarcely affects the appearance, flavour and nutritional value of the meat. The term "quick chilling" refers to a rapid lowering of the carcase temperature within an hour after slaughter. Pathogenic microorganisms like *Salmonella*, *Staphylococcus* spp. and *Clostridium perfringens* are inhibited by cooling. However, *Listeria monocytogenes* can grow at 2°C, some *Salmonella* spp. at 5°C and *Campylobacter* at 7°C. Non-pathogenic microorganisms like some *Pseudomonas* spp. which predominate on the exposed surface of chilled meat and *Lactobacillus* spp. on vacuum-packed meat.

Freezing

Freezing is the process of refrigeration of carcase/meat at temperatures below 0°C. The freezing point of meat lies between -1°C and -1.5°C. Different freezing temperatures are maintained in different countries such as Germany (-6°C), UK (-9°C to -10°C), South America (-8°C to -9°C during sea transport and -18°C for pork) and Australia (-11°C). Freezing of carcase/meat at -15°C diminishes the dangers of most pathogenic organisms and slows the growth of spoilage organisms; growth of some organisms, moulds, virtually ceases at -10°C. Freezing of meat commercially at -29°C and domestically at -18°C is now a standard method of preserving for periods of 1-2 years. However, there is some deterioration of eating quality compared with fresh or chilled meat.

Effect of Freezing on Pathogenic Microorganisms and Parasites

Freezing can destroy some bacteria while inhibits the growth and multiplication of others. Freezing can also inhibit the growth or even destroy certain parasites. However, there is little effect of freezing on viruses. Anthrax bacilli can withstand a temperature of -130°C. *Salmonella* can withstand a temperature of -175°C for 3 days. Tubercle bacilli can remain alive in carcases frozen at a temperature of -10°C for more than 2 years.

Pork affected with *Cysticercus cellulosae* can be rendered safe, if held at -10°C to -8°C for 4 days. *Trichinella* cysts in pork can be destroyed by freezing the carcase at -15°C for 20 days or at -18°C for 24 days. Beef affected with *Cysticercus bovis* can be rendered safe for 14 days if frozen at -10.5°C or below. Foot and mouth disease virus can survive in chilled or frozen carcase for 76 days if animal slaughtered during the incubation period of the virus. Swine fever virus may remain infective in the bone marrow of the frozen carcase for at least 73 days.

Problems Associated with Chilling and Freezing

1. **Cold Shortening:** There are some problems in chilling and freezing of meat. A condition "cold shortening" may develop which is characterized by contraction of muscle fibres and toughening of meat after its cooling. This condition may develop due to rapid cooling of carcase/meat (below 10°C) before falling of muscle pH below 6.0. This problem is more common in small animals, *e.g.*, lamb, which cool down rapidly, while lesser extent in beef but pork is almost unaffected. Cold shortening can be prevented by a modern procedure involving cooling of the carcases to 10-15°C (conditioning) and holding that temperature for a few hours until the pH has fallen to 6.0. Cold shortening in beef carcases can be prevented by suspending them in such a way as to exert a pull on certain muscles to prevent contraction. Another method of preventing this condition is by applying electrical stimulation to the carcase after slaughter (at low volt) or after evisceration (at high volt) for 2-4 minutes to bring down the pH rapidly.
2. **Thaw Rigor:** Another problem like "thaw rigor" can arise during thawing of pre-rigor frozen meat when the muscle contracts and exudes a substantial part of its weight as tissue fluids. After slaughter, direct and immediate freezing of meat is not recommended. Only post-rigor meat/ carcase should be frozen.
3. **Dripthaw:** There may be another problem "dripthaw" which is associated with nutritional losses in frozen meat. These losses incurred in frozen meat when it is thawed, and juices are exuded containing soluble proteins, vitamins and minerals. The amount of losses depends on the length of time of ageing (time between slaughter and freezing), whether frozen as carcase or meat cuts, conditions of freezing and speed of thawing. The

amount of losses varies between 1 per cent and 10 per cent of the weight of the meat and is usually about 5 per cent.

Heat Treatment

High-temperature heat treatment of food can destroy bacteria, depending on the cooking temperature and duration. Sterilization (mainly used in canneries) involves submitting the already canned product to the action of steam, generally in a closed container such as an autoclave or continuous cooker. Pasteurization is particularly reserved for liquids such as fruit juice, beer, milk or cream.

Ozonization

Ozonization can restrict the growth of moulds and yeasts on the surface of meat. This is effective on bacteria only when the moisture content of the superficial tissues has been considerably reduced. This is valuable in the treatment of various types of taints such as oil taint, fruit taint *etc.* in the frozen imported meat. However, ozonization is ineffective in removing ammonia taint.

Irradiation

Irradiation is a process of destruction of microorganisms and parasites by using radioactive substances. The process has some harmful effects such as loss of nutritive value of the treated food and production of some carcinogenic substances in it. Irradiation can be carried out by using several rays. Infrared rays have been mainly used for treatment of dry fruits and vegetables. Infrared rays are also used for heat blanching of fruits and vegetables in high frequency radiation. Ultraviolet rays are bactericidal and can be used to destroy the atmospheric bacteria. Ultraviolet rays are also used for aging of meat at relatively higher temperatures to control the growth of the surface organisms. Gamma rays and electrons cause ionizing radiation and these can be used for irradiation of food. Gamma rays are produced by radioactive substance like cobalt 60 or high energy electrons generated by machines. Gamma rays and electrons cause chemical changes in the food which leads to destruction of microorganisms.

Ionizing radiation is used for various purposes such as decontamination of food ingredients like spices; reduction in the number of pathogenic microorganisms like *Salmonella* spp., *Campylobacter* spp. and *Listeria* spp.; and extension of shelflife of meat, meat products, fruits and vegetables. "Radiation pasteurization" using much lower doses enables the refrigerated shelflife of many foods to be considerably extended. However, sterilizing canned foods with radiation requires such high dosage that unacceptable flavours and odours result. Ionizing radiation is very useful in food industry for two main purposes that are the screening of food packs for foreign matter and monitoring to detect underfilling.

Microwave Sterilization

Microwave sterilization is another type of electromagnetic emission that is currently being used in food industry. It is used for rapidly thawing raw frozen ingredients before further processing, as well as for heating frozen cooked foods in 2 to 3 minutes. This is a very useful method of food preservation as it causes low moisture content loss. This method also preserves the appearance and flavour of the food.

Canning

Canning is being used since early in the 19th century. This is useful for storage of food for many years because the food is sterilized and protected from recontamination. Canning operations involve two main processes that are exhausting and processing. Exhausting is the removal of oxygen content from the can which prevents discolouration of the surface of the food. In fruit packs, it also reduces the chemical action between the food product and container. Exhaustion of a can may be carried out by heat exhausting which involves passing of the filled cold can through steam-heated chamber and vacuumizing which involves closing of the cold material filled can in a vacuum closing machine. The processing is the heating of the can and foodstuff after the can has been hermetically sealed. Prompt cooling is important in order to prevent the action of heat on the food and undue changes in its texture and colour.

Types of Spoilage in Canned Foods

Spoilage in canned foods may occur due the action of spore-forming microorganisms which are resistant to commercial processing or due to the action of non-spore forming microorganisms like *Proteus* spp. and *E. coli* which gain access to the can by leakages after processing. Canned foods may be spoiled due to deleterious changes in foods and changes in the cans as follows:

1. **Swell or Blower:** This condition is characterized by swelling or bulging of the can at the ends due to accumulation of gas generated from microbial or chemical spoilage of the food.
2. **Flipper:** This condition is characterized by normal appearance of the can but flipping out one end of the can when struck against a solid object and regaining its normal position after applying light pressure. This condition indicates early spoilage in the canned food.
3. **Springer:** This condition is characterized by bulging of one end of the can when struck against a solid object and regaining of its normal position on applying pressure but bulging of the other end. Like flipper the springer also indicates early spoilage in the canned food. All blown cans pass successively through the flipper and springer stages. These conditions occur due to improper canning operations such as inadequate exhaustion of air before sealing, and overfilling.
4. **Leaker:** This condition is characterized by the hole in the can which may be contaminated by air and escaping of food content through it.
5. **Overfilled Can:** This is characterized by bulging of both the ends of the can due to overfilling of food.

Microbial Spoilage in Canned Foods

Flat souring may develop in canned foods due to the action of mainly thermophilic microorganisms such as *Bacillus stearothermophilus*, *B. coagulans* and *B. circulans*. The can is not blown in canned food subjected to flat souring. Canned foods containing sugar or starch and canned meat products containing cereal are more liable to development of flat souring.

Chemical Spoilage in Canned Foods

1. **Hydrogen Swell:** This is a type of chemical spoilage in the canned foods which occurs due to internal corrosion in the acidic medium. An acidic pH range of 3.5-4.5 of a canned food is most favourable for such spoilage.
2. **Purple Staining:** This is a type of chemical spoilage which involves breakdown of sulphur containing proteins by *Clostridium nigrificans* 'sulphur stinker', which liberates hydrogen sulphide that leads to the formation of a thin layer of tin sulphide on the inside of the can. This is found on the inner surface of the can in which sulphur containing foods such as fish, liver, kidneys and tongue are packed. Purple staining does not involve the foodstuff itself. This discolouration varies from light pink to dark purple which may be accompanied by blackening of both the inside of the can and surface of the foodstuff. Both types of discolouration can be prevented by a sulphur-resisting lacquer.
3. **Rust or Damage:** Rust on the external surface of the can may occur beneath the can labels when the adhesive contains hygroscopic substances. This may cause minute perforation of the tin plate (pinholing) which permit the entrance of air and lead to the spoilage of the canned product.

Public Health Aspects of Canned Foods

Nowadays the hygienic processing in canning industries has been significantly improved. Salmonellae are destroyed during commercial thermal processing. *Clostridium botulinum* spores are destroyed during processing in low-acid and medium acid foods. Staphylococci and streptococci are ubiquitous in nature that may contaminate the foods if not adequately heat treated. Staphylococci produce heat-stable enterotoxin which is not destroyed on ordinary cooking temperatures. However, it is destroyed during commercial processing. The cans may occasionally become contaminated by these organisms through a leak.

Use of Preservatives

Preservatives are the substances capable of retarding or inhibiting the growth of microorganisms and thereby the process of fermentation, acidification or other deteriorative changes (*e.g.*, putrefaction) in the food. There are certain preservatives that can be used safely in their optimum concentrations. The permitted preservatives include sodium nitrate, sodium nitrite, sulphur dioxide, benzoic acid, propionic acid, sorbic acid, methyl para-hydroxybenzoate, propyl para-hydroxybenzoate, copper carbonate, ortho-phenylphenol, tetracycline *etc.* The permitted preservatives for cured meat; sausages and sausages meat; and bacon and ham are sodium nitrite (200 ppm); sulphur dioxide (450 ppm); and sodium nitrate (500 ppm), respectively.

Use of Antioxidants

Antioxidants are substances that prevent or retard the development of rancidity in food or other flavour deterioration due to oxidation. The permitted oxidants used in food include butylated hydroxyanisole (BHA), butylated hydroxytoluene (BHT), propyl gallates, octyl gallates, dodecyl gallates and ethoxyquin. These are mainly used in fats, oils, or foods containing fats and oils. The maximum permissible limit of gallates, BHA and BHT in fats and oils are 100 ppm, 200 ppm and 200 ppm, respectively.

Physical Changes in Stored Meat

1. **Shrinkage:** Shrinkage in carcase/meat occurs due to loss of water content from their surfaces and lead to weight loss. The rate of weight loss during storage depends on the humidity of the storage room. At chilling temperature, the optimum relative humidity level is 83-90 per cent. The rate of weight loss is higher when the air of the storage room is drier. Carcase weight loss may be from 1.5 to 2.0 per cent in the first 24 hours of storage. Water evaporation from the meat surface can be prevented or reduced by using membranes like pleura and peritoneum. Low humidity of air leads to rapid evaporation of water from the meat surface, while high humidity facilitates the growth of moulds. Therefore, the relative humidity of the storage room should be optimum.
2. **Sweating:** Sweating is the condensation of water vapour on the surface of the meat. The condition develops when the meat is brought from cold store into ordinary room temperature. Sweating occurs due to condensation of air on the meat surface due to lowering of temperature of air below the dew point.
3. **Loss of Bloom:** Bloom is defined as "the colour and general appearance of carcase surface when viewed through the semitransparent layers of connective tissues, muscles and fat which form the carcase surface". Moistening of these connective tissues causes swelling of the collagen fibres in the connective tissues and opacity and dullness of meat. Loss of surface bloom in beef carcases may also be caused by dehydration and excessive oxidation. Loss of bloom can be prevented by avoiding temperature fluctuations, keeping the high relative humidity of the cooling chamber and ensuring proper circulation of air.

Chemical Changes in Stored Meat

The chemical changes in the stored meat take place due to breakdown of protein and fat. Protein breakdown may occur either due to endogenous enzymes or microbial enzymes. The storage life of meat dependents upon chemical changes in fat rather than muscles. The following chemical changes are seen in stored meat:

Weeping or Drip

Weeping refers to the presence of a watery, blood stained fluid which escapes from frozen meat when thawed. Drip consists mainly of water, salts, proteins, damaged blood cells and extractives. Weeping occurs after thawing of frozen meat which is an undesirable feature. This occurs due to rupture of muscle fibres by ice crystals and irreversible changes in muscle plasm. Slow freezing produces large ice crystals which damage the muscle fibres, while rapid freezing produces small ice crystals which cause little or no damage. The amount of drip is greater in beef than in mutton or pork.

Affections of Imported Meat

The conditions commonly encountered in imported meat include brine staining, freezer burn, bone taint, abnormal odours *etc.*

1. **Brine Staining:** Brine staining refers to the contamination of meat with calcium chloride which is commonly used in the brine circulating system for chilling or freezing of meat. Brine staining may occur in meat stored in refrigerating chamber due to leakage of brine from brines circulating system. Brine staining is characterized by dull or pale greenish colour on the carcase surface. Contamination with sea water differs from brine staining in that it gives rise to salty taste but not bitter taste and pale greenish colouration. Carcase/meat contaminated with brine can be safely removed by trimming.
2. **Freezer Burn:** Freezer burn refers to the damage of the surface of imported frozen offal usually liver and kidney due to loss of moisture from the outer tissues. The condition is characterized by development of soft and spongy with withered patches on the surface of the offal. Offal affected with freezer burn is insipid and unpalatable when cooked. The freezer burn can be prevented by enclosing the offal in a parchment paper before packaging in the boxes.

26

Utilization of Slaughterhouse By-Products

The huge quantities of offal and digesta from the slaughtered animals can be profitably utilized through rendering plants for production of value added products, like meat-cum-bone meal, pet foods, bone chips, tallow and methane as a source of energy for value addition in most of the modern plants. Meat-cum-bone meal contains about 50 per cent of good quality protein and is a cheap source of protein for poultry feeding. It is a good source of lysine and other sulphur containing amino acids. Tallow is a cheap source of energy for broiler production. This is also used for soap manufacturing. Meat-cum-bone meal obtained from rendering plants is completely sterilized and free from the risk of prion causing bovine spongiform encephalopathy (BSE) and can be safely used for animal feeding.

Efficient Utilization of Slaughterhouse By-products

Efficient utilization of slaughterhouse by-products has direct impact on the economy and environmental pollution of the country. Non-utilization or underutilization of by-products not only lead to loss of potential revenues but also lead to the added and increasing cost of disposal of these products. Non-utilization of animal by-products in a proper way may create major aesthetic and catastrophic health problems. Besides pollution and hazard aspects, in many cases meat, poultry and fish processing wastes have a potential for recycling raw materials or for conversion into useful products of higher value. In some parts of the world, blood is also utilized as an edible product for human beings.

Traditions, culture and religion are often important when a meat by-product is being utilized for food. Regulatory requirements are also important because many countries restrict the use of meat by-products for reasons of food safety and quality. By-products such as blood, liver, lung, kidney, brain, spleen and tripe have good nutritive value. Slaughterhouse by-products have also some medicinal and pharmaceutical properties. Waste products from the poultry processing and egg production industries must be efficiently dealt with as the growth of these industries depends largely on waste management. Treated fish waste has been useful in many applications among which the most important are animal feed, biodiesel/biogas, dietetic products (chitosan), natural pigments (after extraction) and cosmetics (collagen).

Many types of waste material either already contain large numbers of microbes and/or may be altered quickly through microbial activity. If regulations concerning infections/diseases are not properly observed, then hygienically unacceptable condition can arise, *e.g.*, maggots or molds. The breakdown of protein is always characterized by the evolution of strong odours. The majority of the waste in the meat industry is produced during slaughter. Slaughterhouse waste consists of the portion of a slaughtered animal that cannot be sold as meat or used in meat products. Such waste includes bones, tendons, skin, contents of the gastro-intestinal tract, blood and internal organs.

In the past, by-products were a favourate food in Asia, but health concerns have led to an increased focus on non-food uses, such as pet foods, pharmaceuticals, cosmetics and animal feed. Meat by-products are produced by slaughterhouses, meat processors, wholesalers and rendering plant. Traditional markets for edible meat by-products have gradually been disappearing because of low prices and health concerns. In response to these problems, meat processors have directed their marketing and research efforts towards non-food uses.

Some items may not be used in uncooked products. These include mammalian parts such as blood, blood plasma, feet, large intestines, small intestines, lungs, oesophagus meat, rectum, stomach (non-ruminant), first stomach (tripe, after cooking), second stomach (tripe, after cooking), fourth stomach, testicles and udder. It also includes poultry part such as gizzards and necks.

The yield of edible meat by-products from pigs is around 6.7 per cent of the carcases weight. The world production of edible by-products from pigs is about 625 million MT. Asia is the largest producer (50.4 per cent) of edible by-products from pigs. Europe is the second largest producer, with 37.1 per cent of the world total. Asia and Europe are also the two major consumers of meat by-products, including beef and lamb. Usage of meat by-products often requires treatments such as collection, washing, trimming, packaging and chilling. Whether these products are widely accepted by consumers depends on various factors such as the nutrient content, the price and whether there are comparable competing products.

Nutritional Importance of Meat By-products

Edible meat by-products contain many essential nutrients. Some are used as medicines because they contain special nutrients such as amino acids, hormones, minerals, vitamins, fatty acids *etc.* Blood and several other meat by-products have a higher level of moisture than meat, *e.g.*, lung, kidney, brains, spleen and tripe. Some organ meat, including liver and kidney, contains a higher level of carbohydrate than other meat materials.

Pork tail has the highest fat content and the lowest moisture content of all meat by-products. The liver, tail, ears and feet of cattle have a protein level which is close to that of lean meat tissue, but a large amount of collagen is found in the ears and feet. The lowest protein level is found in the brain, chitterlings and fatty tissue. The mechanically deboned beef and pork contain at least 14 per cent protein and a maximum of 30 per cent fat. The amino acid composition of meat by-products is different from that of lean tissue because of the large amount of connective tissue. As a result, by-products such as ears, feet, lungs, stomach and tripe contain a larger amount of proline, hydroxyproline and glycine, and a lower level of tryptophan and tyrosine.

The vitamin content of organ meats is usually higher than that of lean meat tissue. Kidney and liver contain the largest amount of riboflavin (1.697-3.630 mg/100 g), and have 5-10 times more than lean meat. Liver is the best source of vitamin A, vitamin B6, vitamin B12, folacin, niacin and ascorbic acid. Kidney is also a good source of vitamin B6, B12, and folacin. Lamb kidneys, pork, liver, lungs, and spleen are an excellent source of iron as well as vitamins. The copper content is highest in the livers of cattle, lamb and calf. Liver also contains the highest amount of manganese (0.128-0.344 mg/100 g). However, the highest level of phosphorus (393–558 mg/100 g) and potassium (360–433 mg/100 g) in meat by-products is found in the thymus and sweetbreads. With the exception of brain, kidney, lungs, spleen and ears, most other by-products contain sodium at or below the levels found in lean tissue. Mechanically deboned meat has the highest calcium content (315-485 mg/100 g).

Many organ meats contain more polyunsaturated fatty acids than lean tissue. Brain, chitterlings, heart, kidney, liver and lungs have the lowest level of monounsaturated fatty acids and the highest level of polyunsaturated fatty acids. The organ meats contain 3-5 times higher (260-410 mg/100 g) cholesterol than the lean meat and also have large quantities of phospholipids. Brain has the highest level of cholesterol (1352-2195 mg/100 g) and also has the highest amount of phospholipids compared to other meat by-products.

Utilization of Blood

Blood is used in food as an emulsifier, a stabilizer, a clarifier, a colour additive, and as a nutritional component. Most blood is used in livestock feed in the form of blood meal. It is used as a protein supplement, a milk substitute, a lysine supplement or a vitamin stabilizer, and is an excellent source of most of the trace minerals. Blood plasma has ability to form a gel, because it contains 60.0 per cent albumin. Blood plasma is the best water and fat binder of the blood fraction. Plasma gels appear very similar to cooked egg whites. Plasma forms a gel at a protein concentration of 4.0-5.0 per cent. Blood plasma also has an excellent foaming capacity, and can be used to replace egg whites in the baking industry. Transglutaminase from animal blood and organs is used in meat products for improving the binding ability of fresh meat products at chilling temperature. It occurs due to myosin cross-linkage by transglutaminase.

Blood can be separated into several fractions that have therapeutic properties. Blood plasma contains albumin, globulin and fibrinogen. In the laboratory, many blood products are used as a nutrient for tissue culture media. Blood product is used as a necessary ingredient in blood agar, and as peptones for microbial use. Glycerophosphates, albumins, globulins, sphingomyelins, and catalase are also used for biological assay. Many blood components such as fibrinogen, fibrinolysin, serotonin, kalikreninsa, immunoglobulins and plasminogen are isolated for chemical or medical uses. Purified bovine albumin is used to help replenish blood or fluid loss in animals. It is used in testing for the Rh factor in human beings, and as a stabilizer for vaccines. It is also used in antibiotic sensitivity tests.

Utilization of Hides and Skins

Animal hides have been used for shelters, clothing and as containers by human beings since prehistoric times. The hides represent a remarkable portion of the weight of the live animal. The weight of hide varies in different animal species. The approximate weight of hide in cattle, sheep and swine is 7 per cent, 11 per cent and 6 per cent, respectively. Hides and skins are generally one of the most valuable by-products from animals. Several finished products such as leather shoes and bags, rawhide, athletic equipment, reformed sausage casing and cosmetic products, sausage skins, edible gelatin and glue are obtained from the hides of cattle and pigs, and from sheep pelts. A product made from extracted collagen can stimulate blood clotting during surgery.

Pork skin is similar to human skin, and can be converted into a dressing for burns or skin-ulcers. Pork skin used as a dressing needs to be cut into strips or into a patch, shaved of hair, split to a thickness of 0.2-0.5 mm, cleansed, sanitized and packaged. It can be used for skin grafting. When used for skin grafting, it is removed from the carcase within 24 h of the slaughtered/death of the pig.

Gelatin is produced by the controlled hydrolysis of a water-insoluble collagen derived from protein. It is made from fresh raw materials such as hides and bone that are in an edible condition. Both hides and bones contain large quantities of collagen. The processing of gelatin from hide consists of three major steps. The first step is the elimination of non-collagenous material from the raw material. This is followed by controlled hydrolysis of collagen to gelatin. The final step is recovery and drying of the final product.

Gelatin extracted from animal skins and hides can be used for food. The raw material can also be rendered into lard. In the United States, Latin America, Europe and some Asian countries, pork skin is immersed, boiled, dried and then fried to make a snack food (pork rinds) and in UK they are called "pork scratching". Collagen from hides and skins also has a role as an emulsifier in manufacturing of meat products because it can bind large quantities of fat. This makes it a useful additive or filler for meat products. Collagen can also be extracted from cattle hides to make the collagen sausage used in the meat industry.

Gelatin is added to a wide range of foods, as well as forming a major ingredient in jellies and aspic. Its main use is the production of jellied desserts, because of its "melt in the mouth" properties, but is also added to a range of meat products, in particular to meat pies. Gelatin is also widely used as a stabilizer for ice cream and other frozen desserts. High-bloom gelatin is added as a protective colloid to ice cream, yoghurt and cream pies. The gelatin inhibits the formation of ice crystals and the recrystallization of lactose during storage.

Utilization of Bones

The marrow inside some of the bones can also be used as food. For centuries, bones have been used to make soup and gelatin. Meat-cum-bone meal (MBM) is widely recommended and used in animal nutrition as a protein source in place of proteinaseous feeds because of its content of available essential amino acids, minerals and vitamin B12.

27

Occupational Health Hazards in Meat Industries

The increased mechanization of work in the slaughterhouses and meat processing plants which occurred in the 1980s resulted in the most dangerous aspect of the work changing to the day-to-day cumulative grind that generated crippling occupational illnesses. The most common machinery accidents in food/drink manufacture involve conveyors. Workplace transport accidents in the food and drink industries are the second highest cause of fatal injury, comprising almost 30 per cent of fatal accidents. Cleaning materials used in meat processing can be irritant or toxic to the exposed personnel of the slaughterhouses and meat processing plants.

In the food and drink industries, there are reports of various occupational diseases such as bronchitis, asthma, other respiratory disorders, musculoskeletal disorders, physical coordination disorders, repeated trauma disorders, strains in various parts of body such as knees and elbows, hearing impairment, skin diseases, allergies, infections *etc.* The occupational health hazards such as physical, chemical, biological, electrical, thermal and noise hazards in slaughterhouses and meat processing plants are described as follows:

Physical Hazards

Occupational injuries may occur in slaughterhouses and meat processing plants during handling and slaughter of animals and meat processing. The most common causes of injuries in the meat industry are hand tools especially knives, equipment, instruments, operation of machinery, collisions with moving or stationary objects, falls or slips and burns. Injuries may also occur on the slippery floor which may be due to deposition of fat, blood and spread of water. The commonly occurring occupational physical hazards are described as follows:

Injuries from Animals

Slaughterhouse personnel may get injured from animals during their unloading from the transport vehicles, care and handling in the lairages, restraining at the time of stunning and slaughter. Accidental injuries may occur from stunner if prematurely or inadvertently discharge while workers try to still the animals. Falling animals and nervous system reactions in stunned cattle may cause jerking and thereby hazards to workers in the area. Such injuries in the slaughterhouse personnel can be prevented or minimized by adequate training to them about safety precautions, and importance and use of personal protective equipment (PPE).

Injuries from Falling of Carcases and Equipment

Many operations involve a series of hooks, chains and conveyor tram rails to move the product between processing steps, posing the hazard of falling carcases, products and equipment. Here PPE is very important in preventing the injuries from falling of carcases and equipment.

Several operations require protective equipment, including protective helmets, footwear, mesh gloves and aprons, wrist and forearm guards and waterproof aprons. Protective goggles may be required during boning, trimming and cutting operations to prevent foreign objects from entering eyes of meat plant personnel. Hand knives should also have guards to prevent the hand holding the knife from slipping over the blade. Training and adequate spacing between workers is also necessary to conduct operations safely.

Falling Accidents

Falling accidents are most often caused due to poor condition of the floor such as uneven floor, wet and slippery floor due to spread of water, fats, oils, grease *etc.* Falling accidents can be prevented by designing of anti-slip floors. Falling accidents can also be avoided by proper surface cleaning along with good housekeeping and by using proper footwear. There should be provision of good drainage system to avoid accumulation of water or spillage of other matters. Pits or drainage channels should be covered or barricaded in order to prevent the falling accidents.

Wet and treacherously slippery floors and stairs throughout the meat plant pose a serious hazard to personnel. Elevated work platforms also pose a falling hazard. Personnel must be provided with safety shoes with non-slip soles. Non-slip floor surfaces should be used on floors and stairways. Adequate drainage in wet areas must be provided, along with proper and adequate housekeeping of floors during production hours to minimize wet and slippery surfaces. All elevated surfaces must also be properly equipped with guard rails both to prevent workers from falling accidents and to prevent worker contact and materials falling from conveyors. Guardrails should also be used on stairways on the production floor to prevent falling accidents.

There should be provision of safe means of access to equipment and storage areas and use of sound ladders while working at heights to prevent hazards. The chance of hazard is increased when equipment to be cleaned is located in areas that are not easily accessible. There should be provision to control on operations that generate steam or dust as these make the floor slippery and also interfere with visibility in the area. Insufficient or inconsistent lighting may cause falling accidents. Therefore, the light source should be at appropriate locations and in enough in numbers with the adequate intensity of light so that the employees can clearly observe the processes.

Cuts

Cuts are the most common form of occupational injuries in slaughterhouses and meat processing plants. Cuts may occur due to careless use of sharp knife. Application of excessive and imbalanced force on the sharp knife during making an incision may also lead to severe cut injuries. Injuries caused by knives in meat and fish preparation can be minimized by design and maintenance, adequate work areas, selection of the right knife for the job, provision of tough protective gloves and aprons and correct training of workers on both the knife sharpening and the use of the knife. Mechanical cutting devices also pose hazards which can be prevented by good maintenance and adequate training of personnel.

Abrasions and Contusion

These conditions may be found through knocks from equipment and falling of gambrel and trolley. Although, accidents involving transmission machinery are relatively infrequent, but these accidents are usually very serious. Risks associated with machines and handling systems must be studied individually in each industry. Handling problems can be addressed by close examination of injury history for each particular process and by use of appropriate personal protection for head, foot, leg, hand, arm, eyes and face. Risks from machinery can be prevented by secure machinery guarding.

Mechanical handling equipment, especially conveyors, is widely employed, and particular attention should be paid to in-running nips on such equipment. The conveyor belts as well as pulleys and gearing, should be securely protected. To prevent cuts in canning, for example, effective arrangements for clearing up sharp tin or broken glass are required. Serious injury due to the inadvertent start-up of transmission machinery during cleaning or maintenance can be avoided by strict lockout/tagout procedures. Meat packing has one of the highest rates of injury of all industries.

Puncture and Laceration

These conditions may be found through bone splints and cause severe injuries.

Vibration Disorders

Vibration disorders may occur in slaughterhouse and meat processing personnel due to exposure of vibrations from instruments and equipment. The problem of vibration disorders increase with increasing in the level of mechanization. Cold temperatures increase the probability of vibration disorders to the fingers of the hand. Excessive exposure to vibration leads to musculoskeletal disorders in the wrists, elbows and shoulders. The type and degree of disorder depend on the type of machine and degree of vibration generated from it. High levels of vibration exposure can result in growth of a protuberance on the bone or the gradual destruction of the bone in the joint. This causes severe pain and limited mobility.

Musculoskeletal Problems

Many of the meat, poultry, fish and other food processing plants involve highly repetitive and forceful work. Handling of boxes for shipping can cause back injuries. There are three main factors such as extreme postures, high forces or high levels of repetition responsible for cumulative trauma disorders. Involvement of more than one factor makes the problem more critical. Early detection and treatment of affected workers is important.

Repetitive strain injuries (RSIs) generally occur due to repetitive exposure of the part of the body. The condition occurs mostly in modern abattoirs due to handling of automatic equipment, instruments and performing repetitive tasks. Many workers of the slaughterhouse and meat processing plants report the occupational musculoskeletal disorders. Plant automation and work whose pacing is regulated by a machine or conveyor belt occurs today for more workers in the food industry than ever before. Tasks in automated plants tend to be monotonous, with workers performing the same movement for a long time period. Packaging operations may involve repetitive hand and wrist motions to place finished items undamaged into trays or packs.

RSIs may occur as inflammation of the tendon (tendinitis) and inflammation of the tendon sheath (tenosynovitis). These are prevalent among workers whose jobs require repetitive hand movements, like meat packing workers. Tasks that repeatedly combine the bending of the wrist with gripping, squeezing and twisting motion can cause carpal tunnel syndrome (CTS) which is characterized by a tingling sensation in the thumb and first three index fingers, is caused by inflammation in the wrist joint creating pressure on the nerve system in the wrist.

Lugging of heavy carcases and repetitive lifting of heavy boxes of meat ready for shipping can cause back injuries. Cumulative trauma disorders such as carpal tunnel syndrome, tendinitis and tenosynovitis are widespread in the industry. The wrist, elbow and shoulder are all affected. These disorders can arise from the highly repetitive and forceful nature of the assembly line work in the plants, the use of vibrating equipment in some jobs, the use of dull knives, the cutting of frozen meat and the use of high-pressure hoses in cleaning operations. These disorders can be prevented by ergonomic designing of equipment, use of mechanical assists, and vigilant maintenance of vibrating equipment to minimize vibration, and advanced worker training and medical programmes.

Prevention and Control of Musculoskeletal Problems

1. **Rotation of Workers:** Rotation of workers with a view to avoid repetitive motions may reduce the risk by sharing the critical task across the team.
2. **Engineering Controls:** Engineering controls focus on reduction or elimination of the three primary causes of musculoskeletal problems that are force, position and repetition. The workplace should be analyzed to

identify the need of changes, including workstation design that favours adjustability, working methods, task automation/mechanical assists and ergonomically sound hand tools.

3. **Use of Mechanical Lifting Devices:** Vacuum hoists and other mechanical lifting devices can be used to reduce lifting of boxes.
4. **Adequate Training:** Adequate training should be provided to personnel using knives. Meat plants must provide adequate knife-sharpening facilities and avoid the cutting of frozen meat. Training encourages workers to understand the cause and prevention of musculoskeletal disorders. It reinforces the need to correct use of the tools and machines specified for the task. It should also encourage personnel to report medical symptoms as soon as possible. Elimination of more invasive medical intervention by restriction of duties and other conservative care is effective treatment of these disorders.

Chemical Hazards

Slaughterhouse personnel may be exposed to toxic chemicals like acids and alkalis during the cleaning and sanitizing of equipment. These can cause dryness, allergic rashes and other skin problems. Liquids can splash up and burn the eyes. PPE such as eye, face and arm coverings, aprons and protective footwear must be provided. Hand and eye washing facilities should also be available. Some chemical substances such as anhydrous ammonia, methyl chloride and other halogenated aliphatic hydrocarbons are used as refrigerants in freezing and cold storage which may pose the risks of poisoning and chemical burns. Therefore, emergency planning in addition to the normal fire planning is important. Training of workers in evacuation procedures is also necessary.

Chlorine in the water used to wash the carcases can also cause eye, throat and skin irritation. Adequate ventilation must be provided to escape of gaseous chemicals from the building. Special care must be taken to follow the manufacturers' instructions for storage, handling and use of chemicals. Carbon dioxide is used in refrigeration of products that are to be shipped. This must also be kept under strict controls. Carbon dioxide in the form of dry ice is used in the packaging area. During this process, the gas may escape from these vats and spread throughout the room. Exposure to this gas causes headache, dizziness, nausea, vomiting, and death at high levels. Adequate ventilation must be provided over dry ice bins to prevent the gas from causing health hazards.

Ammonia is used as a refrigerant in many industries and leakage of ammonia from pipes are common. Ammonia gas is irritating to the eyes and skin. Exposure of personnel to ammonia causes headache, burning in the throat, perspiration, nausea and vomiting in them. If escape is not possible, there may be severe irritation of the respiratory tract, producing cough, pulmonary oedema or respiratory arrest. Adequate maintenance of refrigeration lines is the key to prevent the leakage of ammonia. In addition, once an ammonia leak is detected, monitoring and evacuation procedures must be carried out to prevent dangerous exposures. Anhydrous ammonia and the less frequently used refrigerants such as ethane, ethylene, propane and butane are also flammable and explosive. Leakage from pipes may occur due to inadequate maintenance. Therefore, adequate measures should be taken to prevent the leakage, explosion and fire accidents.

Pesticides, fumigants and other hazardous materials must be kept under strict control and used only according to the manufacturer's guidelines. Organophosphate pesticides should only be used when accompanied with biological monitoring to assure the control of exposure. Early warning system through installation of chemicals sensors with a central alarm system in the building to signal the need to evacuate the personnel may be very useful to prevent such hazards.

Biological Hazards

Biological hazards are those which occur due to exposure of slaughterhouse personnel to the biological agents (bacteria, viruses, fungi, parasites, arthropods *etc.*) or substances in meat plant possessing these agents. Biological hazards may occur in several forms which are discussed as follows:

Allergies

Slaughterhouse and meat processing personnel may be susceptible to animal and their environment. The common allergies are asthma, rhinitis, skin rashes *etc.* Asthma and rhinitis may occur due to contact with animals or to mouldy hay and straw which may lead to a condition known as "farmer's lung". The condition is caused by inhalation of mouldy hay and straw containing large amount of spores of a bacterium *Micropolysporum faeni*. The condition is characterized by dyspnea, cough, fatigue and chills.

Respiratory Diseases

The common respiratory problems among slaughterhouse workers reported from various countries include bronchitis, chronic cough, chronic rhinitis, nasal congestion, pneumonia, dyspnoea *etc.*

Skin Diseases

In meat industries, the skin problems may occur as dermatitis and contact allergies like eczema. Slaughterhouse and meat processing personnel frequently wash their hands with soap and using hand-dip stations that contain quaternary ammonium solutions, for maintaining sanitary and hygienic conditions. This constant wetting of the hands can reduce the lipid content of the skin and lead to dermatitis. Personnel may become highly sensitized to microbial proteins and peptides generated by fermentation and extraction, which can lead to eczema and other allergies.

Occupational Zoonotic Diseases

Slaughterhouse and meat processing occupational groups may be exposed to zoonotic pathogens through direct contact with infected animals and affected with several zoonotic diseases such as anthrax, brucellosis, clostridial cellulitis, erysipeloid, leptospirosis, listeriosis, lyme borreliosis, Q fever, psittacosis, skin sepsis (caused by streptococci and staphylococci), cutaneous salmonellosis (caused by *Salmonella* Dublin, *S.* Typhimurium and *S.* Virchow), contagious pustular dermatitis or contagious ecthyma, looping ill, ringworm *etc.*

Brucellosis may be transmitted due to contact with infected cattle, goats and sheep. Erysipeloid may be transmitted due to contact of open wounds with infected pigs and fish. Leptospirosis may be transmitted due to direct contact with infected animals or their urine. Dermatophytosis (ringworm) may be transmitted in slaughterhouse personnel through contact with skin and hair of infected animals. Toxoplasmosis in slaughterhouse personnel may be transmitted through contact with infected cattle, sheep, goats, pigs and poultry.

Prevention and Control of Occupational Zoonotic Diseases

Occupational diseases in slaughterhouse and meat processing plant staff can be prevented and controlled by adopting following measures:

1. Health education to the slaughterhouse and meat processing plant personnel.
2. High standards of personal and environmental hygiene in order to prevent the transmission of infectious diseases.
3. Provision of uniforms, personal protective equipment, hand towels *etc.*
4. Adequate facilities of clean washrooms, toilets, and shower.
5. Carefulness to prevent the incidence of cuts, wounds and abrasions, and prompt and efficient treatment.
6. Adequate first-aid facilities like proper dressing of wounds.
7. Efficient ante-mortem inspection of animals and post-mortem inspection of slaughtered animals.
8. Close liaison with medical expertise.
9. Vaccination in animals and slaughterhouse personnel, early detection and treatment of diseases with antibacterial/antiparasitic drugs for prevention and control of zoonotic diseases.
10. The control and extermination of rodents and the exclusion of dogs, cats and other animals from the slaughterhouse.

Electrical Hazards

Electrical installations, especially in wet or damp places, require proper grounding and good maintenance to prevent electrical hazards. In addition to proper grounds, outlets protected with ground fault interrupters (GFIs) are effective in protecting from electrical shock. Fires may also occur if welding is done around explosive/combustible matters. Explosions may also occur in gas or oil-fired ovens or cooking processes if they are not installed, operated or maintained correctly; provided with the essential safety devices; or if proper safety procedures are not followed. The combination of wet working conditions and elaborate electrical wiring also poses a hazard of electrocution to workers. Therefore, all electrical equipment should be grounded to prevent electrical hazards. Electrical outlet boxes should be provided with covers which effectively protect against accidental contact. All electrical wiring should be checked periodically for defects like cracking, fraying or other defects. GFIs should be used where possible.

Thermal Hazards

The hazards caused due to exposure of extremes of temperatures (low and high) to slaughterhouse and meat processing plant personnel is known as thermal hazards. Thermal hazards may occur as cold injuries as well as heat injuries which are discussed as follows:

Cold Injuries

Thermal extremes may exist in the food work area. People may work in freezers with temperatures of -18 °C or below. The use of freezer clothing helps to insulate the worker from the cold, but warm break rooms with access to warm liquids must be provided. Meat processing plants are kept at 7 to 10 °C which is below the comfort zone and workers may need to wear additional clothing layers. Areas such as freezers may require temperatures below -40 °C. The most common cold-related injuries are frostnip, frostbite, immersion foot and trenchfoot, which occur in localized areas of the body. A serious consequence of cold stress is hypothermia. The respiratory system, the circulatory system and the osteoarticular system can also be affected by overexposure to the cold.

Slaughterhouse and meat processing plant personnel should wear appropriate clothing to prevent the consequences of cold stress and reduce the hazards of cold working conditions. Moreover, the workplace should have proper equipment, administrative controls and engineering controls. Multiple layers of clothing provide better protection than single thick garments. Cooling equipment and air distribution systems should minimize air velocity. Unit coolers should be placed as far away from workers as possible, and wind deflectors and barriers should be used to protect personnel from windchill. Job rotation to more active or warmer locations is advisable. In large tunnel freezing plants, it may be fatal for workers to stay in the rapidly moving stream of air, even if dressed in polar clothing. It is important to prohibit entry into a tunnel freezer in operation and to make effective interlocking arrangements or use confined-space entry protocol to ensure that freezers cannot be started up while workers are still inside them.

Heat Injuries

Heat injuries may occur due to fire accidents in the slaughterhouses and meat processing plants. Burns and scalds from hot liquors and cooking equipment are common. These injuries arise from steam and hot water used in equipment cleaning. Even more serious accidents can occur due to explosion of boilers or autoclaves due to lack of regular examination, poor employee training, poor procedures or poor maintenance. Therefore, all steam equipment should be regularly and carefully maintained in order to prevent major explosion or minor leaks. Ovens and steam cookers have radiant and moist heat.

Heat stress can occur during season changes and heat waves. High-pressure hoses used to transport hot water for disinfecting equipment can also cause burns. Adequate worker training on the use of such hoses is important. There may be some other problems due to heat, which often combined with high humidity in cooking and sterilization process, can cause heat stroke and heat exhaustion. These conditions are found especially in processing that entails evaporation of solutions, *e.g.*, kill floors of slaughterhouses.

Effective ventilation systems are essential, with special attention to avoid condensation problems. Air conditioning may be necessary in some areas. Inflammable substances must be kept or stored in a very protective condition. Care must be taken to avoid any fire work near the inflammable chemicals. Installation of fire alarming system with provision of adequate exits for safe evacuation of personnel in case of fire accidents is very important to prevent such hazards. Moreover, proper training of personnel regarding safety precautions and safe handling of situation are also important to prevent the happening of such hazards.

Noise Hazards

Machines and equipment in slaughterhouses and meat processing and packing plants produce a high level of noise. Safety and health codes vary from country to country, but personnel exposure to noise is usually limited to 85 dB for 8 continuous hours. Hearing impairment may occur as a result of continuous and prolonged exposure to noise above threshold levels. This impairment causes communication disorders. This can also lead to deterioration in psychological and physiological performance. Exposure to high level of noise may cause abnormal blood pressure, heartbeat, respiration rate and nervous disorders.

Health and safety committees that effectively involve the entire organization, including production operators, in the development of plant programmes is the key to a safe operation. Annual audiometric testing is recommended. One of the most important factors in noise reduction is the design of the workplace. Equipment should be designed that produce low noise and low vibration. The noise level can be reduced by replacing metal parts with softer materials like rubber. Silencers should be installed at air valves and exhaust pipes. Noise-producing machines and processes should be enclosed to reduce to a minimum number of workers exposed to high noise levels. Noise-proof partitions and noise-absorbing ceilings should be installed. PPE should be used to protect the ears of the worker exposed to high level of noise.

Machines and equipment found in meat packing plants produce a high level of noise. Workers must be provided with ear plugs, as well as hearing examinations to ascertain any potential hearing loss. Further, sound-dampening equipment should be used on machinery where possible. Good maintenance on conveyor systems can prevent unnecessary noise. Certain engineering techniques such as sound-absorbent mounting, magnetic elevators, nylon-coated cables and speed-matching in the cans conveyor systems reduce the level of noise. Nowadays a hearing conservation programme based on audiometric examinations, hearing-protection equipment and education should be instituted.

Identification of Hazards

Identification of hazards in any occupational activity involves characterization of the workplace by identifying hazardous agents and groups of workers potentially exposed to these hazards. Some hazards in the work environment are easy to recognize, *e.g.*, irritants, which have an immediate irritating effect after inhalation or skin exposure. Other hazards are not so easy to recognize, *e.g.*, chemicals which are accidentally formed and have no warning properties. Some agents like metals (*e.g.*, lead, mercury, cadmium, manganese), which may cause hazards after several years of exposure.

Prevention of Accidents in Meat Plants

1. **Maintenance of Machines and Equipment:** Regular and adequate maintenance of all equipment is necessary, especially equipment used to move meat. Such equipment must be checked regularly and repaired as per the need.
2. **Use of Personal Protective Equipment:** Adequate safeguards must be taken for knocking guns, such as safety switches and making sure to avoid blow back. Slaughterhouse workers involved in knocking and sticking operations must be trained on the hazards of this job, as well as provided with guarded knives and protective equipment to prevent injury, *e.g.*, the use of arm guards, mesh gloves and special guarded knives for sticking operations.
3. **Training of Slaughterhouse Personnel:** Slaughter and further processing involve the use of instruments and equipment including hand knives and mechanical cutting devices such as head splitters, bone splitters, snout pullers, electric band and circular saws, electric or air-powered circular-blade knives, grinding machines and bacon processors. These operations have a high rate of injury, from knife cuts to amputations, because of the speed at which workers operate the inherent danger of the tools being used and the often slippery nature of the product from fat and wet processes. Slaughterhouse workers may be injured due to cut by their own knives and by other workers' knives during the butchering process. Therefore, adequate training associated with these accidents should be given to the slaughterhouse personnel to prevent accidents.

28

Meat-Borne Diseases

Meat-borne diseases are defined as diseases usually either infectious or toxic in nature, caused by agents that enter the body through the ingestion of meat and meat products. Most of the food-borne illnesses are caused by bacteria which arise from animal, human or environmental sources. Viral food-borne infections are unlikely to be from animals and may be due to direct contamination from human or indirectly through the environment, *e.g.*, from shellfish contaminated by discharged human sewage.

Food-Borne Illnesses

Pathogenic organisms may cause food-borne illnesses which may be of the following type:

1. **Food Infection:** The ingestion of viable pathogenic organisms along with the food leads to their lodgment and establishment in organs of consumers.
2. **Food Intoxication:** Ingestion of toxins already produced by microorganisms in the food (that is preformed toxins in the food) brings about poisoning syndrome in the consumers. This is called food intoxication and the toxins affecting the gastrointestinal tract are called enterotoxin.
3. **Toxico-Infection:** This is a type of infection in which a certain group of organisms infect the intestine when ingested along with the food and produce toxins in situ to bring about symptoms of poisoning. This situation is called toxico-infection. *Clostridium perfringens* causes such type of food-borne illness.

Organisms Causing Meat-Borne Infections

The important organisms causing meat-borne infections include *Campylobacter* spp., *Clostridium perfringens, Escherichia coli, Salmonella* spp., *Vibrio parahaemolyticus, Yersinia* spp. *etc.*

Campylobacter spp.

Campylobacter jejuni causes most of the campylobacteriosis. Poultry and cattle are the main reservoirs of *Campylobacter* spp. and act as source of infection for humans. The disease may occur due to consumption of undercooked chicken and contaminated raw milk. The incubation period varies from 1 to 10 days. The disease is characterized by acute onset of fever, abdominal pain and diarrhoea which may be blood stained.

Clostridium perfringens

Clostridium perfringens is a common inhabitant of the intestinal tract of animals and humans. The spores of the organisms are very resistant and can survive the usual cooking temperatures to produce toxin in the intestine after germination into vegetative forms. The incubation period varies from 8 to 24 hours. The disease is characterized by nausea, abdominal pain, diarrhoea *etc.*

Escherichia coli

The main source of *Escherichia coli* infection is improperly cooked meat. The disease may be transmitted due to consumption of infected or contaminated raw meat. The incubation period varies from 1 to 12 days. The disease is characterized by fever, abdominal pain and diarrhoea. The infection caused by enterohaemorrhagic *Escherichia coli* (EHEC) strain O157:H7 may lead to hemorrhagic colitis (HC), haemolytic uremic syndrome (HUS) and thrombotic thrombocytopenic perpura (TTP).

Salmonella spp.

The meat may be contaminated by *Salmonella* organisms either directly during slaughter from animal excreta or from human excreta transmitted to meat by hands, utensils, equipment, flies *etc. Salmonella* infection is caused by various *Salmonella* serovars and characterized by gastrointestinal upset including diarrhoea following about 12 hours on consumption of contaminated meat. The incubation period may vary from 6 to 36 hours. *Salmonella* food poisoning outbreaks occur as sporadic, smaller outbreaks and larger outbreaks. *Salmonella* infections can be prevented and controlled by the following measures:

1. Hygienic standards of animal husbandry.
2. Hygienic conditions of transport and lairaging with avoidance of stress.
3. Hygienic method of slaughter and dressing.
4. Hygienic method of processing and handling of meat and meat products.
5. Adequate cooling of meat and meat products.

Vibrio parahaemolyticus

Vibrio parahaemolyticus food poisoning occurs frequently in Japan and other eastern countries where it is commonly found in sea creatures (fish and shellfish) and sea foods. The organism is found in sea waters, especially at temperature above 10°C. The infection may be transmitted through fish and shellfish from contaminated water. The incubation period varies from 12-24 hours. The illness is characterized by profuse diarrhoea, abdominal pain, and sometimes fever and vomiting.

Yersinia spp.

Yersinia enterocolitica and *Y. pseudotuberculosis* cause yersiniosis. The infection occurs due to consumption of infected raw pork, unpasteurized milk, cheese *etc.* The disease can also be transmitted from person to person. The incubation period varies from 3 to 7 days. The disease is characterized by an acute onset of fever followed by abdominal pain and diarrhoea.

Organisms Causing Meat-Borne Intoxications

Staphylococcus aureus

Staphylococci are found on the skin and in the upper respiratory tract of animals and humans and can easily contaminate different types of foods. Staphylococcal intoxication occurs due to consumption of meat and meat products already containing staphylococcal toxins (intradietetic). The incubation period is short (2-3 hours) as the toxins are already formed in the meat and meat products (that is the toxins are ready to act on the body of consumer). Staphylococcal intoxication is characterized by vomiting, severe abdominal cramps and diarrhoea. The symptoms usually disappear within 24 hours.

Clostridium botulinum

Clostridium botulinum causes botulism. This is highly fatal food intoxication in animals and humans. Botulism is transmitted through ingestion of food containing powerful toxin of the organism. The toxins are usually produced by the organism in decaying materials like carcases and vegetables in which spores multiply rapidly under favourable conditions of warmth and moisture. The organisms also present in the soil and are worldwide in distribution.

On the basis of toxin production, the organism has been classified into various types such as type A, B, C (C-α and C-β), D, E, F and G. Botulism affects mainly the nervous system, while the gastrointestinal symptoms are minimal. Type A, B and E are most important in causing human botulism. Type C-α causes disease in poultry, ducks and pheasants (limber neck). Type C-β causes disease in cattle, horses and mink. Type D causes Lamziekte in cattle and sheep.

Botulism in animals is similar to that of humans and the disease is associated with a paralysis of muscles and progressive motor paralysis which may lead to death due to respiratory and cardiac failure. Human botulism usually occurs as sporadic outbreaks. Animals suspected of botulism can be subjected to casualty slaughter only after serological examination revealing negative sera for circulating botulinum toxins. Lesions may be absent. Hemorrhages may be present in the intestine, ear and brain. The presence of carrion materials may be suggestive of botulism. Carcases affected with botulism are totally condemned. Botulism can be prevented and controlled by the following measures:

1. The spores of *Clostridium botulinum* can be destroyed by subjecting them at higher temperature (120°C for 20 minutes). More acidic food can also destroy them.
2. Pickled foods can be rendered safe by using high amount of common salt (10 per cent or more) in brine.
3. The care should be taken about the manufacturing of cans, their transport, handling and storage.
4. Preserved foods possessing rancid or other odours should be rejected.

Table 28.1. Bacterial Meat-Borne Diseases

Disease	*Causative Agent*	*Transmission*	*Characteristics of Disease in Humans*
Aeromonas hydrophila infection	*Aeromonas hydrophila*	Ingestion of contaminated seafood, chicken, and red meat.	Gastroenteritis, diarrhoea *etc.*
Bacillus cereus food poisoning	*Bacillus cereus*	Ingestion of contaminated meat.	Nausea, vomiting, diarrhoea, abdominal pain *etc.*
Botulism	*Clostridium botulinum*	Ingestion of meat containing preformed toxins.	Dysphagia, quadriplegia, dysarthria, diplopia *etc.*
Campylo-bacteriosis	*Campylobacter jejuni, C. coli* and *C. lari*	Ingestion of contaminated poultry meat and pork.	Acute enteritis, diarrhoea, colic *etc.*
Clostridium perfringens gastroenteritis	*Clostridium perfringens*	Ingestion of contaminated meat and meat products.	Diarrhea, abdominal cramps *etc.*
Escherichia coli infection	*Escherichia coli*	Ingestion of contaminated ground meat products, hamburgers and salami.	Fever, vomiting, diarrhoea, haemorrhagic colitis, haemolytic uraemic syndrome *etc.*
Salmonellosis	*Salmonella* serovars	Ingestion of contaminated poultry meat, eggs and pork.	Fever, nausea, vomiting, diarrhea *etc.*
Staphylococcal food intoxication	*Staphylococcus aureus*	Ingestion of toxins pre-formed in the meat.	Vomiting, abdominal cramps, diarrhea *etc.*
Vibrio cholerae infection (Cholera)	*Vibrio cholerae*	Ingestion of infected raw fish, crabs, seafood and oysters.	Watery diarrhoea, vomiting *etc.*
Vibrio parahaemolyticus infection	*Vibrio parahaemolyticus*	Ingestion of raw or insufficiently cooked seafood like shellfish.	Diarrhea, vomiting, fever *etc.*
Yersinioses	*Yersinia enterocolitica* and *Y. pseudot-uberculosis*	Ingestion of infected pork.	Enterocolitis, mesenterial and ileocecal lymphadenitis *etc.*

Prevention and Control of Meat-Borne Intoxications

Identification of source of infection is very important to prevent the further spread of disease. In meat-borne intoxications, the great care must be taken to avoid contamination of meat and meat products. Once the meat or meat product becomes contaminated, it cannot be rendered safe by cooking as the toxins are not destroyed on ordinary cooking temperatures. Therefore, the prevention of contamination of meat or meat product is the most important step for prevention and control of meat-borne intoxication. Another method to render the meat and meat products safe from meat-borne intoxication is to store them at refrigeration temperature in order to prevent the multiplication of the organisms and thereby toxins production. Meat-borne intoxication can be managed by the following measures:

1. Creation of outbreak control group (normally include public health, environmental health, medical microbiologists and veterinarian)
2. Practicing outbreak control plan
3. Formulation of hypothesis
4. Testing of hypothesis (microbiologically or epidemiologically)

Food Poisoning Outbreak Investigation

Surveillance of disease outbreak can provide valuable information on organisms involved, the food vehicles and factors contributing to the cause of the outbreak. There are three main sources of data such as notification of food poisoning, surveillance of laboratory-confirmed infections and investigation of food poisoning outbreaks. Food poisoning outbreaks involves informing to higher health authorities in order to prevent its further spread and also important in making strategies for prevention and control of disease.

Objectives of Food Poisoning Outbreak Investigations

Food poisoning outbreak investigations have the following objectives:

1. To prevent any further cases arising from the specific food poisoning outbreak by identifying and removing the food source.
2. To control further spread of food poisoning from infected persons to healthy person.
3. To use the information to reduce the further outbreaks.

Important Steps in an Outbreak Investigation

1. **Preliminary Assessment in Outbreak Investigation:** It is important to confirm the existence of outbreak.
2. **Case Definition and Identification:** Define cases in terms of time/place/present symptoms/lab results. Define population at risk and case identification.
3. **Descriptive Study:** It includes collection and analysis of data. Draw epidemic curve and generate hypothesis. Try link in some way to source *e.g.*, questionnaires.
4. **Analytical Study:** The purpose of analytical study is to test hypothesis. Cohort study may be conducted for food-borne outbreak where the

population is known. Case-control study may be conducted in case of exposure of large population.

5. **Verification of Hypothesis:** It can be done by microbiological analysis of food and environment samples.
6. **Control Measures:** It involves closing the outlet, isolate and treat cases, destroy food and protect person at risk by improving hygiene and prophylaxis. Measures should be taken to prevent recurrence by following the recommendations and guidelines.
7. **Communication:** Communicate the public and professionals.

Organisms Causing Meat-Borne Zoonotic Diseases

Anthrax

Anthrax is primarily a disease of herbivore and secondarily transmissible to humans. The disease is caused by *Bacillus anthracis*. The disease occurs worldwide. The infection in the cattle, sheep and horses usually occurs from ingestion of food containing spores of the organism. The meat from anthrax infected animals is dangerous to humans. Although the vegetative form of the organism is readily destroyed by gastric juice but the spores are resistant and may gain access to the system through abrasions of the mucosa. In humans, anthrax is found in three main forms that are cutaneous, pulmonary and intestinal anthrax.

The type of anthrax depends on the route of entry of the organism in the host. The cutaneous form is the most common form of anthrax which results from the handling of infected materials. This form is characterized by development of papules which later on changed into vesicles. These vesicles rupture and appear as black scar. The pulmonary form (wool sorter's disease) is characterized by bronchitis, pneumonia and pleurisy and high mortality. The intestinal form is characterized by gastroenteritis and ulceration in the intestine. The disease can be prevented and controlled by the following measures:

1. Proper disposal of infected carcases and materials such as litter, blood, hide, fat, internal organs *etc.* by burial or incineration.
2. Disinfection of slaughterhouse with 5 per cent phenol or 10 per cent formalin before subsequent use.
3. Disinfection of pastures, soil, and incubation areas with 30 per cent per acetic acid to destroy the spores.
4. Vaccination in animals can be carried out with anthrax spore vaccine at the dose rate of 1ml subcutaneous for all age group of cattle, sheep, goat and horse. Booster dose is given after 6 months of primary vaccination.
5. Vaccination in humans can be carried out with alum precipitated cell free filtrate of Sterne strain.

Brucellosis

Brucellosis in humans is known as undulant fever. The disease is caused by *Brucella* spp. The disease may be transmitted by contact of infected materials with conjunctiva or the skin from infected animals and their meat. The disease may be transmitted through inhalation of infected aerosols. The disease can also be transmitted through ingestion of meat and milk from infected animals. The disease is characterized by inclining and declining pattern of fever (undulant fever), sweating, myalgia, arthralgia, constipation, abdominal pain, dysurea, frequent micturition, mental depression *etc.* There is also enlargement of lymph nodes especially cervical and axillary, spleen, liver and testes. The persons at risk are those handling the livestock, dressing the carcase, handling of gut and tripe, skin and rejected materials. All abattoir staff is at risk of hazards, especially when reactor animals are slaughtered as in case of brucellosis eradication programme. The disease can be prevented and controlled by the following measures:

1. High standards of hygiene should be adopted.
2. Great care must be taken by staff dressing *Brucella* infected carcase. Hooks should be used to handle uteri and udders.
3. Udder and uteri of *Brucella* infected animals should not be incised.
4. Protective clothing and equipment such as face mask covering the nose and mouth, goggles to protect the eyes and arm length gloves should be used.

Tuberculosis

Tuberculosis is an important bacterial zoonotic disease which is characterized by development of tubercles in soft as well as hard tissues of the body. The disease is caused by *Mycobacterium tuberculosis, M. bovis* and *M. avium*. Though, majority of human cases are caused by *Mycobacterium tuberculosis.* The disease may be transmitted through ingestion, inhalation, contact, genital and congenital. The main sources of infection are milk, meat and offal from infected animals and contaminated water. In humans, the symptoms may be chest pain, hemoptysis, cough, continuous fever *etc.* During post-mortem inspection of the slaughtered animal, if the lesions are localized, only affected organs need to be condemned. If the lesions are generalized, total condemnation is advised. The disease can be prevented and controlled by the following measures:

1. Adoption of high standards of hygiene.
2. Adequate heat treatment of milk and meat before their consumption.
3. Proper disinfection of premises inhabited by sick animals.
4. Proper disposal of meat and offal.
5. Implementation of test and slaughter policy for eradication of tuberculosis.

Bovine Spongiform Encephalopathy (BSE)

This is also known as mad cow disease. BSE is a highly contagious and fatal disease of bovine. The disease is of great economic importance. BSE affected animals are not allowed for slaughter. In addition to this, animal's trades are not permitted in BSE outbreak regions. It is caused by an infectious protein particle "prion". The infectious agent is extremely resistant to physical and chemical conditions. High-temperature autoclaving is necessary to totally destroy the agent. BSE affected carcases are totally condemned and not allowed for manufacturing of any product through rendering.

The key food safety control is the removal of specified risk material; however there are also controls on animal feed and a requirement to test certain categories of animal for BSE. In addition to these controls, cattle with BSE or suspected of having BSE and the offspring and cohorts of BSE cases are removed from the food chain. The specified risk material must be disposed of and prevented from contamination of food or animal feed. A ban on the feeding of contaminated meat and bone meal to ruminants is important step in prevention and control of BSE.

Rabies

Rabies is a highly fatal viral zoonotic disease. The disease is caused by a virus "*Lyssavirus* type1" which belongs to the family Rhabdoviridae. Rabies is transmitted to humans by inoculation rather than by ingestion. Carcases showing lesions of rabies should be condemned.

Foot and Mouth Disease

Foot and mouth disease is a highly contagious viral zoonotic disease. Though the disease is rare in humans but may occur in a mild form. The lesions may occur on lips, tongue, fingers and at the base of the nail. All infected animals should be destroyed and disposed by burial or burning. Foot and mouth disease infected carcases should be totally condemned.

Taeniasis

Taeniasis is an important parasitic zoonotic disease. The disease is caused by a cestode parasite "*Taenia* spp.". The disease caused by the adult stage of the parasite is known as taeniasis, while the disease caused by the larval stage of the parasite is known as cysticercosis. The important reservoirs of the parasite are pig and cattle. *Taenia solium* is adapted to pig, while *Taenia saginata* to cattle. In the pig, the larvae are found predominantly in the musculature and the brain. In humans, the infection of *Taenia solium* is transmitted through consumption of infected raw pork (containing *Cysticercus cellulosae*). This larval stage of the parasite develops into adult stage in the human.

The clinical manifestations due to adult stage of the parasite are anorexia, abdominal pain, digestive upsets, nervousness *etc.* The adult parasite can survive in the host for many years. Gravid proglottids (containing eggs) are released along the human faeces which can serve as a source of infection for pigs and occasionally for humans. The eggs containing infective larvae hatch in the intestine and develop into onchosphere which penetrate the intestinal mucosa and spread via circulation to various organs such as central nervous system, muscles, subcutaneous tissue *etc.*

The larval stage of *Taenia solium* (*Cysticercus cellulosae*) causes cysticercosis. Human cysticercosis due to *Cysticercus bovis* is extremely rare. Cysticerci that develop in the central nervous system are known as neurocysticercosis, which represents a serious threat to the individual. As a result of mechanical pressure, a variety of pathological changes are produced, leading to intracranial hypertensive syndromes, hydrocephalus, epilepsy and psychiatric diseases or death. Affected carcases and organs must be condemned. The disease can be prevented and controlled by the following measures:

1. Proper inspection of carcases with condemnation of generalized cases.
2. Identification, isolation and treatment of human cases harbouring adult worm with proper disposal of faeces.
3. Prevention of pigs to access human faeces.
4. Health education to the public about personal hygiene; prevention of contamination of soil, water and animal food with human faeces; and avoidance of use of sewage effluent on pastures.
5. Thorough cooking of beef and pork before their consumption.

Hydatid Disease

Hydatid disease is an important parasitic zoonotic disease. The disease is caused by a cestode parasite "*Echinococcus* spp.". The disease occurs primarily in herbivorous animals and incidentally transmissible to humans. The disease in humans is transmitted through consumption of food and water contaminated with faeces of infected dog. The disease can also be transmitted through inhalation of air contaminated with dog faeces. Hand to mouth transmission of infection is also possible through transfer of eggs by handling or playing with infected dogs. Hydatid cysts are found mainly in liver and lungs. Infected organs must be condemned and destroyed. If the carcases of hydatid disease are emaciated or oedematous, they are totally condemned. The disease can be prevented and controlled by the following measures:

1. Adoption of high standards of hygiene.
2. Prohibition of access by dogs to raw offal to interrupt the life cycle of the parasite.
3. Elimination of stray dogs.
4. Inspection of all slaughtered animals and destruction of all organs containing cysts.
5. Health education to the public particularly butchers, dog owners, animal breeders and shepherds is the basis of effective prevention.

Trichinellosis

Trichinellosis is also known as trichiniasis or trichinosis. The disease is caused by a nematode parasite mainly "*Trichinella spiralis*". The adult form of the parasite lives in the intestine of mammals and the larval form lives in the muscular tissues of the same host. *Trichinella spiralis* is found in the intestine and muscular tissues of the pig, wild boar, rat, mouse, dog, cat, bear, human and other mammals. The main source of infection is flesh infected with the parasite. The disease is transmitted through consumption of trichinosed flesh. Trichinellosis in the pig is most likely to be acquired from the flesh of another pig via raw or unboiled garbage. Human acquires the infection from infected pig meat.

In humans, the disease is characterized by abdominal pain and diarrhoea associated with the adult worms penetrating the intestinal mucosa. The migration of the larval stage of the parasite in the blood stream may cause symptoms simulating to influenza or typhoid fever. In the later stage, it is manifested by pain resemble to rheumatism. The larvae may reach to the heart and brain and cause myocarditis and encephalitis. The carcase affected with *Trichinella spiralis* should be condemned. The disease can be prevented and controlled by the following measures:

1. Adequate heat treatment of meat before its consumption.
2. Refrigeration of meat at (-25°C) or below for 10 days. Such lower temperatures can destroy the cysts of the parasite.
3. Irradiation of pork by high voltage of X-rays for few minutes.
4. Salting, smoking and drying of meat to destroy the cysts of the parasite.

Factors Contributing to Increase the Occurrence of Meat-Borne Diseases

1. Intensive animal husbandry practices.
2. Unhygienic conditions of the slaughterhouses.
3. Poor personal hygiene of the butchers, meat handlers and other personnel of the slaughterhouse.
4. Globalization of foods of animal origin.
5. Use of contaminated cattle and poultry feeds.
6. Frequent use of antibiotics in livestock and poultry.
7. Improper rendering of slaughterhouse by-products.
8. Improper disposal of condemned carcases/meat and slaughterhouse wastes.
9. Improper treatment and disposal of slaughterhouse effluents.
10. Improper implementation of the HACCP system in the meat production and processing plants.

Public Health Impact of Meat-Borne Diseases

A large number of populations in industrialized countries are affected by meat-borne illness each year and the problems are likely to be even more serious in developing countries. Microbiological hazards contribute a major role among all kinds of hazards and cause problems in not only in developing countries but also in developed countries. Despite remarkable advances in food science and technology, food-borne illness is a rising cause of morbidity in all countries and the list of potential food-borne microbiological pathogens keeps increasing. Therefore, microbial food safety is an essential public health issue of all countries.

Prevention and Control of Meat-Borne Diseases

1. **Hygiene Implementation:** Implementation of hygienic practices in livestock and meat industries are important in preventing and controlling the meat-borne diseases. Adoption of high standards of personal and environmental hygiene is important in this regard.
2. **Prevention of Contamination:** Contamination must be prevented during all the stages of production and processing of meat.
3. **Meat Inspection:** Efficient ante-mortem inspection, post-mortem inspection and laboratory examination (if necessary) should be carried out to ensure the meat fit for human consumption.
4. **Implementation of HACCP:** The HACCP system should be properly implemented in the slaughterhouses and meat processing plants to achieve the quality meat and preventing the hazards from consumption of meat and meat products.
5. **Heat Treatment of Meat:** Adequate heat treatment of meat is important to render the meat safe for human consumption by destroying the pathogenic organisms in meat and meat products.
6. **Condemnation of Unsound Meat:** Unsound meat must be condemned without any consideration in order to prevent health hazards in consumers.
7. **Elimination of Stray Dogs:** Elimination of stray dogs and prevention of entry of wild animals are important to prevent the access of viscera and slaughterhouse by-products in order to interrupt the transmission chain of infectious diseases, *e.g.*, hydatid disease. Strong fencing can be used to keep out trespassing animals, particularly stray dogs and wild animals.
8. **Refrigeration Storage of Meat:** The meat should be stored at refrigeration temperature in order to prevent the multiplication of the organisms as well as toxins production by them.
9. **Disposal of Condemned Carcase/Meat:** Condemned carcase/meat must be properly disposed in order to prevent the spread of pathogenic organisms in animals, humans and environment.
10. **Disposal of Slaughterhouse Wastes and Effluents:** Slaughterhouse wastes should be properly collected and disposed, and the effluents released from slaughterhouse must be properly treated before its disposal into any water bodies.
11. **Immunization and Treatment:** Immunization and treatment of persons at risk especially butchers and other occupational groups are important in prevention and control of meat-borne diseases.
12. **Surveillance:** Effective surveillance and epidemiology of meat-borne diseases should be carried out regularly to know the current status of diseases and to make strategies for prevention and control of meat-borne pathogenic organisms and diseases caused by them.

29

Food Safety

Food safety refers to limiting the presence of those hazards whether chronic or acute, that may make food injurious to the health of the consumers. Food safety covers all aspects of ensuring that food is safe for a person to eat, whereas, food hygiene usually more specifically concerns food-borne illnesses, which arise because of primary bacterial contaminants, but also physical and chemical hazards. Food safety is about producing, handling, storing and preparing food in such a way as to prevent infection and contamination in the food production chain, and to help ensure that food quality and wholesomeness are maintained to promote a good health.

Ensuring food safety starts with production at the farm level. In this regard, misuse of agrochemicals, including pesticides and growth hormones and veterinary drugs adversely affect the health of consumers. Ensuring food safety requires due attention during harvest, transport, processing, storage, and finally during food preparation and storage by consumers. Microbial and chemical risks could be introduced at farm level, *e.g.*, using water contaminated with industrial wastes or poultry farm waste for irrigation of crops. Application of good agricultural practices can reduce microbial and chemical hazards. Organic farming (without the use of pesticides) has been promoted in many countries of southeast Asia Region.

Every food establishment must start with a written food safety plan that may be implemented into a documented food safety programme. Essential elements of a food safety programme must contain sanitation standard operating procedures (SSOPs), and a Hazard Analysis Critical Control Point (HACCP) Programme along with a tried and true recall programme. In addition, other programmes such as pest control, employee training and supplier certification programmes complete a competent food safety programme.

Principles of Food Safety

The food safety principles are as follows:

1. **Cleaning:** Unclean hand is the most important source of contamination of foods. Therefore, hands must be washed with soap and water.
2. **Chilling:** It must be ensured to prevent the growth of organisms.
3. **Cooking:** It is important to render the food safe for human consumption.
4. **Preventing cross-contamination:** It is important to prevent contamination of food from any other sources of organisms.

Food Safety Hazard

Food safety hazard is any physical, chemical or biological agent or factor in food, or condition of food that may cause an adverse health effect in the consumers. Food safety hazards occur when food is exposed to hazardous agents which result in contamination of that food. Food hazards may be physical, chemical, biological, allergenic, nutritional and/or biotechnology-related. Hazards may be introduced into the food supply at any time during harvesting, formulation and processing, packaging and labeling, transportation, storage, preparation, and serving. Food safety hazards include the following:

Physical/Extraneous Material Hazards

Extraneous material like metal, glass, plastics, wood pieces, hair, insects *etc.* may be found in a food that are foreign to that particular food. These materials are usually introduced during unsanitary conditions of production, processing, handling, storage and distribution of food and due to carelessness. Extraneous material can be considered hazardous due to its sharpness, hardness, size or shape. It may cause lacerations, perforations and wounds or may become a choking hazard. Therefore, the sale of food contaminated with injurious extraneous material is objectionable from public heath point of view.

Chemical Hazards

Chemical hazards in the consumers may occur due consumption of foods containing various types of chemicals such as mycotoxins (aflatoxin, ochratoxin A and fumonisin), natural toxins (glycoalkaloids and other natural toxins), marine toxins, environmental contaminants (arsenic, lead, cadmium and mercury), food additives, processing-induced chemicals (acrylamide, ethyl carbamate or urethane and furan), pesticides, agricultural products and veterinary drugs. Chemical contamination of food may occur through various pathways such as the environment (air, soil and water), intentional use of chemicals (such as pesticides and veterinary drugs), manufacturing processes and addition of food additives.

Biological Hazards

Biological hazards include bacteria, viruses, fungi and parasites of public health significance. Biological hazards may occur when hazardous or pathogenic organisms are introduced to food and thus pose a food safety concern to consumers.

Table 29.1. Chemical Hazards due to Consumption of Seafood Containing Common Marine Toxins

Sl.No.	Toxin	Seafood	Illness
1	Saxitoxin	Oysters, clams, scallops, mussels, cockles and whelks	Paralytic shellfish poisoning (PSP)
2	Domoic acid	Bivalve molluscan shellfish, clams, mussels, oysters and scallops	Amnesic shellfish poisoning (ASP)
3	Ciguatoxin	Tropical fish such as barracuda, amberjack, red snapper and grouper	Ciguatera poisoning
4	Okadaic acid	Various shellfish, cockles, mussels and oysters	Diarrhetic shellfish poisoning (DSP)
5	Tetrodotoxin	Pufferfish, California newt, parrotfish, octopus, starfish, angelfish and xanthidcrabs	Tetrodotoxin poisoning
6	Brevetoxin	Oysters, clams and mussels	Neurotoxic shellfish poisoning

Biological hazards can be introduced to food from the environment (*e.g.*, soil bacteria, agricultural run-off) or from inadequate sanitation practices and cross contamination during transportation, handling, processing and storage (*e.g.*, poor food hygiene practices).

The biological hazards causing bacteria associated with food-borne illnesses include *Bacillus cereus, Campylobacter jejuni, Clostridium botulinum, Clostridium perfringens, Escherichia coli* O157:H7, *Escherichia coli* O104:H4, *Listeria monocytogenes, Salmonella* spp., *Shigella* spp., *Staphylococcus aureus, Vibrio cholerae, Vibrio parahaemolyticus, Vibrio vulnificus* and *Yersinia enterocolitica.* Food safety hazards causing viruses include bacteriophage, enteric virus (other than hepatitis A and Noroviruses), hepatitis A virus, Norovirus, Norwalk virus and Rota virus. Food safety hazards causing other biological agents include prions, also known as proteinaseous infectious particles, which are infectious agents made of protein. They are known to cause a number of diseases that affect both humans and animals. The health hazards caused by these agents include BSE or "Mad Cow Disease" in cattle and Creutzfeldt-Jakob disease in humans. Food safety hazards causing fungi include *Aspergillus* spp. and *Penicillium* spp. Food safety hazards causing parasites include *Cryptosporidium parvum, Entamoeba histolytica, Entamoeba coli, Giardia intestinalis, Taenia* spp., *Toxoplasma gondii* and*Trichinella spiralis.*

Allergenic Hazards

An allergen is a protein substance which is capable of producing an abnormal immune response in sensitive individuals. Some edible items that may cause allergic reactions in sensitive individuals include peanuts, tree nuts (almonds, cashews, hazelnuts, macadamia nuts, pecans, pine nuts, pistachio nuts, and walnuts), soy, wheat, mustard, sesame seeds, milk, eggs, seafood (fish, crustaceans, and shellfish) *etc.* Allergic reactions to food usually involve IgE antibodies. Symptoms of an allergic reaction can range in severity from a skin rash or slight itching of the mouth, to migraine headaches, to anaphylactic shock and death.

The type and severity of an allergic response is determined by many factors including dosage, route of administration, frequency of exposure, and genetic factors. Allergic reaction differs from food intolerance which is an abnormal physiological response to a specific food. Symptoms of food intolerance may include cramps, diarrhoea and bloating. Anaphylactic shock is the most severe adverse reaction to food and can be fatal if left untreated. It generally occurs within minutes of consumption, but occasionally the reaction may be delayed, with symptoms appearing several hours after the initial exposure. Initial symptoms of an IgE mediated allergic reaction are characterized by itching, hives, and/or swelling of the lips, palate, tongue, and throat. When the food enters the stomach and intestine, symptoms like cramping, nausea, pain, and diarrhoea may develop. Subsequent systemic symptoms generally affect the respiratory and cardiovascular systems which cause breathing difficulties and a drop in blood pressure or shock.

Sanitation Standard Operating Procedures (SSOPs)

SSOPs is a documented system for assuring that personnel, facilities, equipment and utensils are clean and where necessary, sanitized to specified levels prior to and during operations. The plant manager/designated personnel are responsible for the implementation, daily monitoring of the SSOP, recording the findings and any corrective actions. Before starting cleaning procedures, all clean up personnel must be wearing proper protective clothing, footwear, face shield, rubber gloves *etc*. The SSOPs include pre-operational sanitation and operational sanitation that are discussed as follows:

I. Pre-Operational Sanitation

This is carried out as follows:

1. **Cleaning of Equipment:** All equipment are properly cleaned and sanitized after production. General equipment cleaning includes established cleaning procedures in a meat plant; implementation, monitoring and record keeping; and corrective actions if necessary. Food debris is removed from equipment and garbage is removed from the production area. An approved cleaning solution is applied to equipment parts/surfaces and scrubbed as need to remove soil. Equipment/parts are rinsed with potable water and inspected for cleanliness, and re-cleaned if necessary. Equipment/parts are sanitized with approved sanitizer solution. The plant manager/designated personnel perform daily sanitation inspection after pre-operational cleaning and sanitizing. The results of this inspection are recorded on pre-operational log sheet.
2. **Cleaning of Facilities, including Floors, Walls and Ceilings:** In this process the debris is swept up and discarded and facilities are rinsed with potable water. Floors and walls are cleaned at the end of each processing day. Ceilings are cleaned as needed, but at least once a month. The plant manager/designated personnel perform daily organoleptic, sanitation inspection after pre-operational cleaning of facilities.

3. **Corrective Actions:** When the plant manager/designated personnel determine that equipment or parts are not properly cleaned, the equipment are placed on hold, and the cleaning procedure and inspection are repeated. The plant manager/designated personnel monitor the cleaning process, and re-train the employees doing the cleaning, if necessary. Corrective actions are recorded in the "action taken" column of the pre-operational inspection sheet.

II. Operational Sanitation

Operational sanitation includes the following:

1. **Food Processing Operations:** Food processing is performed under sanitary conditions to prevent direct and cross-contamination of ingredients.
2. **Personal Hygiene Procedures:** All employees handling food ingredients should wear hairnets, and gloves where required during processing. All employees should clean and sanitize hands, knives, other equipment *etc.*
3. **Food Products Handling and Storage Procedures:** Smocks and gloves are identified and designated specifically for production area. Employee's garments are maintained in a clean and sanitary manner and are changed at least daily, and if necessary, more often.
4. **Insect and Rodent Control Procedures:** Insect and rodent traps are maintained in non-food handling areas by the commercial outfit. All areas are checked for rodent and insects and appropriate actions are taken.
5. **Processing Products Clean-Up Procedures:** A full cleanup is carried out for the equipment and processing area, if different food product is processed on the same day.
6. **Implementation, Monitoring and Record Keeping:** The plant manager/ designated personnel is responsible for ensuring that employee hygiene practices, sanitary product handling procedures, pest and rodent control and cleaning procedures are maintained during processing. The plant manager/designated personnel monitors these operational sanitation procedures once during each change in product and these results are recorded on the log sheet
7. **Corrective Actions:** When the plant manager/designated personnel identify operational sanitation problems, he/she notify employees to take appropriate action to correct the sanitation problems. If necessary, processing is stopped and/or employees are re-trained. Corrective actions are recorded in the "actions taken' column on log sheet and also recorded in corrective actions record book.

Hazard Analysis Critical Control Points (HACCP)

HACCP is an internationally recognized way of ensuring the food safety and protection of consumers. The HACCP approach provides a systematic way of identifying food safety hazards and making sure that they are being controlled.

An important priority in meat production is to minimize contamination during slaughter, dressing and subsequent handling of meat. A recent development to achieve and ensure safety in food production in general is a systematic approach based on an assessment of the various risks associated with each step of the process. The object is to identify the relative seriousness of risk.

The concept of HACCP was introduced by H. E. Bauman and other scientists in the food industry in 1971 in the USA to ensure that there would be effective control of the quality of processed foods. The WHO recommends that this concept also be applied to meat inspection and meat hygiene in particular to control salmonellosis. It can also be used to reduce bacterial contamination during slaughtering and dressing and to ensure quality control in meat inspection. The use of proper sanitation methods in slaughterhouse and meat processing plant is the backbone of an effective HACCP programme in ensuring the safe meat production.

In India, HACCP system is made mandatory to all the export oriented units (EOU) by APEDA in order to ensure the identification and prevention of microbial and other hazards in meat production. The Bureau of Indian Standards (BIS) and the National Standards Authority of India (NSAI) on food hygiene have developed HACCP and its guidelines for its application. ISO 9001:2008 is technically equivalent to International Standard (ISO) of Codex, HACCP. Integrated Management System Certification provides for certification that is Quality Management System against IS/ISO 9001 and HACCP Certification against IS: 15000: 1998 through one audit, under one certificate.

Hazard

As per the International Commission of Microbiological Specification for Foods (ICMSF, 1988), the hazard is an unacceptable contamination, unacceptable growth and/or survival by microorganisms of concern to safety or spoilage and /or the unacceptable production or persistence in foods of products of microbiological metabolism (*e.g.*, toxins, enzymes, biogenic amines *etc.*). Hazard may be a physical, chemical or biological agent or property that may cause a food to be unsafe for consumption, or aesthetically unacceptable to the consumer.

Hazard Analysis

The first step in applying the HACCP system to a food manufacturing operation is to identify and quantify the physical, chemical and microbiological hazards and risks within the operation. Hazard analysis requires the specialist knowledge of a multidisciplinary team and should include food microbiologists, veterinarians, engineers, cleaning experts *etc.* A step-by-step investigation of the process is carried out from the specification required for the raw material through the manufacturing process to the distribution chain.

Epidemiological investigation of historical episodes of premature spoilage or food poisoning in which the product was implicated can provide valuable information about potential hazards. All of the data emerging from this analysis should be collected into a flowchart and all of the hazards identified and evaluated with regard to their severity and likely frequency of occurrence.

Critical Control Points (CCPs)

As per the Codex Alimentarius Commission, Codex 1991, CCP is a location, practice, procedure or process at which control can be exercised over one or more factors which, if controlled, could minimize or prevent a hazard. With regard to meat production the HACCP concept systematically identifies potential hazards in the entire chain from animal production to consumption and ranks them according to severity and likely frequency. This covers facilities, equipment and operation and is intended to augment and refine the various codes of manufacturing practice undertaken by industry. The procedure is intended to enable management to take preventive measures rather than depend on intensive testing of the end-products. CCPs can be categorized into two types:

1. **CCP1:** It is a CCP where the hazard can be prevented or eliminated. It indicates effective CCP.
2. **CCP2:** Where the hazard can be minimized, reduced or delayed. It is not absolute.

CCPs can also be classified as follows:

1. **CCPe:** Where the hazards are **e**liminated and no further problem exists.
2. **CCPp:** Where the hazards are **p**revented but not necessarily eliminated.
3. **CCPr:** Where the hazard is significantly **r**educed, minimized or delayed.

Some Critical Control Points in Meat Industry

To control microbial hazards some CCPs have been described as follows:

1. **Raw Meat:** Contaminated animal feeds, skinning, evisceration, spillage of intestinal contents, chilling, scalding (for pigs and poultry), washing (poultry) and freezing (*Cysticercus cellulosae* and *Trichinella* spp. are killed by freezing).
2. **Sausages:** Chilling, freezing and smoking.
3. **Semi-dry/Dry Sausages:** Inoculation of culture, ripening; salt, water and acid content; and fermentation (time and temperature control).

Aims of HACCP

The main aim of seven principles of HACCP is to focus attention on the identification and control of physical, chemical and microbiological food safety hazards during production. The hazard assessment and the regular monitoring of critical control measures must be documented to provide the basis for audit checks and may provide evidence of due diligence in the event of legal action.

Plan of HACCP

HACCP plan means what needs to be done to maintain food safety. In meat plants, HACCP plans should focus on control measures that can reduce the likelihood of contamination of meat from microbiological hazards, such as *Salmonella*, *E. coli*

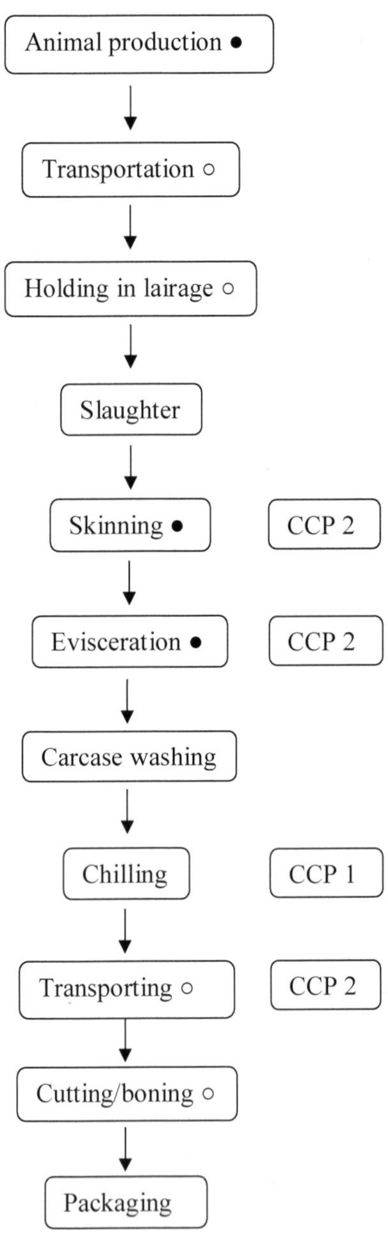

Figure 29.1. Flowchart of Fresh Meat Production and Processing from Cattle, Sheep, Goat and Horses. ● Indicates a site of major contamination ○ Indicates a site of minor contamination. CCP 1: effective CCP. CCP 2: not absolute.

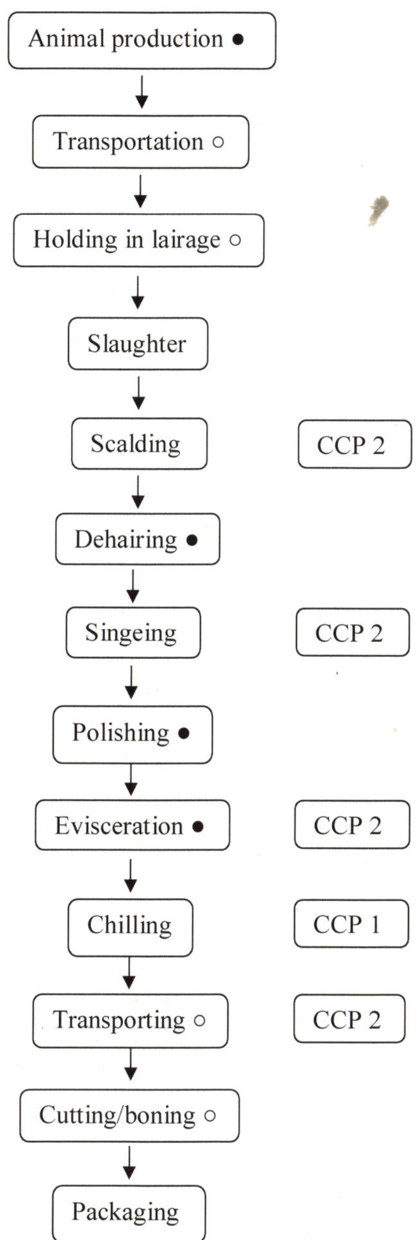

Figure 29.2. Flowchart of Fresh Meat Production and Processing from Pigs. ● Indicates a site of major contamination ○ Indicates a site of minor contamination. CCP 1: effective CCP. CCP 2: not absolute.

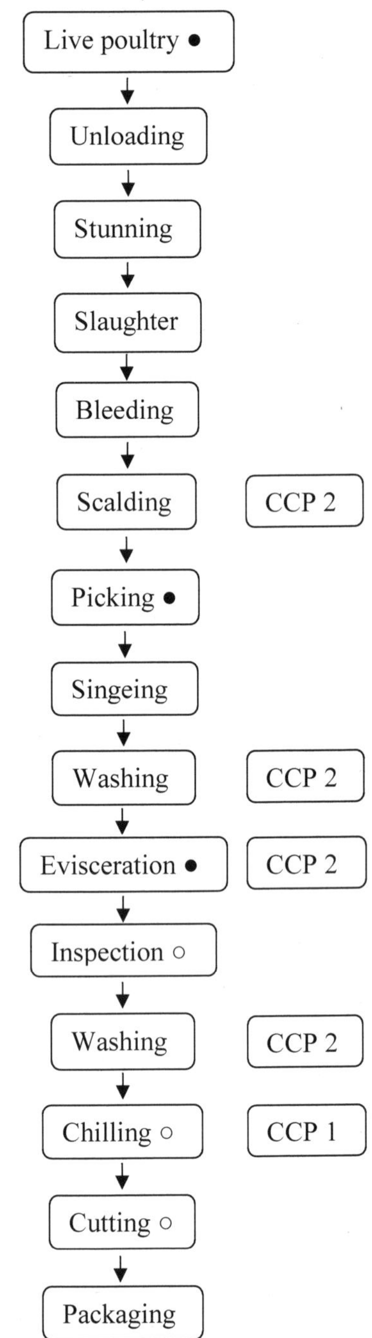

Figure 29.3. Flowchart of Fresh meat Production and Processing from Poultry.
● Indicates a site of major contamination ○ Indicates a site of minor contamination.
CCP 1: effective CCP. CCP 2: not absolute.

O157:H7 and *Campylobacter*, during production. There is a need for retailers and consumers to take precautions, including temperature controls and keeping raw meat and cooked meat and other ready to eat foods separate. The following points must be included in plan:

1. Minimize the likelihood of food poisoning bacteria contaminating meat and meat products.
2. Avoid physical and chemical contamination of meat.
3. Reduce the potential for growth of food poisoning bacteria on meat and meat products.
4. Minimize the potential for cross contamination of ready-to-eat foods by food poisoning bacteria on meat during further processing or in the kitchen.

Steps in the HACCP Plan to Maintain the Food Safety

The HACCP plan involves the following steps to maintain the food safety:

1. **Do:** Do the things that have been planned to maintain the food safety.
2. **Check:** It is important to check the things that have been planned and are being done to maintain the food safety. It is also important to maintain the record that when and what was checked. The checking processes involve supervision, monitoring, verification, microbial testing and review.
3. **Act:** Act to correct any food safety problems and record that what has been done about the problem and when.

Constitution of HACCP Team

A multi-disciplinary HACCP team should be constituted which represents the expertise and experience needed for documenting, verifying and implementing HACCP system. The HACCP team members are trained on basic principles of HACCP/Food Safety Management System. The HACCP team meets once in 6 months to review the status of implementation of HACCP system.

Principles of HACCP

The HACCP system is internationally accepted as the system of choice for food safety management. HACCP is a preventative approach to food safety based on the following seven principles:

1. **Hazard Analysis:** The first step in applying the HACCP system to food manufacturing operation is to identify the microbial hazards and risks within the operations. The hazard may be present in the ingredients, equipment, or introduced during processing. Hazard analysis requires many specialists such as microbiologist, engineers, veterinarians, cleaning experts *etc.* The HACCP team should conduct a hazard analysis and list all the biological, chemical and physical hazards that are reasonably expected to occur at each step and should describe the preventive measures that can be used to control these hazards. The team takes preventive measures

for each hazard to eliminate hazards or reduce their impact or occurrence to acceptable levels. While conducting the hazard analysis, the HACCP team should keep in mind the following things:

i. The likely occurrence of hazards and severity of their adverse health effects.
ii. The qualitative and quantitative evaluation of the presence of hazards.
iii. Survival or multiplication of microorganisms of concern.
iv. Production or persistence of toxins, chemicals or physical agents in foods.
v. Conditions leading to the above.

The team then consider what control measures, if any, exist which can be applied for each hazard. At the end of each product, the basis for determination of occurrence of hazards and severity of their adverse health effects is given.

2. **Identification of Critical Control Points (CCPs):** It is important to identify the CCPs so that the appropriate measures can be taken to prevent or minimize the hazards. The HACCP team can identify CCPs by applying the decision tree, using their experience and on the basis of training provided to them to determine whether a step is a CCP for the identified hazard during hazard analysis. All hazards that may be reasonably expected to occur, or be introduced at each step, have been considered. If a hazard has been identified at a step where control is necessary for food safety, and no control measure exists at that step, or any others, then the process is modified at that step, or at any earlier or later stage, to include a control measure.
3. **Establishing Critical Limits at each CCP:** The criteria are the specific limits of characteristics of a physical (*e.g.*, time, temperature *etc.*), chemical (*e.g.*, salt, acetic acid *etc.*) or biological (*e.g.*, sensorial, microbiological *etc.*) nature. The main emphasis is given on quantitative (*e.g.*, time, temperature, pH and salt concentration of various chemicals) rather than qualitative (*e.g.*, colour, smell *etc.*) criteria which indicate whether or not an operation is under control. Since CCPs define the boundaries between safe and unsafe products, it is vital that they are set at the correct level for each criterion. Therefore, the HACCP team should use criteria governing safety at each CCP in order to set the appropriate critical limits. Criteria used include control of foreign matter, moisture, microbial load and sensory parameters such as taste and flavour and visual appearance and enforcement of strict plant and personnel hygiene. Quantitative criteria such as time, temperature, pH and the concentration of various chemicals should be measured and indicated that whether a system is under control or not.
4. **Monitoring of each CCP:** Monitoring is one of the most important aspects of HACCP system. Monitoring is the systematic checking of the process or procedure at a particular CCP in an attempt to detect any deviation

from the criteria which have been set. It is the scheduled measurement and observation of a CCP relative to its critical limits. The system also includes calibration of equipment. The HACCP team should establish a monitoring mechanism for control of CCPs. Best monitoring can be done by the operatives themselves or their supervisors rather than by special team of quality assurance staff.

5. **Corrective Actions at each CCP:** A system of corrective action must be established so that immediate action can be taken for detection of deviation from the established criteria. Corrective action may involve rectifying an out-of-control situation before an operation is allowed to commence, or halting the manufacturing process. Corrective actions taken also include proper disposition of the affected product. Corrective action also occurs when monitoring results indicate a trend towards loss of control at a CCP. Remedial action is taken to bring the process back into control before the deviation leads to a safety hazard.
6. **HACCP Verification/Validation:** It is important to establish the procedures to verify whether these procedures are working effectively. Verification checks are carried out systematically by the quality assurance staff and the veterinary inspection team. Verification may involve some microbiological checks of end product, structures and equipment. All checks should be carried out at a statistically-significant frequency and in a systematic targeted manner. Monitoring and auditing methods, procedures and tests, including random sampling and analysis are used to determine if the HACCP system is working correctly. The frequency of audit and verification are kept at appropriate intervals to validate the HACCP system. Verification activities include the following:
   i. Review of the HACCP system and its records (during internal audit once in six months).
   ii. Review of deviations and product dispositions (immediately after occurrence of deviations).
   iii. Verification of operations to determine if CCPs are under control during internal audit.
   iv. Validation of established critical limits should be once in a year.
7. **HACCP Documentation:** Establishment of the documentation is essential for working of HACCP system in a proper way. Records should be easy to keep up to date because they provide evidence of the operator's thinking and decisions. The hazard assessment and the regular monitoring of critical control measures must be documented to provide the basis for audit checks and may provide evidence of due diligence in the event of legal action.

Implementation of HACCP System

For implementation of a HACCP system, it is essential to first assemble a HACCP team. This is followed by the description of the product with the intended use of the product, thereafter; a flowdiagram is drawn with an on-site verification of the flow diagram. The hazards associated with each step are determined and the preventive measures to control them are taken.

HACCP Training

Food business operators need to understand the HACCP principles so that they can set up and maintain HACCP-based procedures. Staff should also be aware of the importance of the HACCP procedures for minimizing the spread of food-borne hazards, which may be physical (wool, metal), chemical (veterinary drugs or residues of cleaning and disinfecting agents) and biological hazards (*Campylobacter* spp., *E. coli* O157, *Salmonella* spp. *etc*.).

Employees Training

In smaller public slaughterhouses, owned by the municipalities, there is often a manager and no specific staff, slaughtering being carried out by local butchers and their working teams. In these situations, it is difficult to give any training to workers on hygiene and cleaning. Large slaughterhouses have permanent staff which performs all kinds of work. When the staff is permanent, they can receive some training in process hygiene, personal hygiene and cleaning and disinfection.

Supplier Certification Programmes

Supplier certification is an important component of a total quality management system that assures that a supplier's product is produced, packaged, and shipped under a controlled process that results in consistent conformance to our requirements.

Meat Quality and Safety Programmes

Accreditation with HACCP and ISO Standards is mandatory requirements for all the export oriented units (EOU) to ensure the safety and quality of meat and to develop confidence in consumers. International acceptance of the HACCP approach was underlined during the Uruguay Round of the General Agreement on Tariffs and Trade (GATT), when the inclusion of the Codex Alimentarius Commission's recommendations on the application of HACCP were specifically identified as the baseline for consumer protection under the Agreement on the Application of Sanitary and Phytosanitary (SPS) measures.

Pest Control

Meat traders/exporters should be conscious about the menace of the flies, mosquito, rats and other pests. Pest not only contaminates the food, food contact surface, food packaging materials but they themselves are carriers of different diseases. It is to be ensured that the utility building and its surroundings must be insect and pest free. Pest control measures should be taken both outside and inside the meat plan. Rat traps can be placed in switch room. Cartons and packing material should be stored hygienically by sealing crevices to prevent entry of rodent and pests through the drain.

Air curtains can be placed at each entry point in the utility building. Insect catchers can also be fitted in canteen and at halal point in the slaughterhouse to eliminate insects. Automatic closing windows (chutes) can be provided in the slaughterhall to deliver the rejected carcases for rendering. Chutes are also provided in the bone room. This prevents the entry of birds and other flying objects to enter the deboning/slaughterhall. Regular cleaning of toilets and keeping of naphthalene balls can prevent the flies and mosquito entry in the meat plant. Pesticides can be sprayed outside the building to reduce insects from immediate vicinity.

Quality Management System- ISO 9001: 2008

The Bureau of Indian Standards (BIS) provides guidance to the Indian food industries by formulating the national standards and certification system. The BIS is operating Quality Management System Certification Scheme (QMSCS) according to IS/ISO 9001:2008 as one of the scheme under Management Systems Certification. This standard is identical to ISO 9001:2000.

Control of Hygiene and Safety

The Codex Alimentarius Commission (CAC) has elaborated (besides meat inspection Codes) the recommended International Code of Hygiene Practice for Fresh Meat (CAC/RCP 11-1976) and the recommended International Code of Hygienic Practice for Poultry Processing (CAC/RCP 14-1976) which describe the minimum requirements of hygiene for meat and poultry production. The application of these Codes can be an important step towards the targets as follows:

1. The food will not cause infection or intoxication when properly prepared.
2. The food does not contain residues of pesticides, veterinary drugs and heavy metals in excess of established limits.
3. The food is free from disease.
4. The food is free from obvious contamination.
5. The food is free from defects generally recognized as objectionable.
6. The food has been produced under adequate hygienic control.
7. The food fulfills the expectation of the consumer in regard to composition.

The CAC guidelines include advice on the construction of abattoirs and the facilities required; control of pests, quality of water for cleaning and disinfection; rules of meat inspection and hygienic practices including supervision by a veterinary inspector.

Food Safety Management Procedures for Meat Shops

Food safety management can be achieved by a combination of good hygiene practices (GHPs) and operational hygiene procedures. Food safety management procedures based on HACCP should be documented for all food businesses. In premises that have higher risks, such as butchers handling raw and cooked food, more rigorous procedures are required with more in depth monitoring and recording requirements. Butchers' shops require more in depth record keeping than lower risk food businesses.

All activities that can affect food safety should be considered when developing food safety management procedures such as methods of carrying out in the premises and management method to ensure food safety. These activities include cooling and subsequent storage, times and temperatures, controlling cross-contamination from slicing, pests and wastes, cleaning and disinfection arrangement, personal hygiene requirements, water quality, management of food safety activities such as details of suppliers and other traceability arrangements, staff training, complaint and maintenance of equipment including calibration (accuracy) of temperature monitors.

Risk Analysis

Risk is a function of the probability of an adverse event and the magnitude of that event, consequential to a hazard(s) in food. Risk analysis is a process which consists of three components, *viz.*, risk assessment, risk management and risk communication.

Risk Assessment

Risk assessment is a scientific process of identifying hazards, and estimating risk in qualitative or quantitative terms. The aims of risk assessment are to characterize the types of health effects expected as a result of a certain exposure to a given agent, as well as providing estimates on the probability of occurrence of these health effects, at different levels of exposure. It involves hazard identification, the establishment of exposure-effect relationships, and exposure assessment, leading to risk characterization. Risk assessment provides a basis for risk management. Risk assessment involves the following four analytical steps:

1. **Hazard Identification:** The qualitative indication that a hazard(s) could be present in a particular food.
2. **Hazard Characterization:** The quantitative and/or qualitative evaluation of the nature of the adverse effects which may include a dose-response assessment. Dose-response assessment is the determination of the relationship between the magnitude of exposure and adverse effects.
3. **Exposure Characterization:** The qualitative and/or quantitative evaluation of the degree of human exposure likely to occur.
4. **Risk characterization:** Integration of the above steps into an estimation of the adverse effects likely to occur in a given population, including attendant uncertainty.

Risk Management

Risk management is the process, distinct from risk assessment, of weighing policy alternatives, in consultation with all interested parties, considering risk assessment and other factors relevant for the health protection of consumers and for the promotion of fair trade practices and, if needed, selecting appropriate prevention and control options. Risk management involves decisions and actions that aim at preventing, or reducing to acceptable levels, the occurrence of agents which may pose hazards to the health of workers, surrounding communities and the environment, also accounting for the socio-economic and public health context.

Risk Communication

Risk communication is the interactive exchange of information and opinions throughout the risk analysis process, concerning risk, risk-related factors and risk perceptions, among risk assessors, risk managers, consumers, industry, the academic community and other interested parties, including the explanation of risk assessment findings and the basis of risk management decisions. Food safety risk communication is important to the protection of public, animal, plant and environmental health, and people's quality of life, including socio-economic factors such as livelihoods. Food safety risk communication may involve communication of both risks and benefits. Effective food safety risk communication can improve the following:

1. Physical well-being of people.
2. Trust of consumers in the food supply and in the regulatory systems.
3. The environment in which we live.
4. Overall quality of life of people, including socio-economic factors such as livelihoods and psychological factors.

National Food Safety Agencies/Organizations

In India, the national agencies/organizations such as Bureau of Indian Standards (BIS), Food Safety and Standards Authority of India (FSSAI), Agricultural and Processed Food Products Export Development Authority (APEDA), Directorate of Marketing and Inspection (DMI), State and Local Self Government *etc.* associated with the food safety are discussed as follows:

Bureau of Indian Standards (BIS)

BIS is the National Standards body of India which works under the aegis of Ministry of Human Affairs, Food and Public Distribution, Government of India. The BIS was established under the BIS Act, 1986. The BIS formerly known as ISI (Indian Standards Institution) which was set up under the resolution of the Department of Industries and Supplies No. Std. (4)/45, dated 33 Sept., 1946. The ISI was registered under the Societies Registration Act, 1960. The BIS has its headquarters in New Delhi, with Regional Offices located in Kolkata, Mumbai, Chennai, Chandigarh and Delhi also. The BIS has 20 branch offices also. The BIS is a very important department in every department because it makes sure every standard is followed perfectly without any problem.

Presently, BIS has 14 Division Councils which are formulating Indian Standards covering all areas other than those mentioned in Agricultural Produce (Grading and Marking) Act, 1937 and the Drugs and Cosmetics Act, 1940. BIS does not make technical regulations. However there are technical regulations which make compliance to BIS standards mandatory. Technical regulations are issued by various departments under different ministries of Government of India or by different regulators empowered under different states.

The BIS is empowered by the BIS Act, 2016 operates product certification schemes by which it grants licenses to manufacturers covering particularly every industrial discipline from agriculture and textile to electronics. The BIS and FSSAI are complementary to each other. The BIS has a lot of significance pertaining to the FSSAI as BIS is the one which has provided the right of safety and standards to the FSSAI license. The BIS is important for the FSSAI Registration as it helps it to do all the safety work. The certification allows licensees to use the popular ISI mark, which has become synonymous with quality products for the Indian and neighbouring market for over 55 years.

The Food Safety and Standards (Prohibition and Restriction Sales) Regulations, 2011 has prescribed mandatory certification under the BIS Act for several products such as milk powder (IS 1165), skimmed milk powder (IS 13334), packaged mineral water (IS 13428) *etc.* The Indian Government has, in public interest, enforced mandatory certification on various products through various quality control orders issued from time to time, under various Acts. Quality control orders issued under Essential Commodity Act, 1955 include Fruit Product Order (FPO) 1955, Meat Food Product Order (MFPO) 1973, Milk and Milk Product Order (MMPO) 1992 *etc.*

As per the Meat Food Product Order (MFPO) 1973, the meat food product shall be obtained from animal species like bovine, ovine, caprines, suillines, poultry and fish. No meat food product shall contain any poisonous metal in excess of their permissible limits such as lead (2.5 ppm), copper (20 ppm), zinc (50 ppm) and tin (250 ppm). No meat food product shall contain any preservatives in excess of their permissible limits such as potassium nitrate (200 ppm), sulfur dioxide ((450 ppm) and commercial saltpetre (500 ppm). No meat food product shall contain any insecticide in excess of their permissible limits such as monocrotofos (0.02 ppm), fentrothion (0.03 ppm), pholate (0.05 ppm), 2,4-D (0.05 ppm), chlorpyrifos (0.1 ppm), carbafuron (0.1 ppm), benomyl (0.1 ppm), carbendazim (0.1 ppm), trichlorfon (0.1 ppm), aldrin/dialdrin (0.2 ppm), chlorfenvinphos (0.2 ppm), lindane (2 ppm), fenthion (2 ppm) and DDT (7 ppm). No meat food product shall contain sequestering and buffering agents in excess of their permissible limits such as gluconodelta (5000 ppm, by weight) and mono-sodium glutamate (1 per cent, by weight).

Food Safety and Standards Authority of India (FSSAI)

FSSAI comes under the Ministry of Health and Family Welfare. FSSAI has been established under the Food Safety and Standards (FSS) Act, 2006 as a statutory body for laying down science based standards for articles of food and regulating manufacturing, processing, distribution, sale and import of food so as to ensure safe and wholesome food for human consumption. Food Safety and Standards Act, 2006 consolidates various Acts and Orders that have hitherto handled food related issues in various Ministries and Departments. The Act also aims to establish a single reference point for all matters relating to food safety and standards by moving from multi-level and multi-departmental control to a single line of command. To this effect, the Act establishes independent statuary Authority (FSSAI) with head office at Delhi. FSSAI and State Food Safety enforce various provisions of the Act. FSS Act, 2006 highlights various central Acts like Prevention of Food Adulteration Act, 1954; Vegetable Oil Products (Control) Order, 1947; Fruit Products Order, 1955; De- Oiled Meal and Edible Flour (Control) Order, 1967; Meat Food Products Order, 1973; Milk and Milk Products Order, 1992; Solvent Extracted Oil, Edible Oils Packaging (Regulation) Order 1998 *etc.* FSSAI has been mandated by the FSS Act, 2006 for performing the following functions:

1. Framing of regulations to lay down the standards and guidelines in relation to articles of food and specifying appropriate system of enforcing various standards thus notified.
2. Laying down mechanisms and guidelines and for accreditation of certification bodies engaged in certification of food safety management system for food business.
3. Laying down procedures and guidelines for certification of laboratories and notification of the accredited laboratories.
4. Provides scientific advice and technical support to central government and state governments in the matter of framing the policy and rules in areas which have a direct or indirect bearing of food safety and nutrition.
5. Collect and collate data regarding food consumption, incidence, and prevalence of biological risk, contaminants in food, residues of contaminants in foods products, identification of emerging risks and introduction of rapid alert system.
6. Creating an information network across the country so that the public, consumers Panchayats *etc.* receive rapid, reliable and objective information about food safety and issues of concerns.
7. Provide training programme for persons who are involved or intended to get involved in food business.
8. Contribute to development of international technical standards for food, sanitary and phytosanitary standards.
9. Promote general awareness about food safety and food standards.

Agricultural and Processed Food Products Export Development Authority (APEDA)

APEDA comes under Ministry of Commerce, Government of India. APEDA was established by the Government of India under the Agricultural and Processed Food Products Export Development Authority (APEDA) Act passed by the Parliament in December, 1985. The Act (2 of 1986) came into effect from 13th February, 1986 by a notification in the Gazette of India: Extraordinary: Part-II [Sec. 3(ii):13.2.1886]. The Authority replaced the Processed Food Export Promotion Council (PFEPC). APEDA has its Head Office in New Delhi and 5 regional offices at Mumbai, Hyderabad, Bengaluru, Kolkata and Guwahati. The authority also has 14 virtual offices in different States of India.

Mandate of APEDA

As per the APEDA Act, the authority is responsible for the development and promotion of the export of the following 14 products mentioned in the first schedule to the APEDA Act:

1. Meat and Meat Products
2. Poultry Products
3. Dairy Products
4. Fruits and Vegetables and their Products
5. Confectionary, Biscuits and Bakery Products
6. Honey, Jaggery and Sugar Products
7. Cocoa and its Products, Chocolates of all kinds
8. Alcoholic and Non-Alcoholic Beverages
9. Cereals and Cereal Products
10. Groundnuts, Peanuts and Walnuts
11. Pickles, Papads and Chutney
12. Guar Gum
13. Horticulture and Floriculture Products
14. Herbal and Medicinal Plants

Functions of APEDA

The APEDA has been assigned the following functions under APEDA Act 1985, (2 of 1986) :

1. Development of industries by providing financial assistance.
2. Registration of persons as exporters of the scheduled products.
3. Fixing of standards and specifications for the scheduled products.
4. Carrying out inspection of meat and meat products in slaughterhouses, processing plants, storage premises, and other places where such products are kept or handled and conveyances.
5. Improving of packaging of the scheduled products.
6. Improving of marketing of the scheduled products outside India.
7. Promotion of export oriented production and development of the scheduled products.
8. Collection of statistics from the owners of factories or establishments engaged in the production, processing, packaging, marketing or export of the scheduled products and publication of the statistics so collected.
9. Training in various aspects of the industries connected with the scheduled products.

Directorate of Marketing and Inspection (DMI)

The DMI comes under Ministry of Agriculture. DMI headed by the Agricultural Marketing Advisor, Department of Agriculture and Co-operation (DAC) implements agricultural marketing policies and programmes of the Government of India. The functions of DMI include Standardization, Grading and Quality control of agricultural and allied produce.

State and Local Self Government

In India, every state has a system of food inspection and mechanism of enforcing the laws. These are generally under Directorate of Health Services. Municipal Corporation or town areas have authority to inspect and enforce public health laws through sanitary inspectors and veterinarians. These include ante-mortem and post-mortem inspections. The authorized veterinary officer certifies meat only from healthy animals.

International Food Safety Agencies/Organizations

The following international agencies/organizations associated with the food safety worldwide are discussed as follows:

Food and Agriculture Organization (FAO) of the United Nations

FAO's work on food safety and quality is very important. FAO plays a key global role in food safety governance and capacity development. The Food Safety and Quality Unit of the Agriculture and Consumer Protection Department supports the strengthening of systems of food safety and quality control at national, regional and international levels. This involves the following:

1. Strengthening national food control regulatory capacities and facilitation of global trade.
2. Supporting development of institutional and individual capacities for food control and food safety management, including the management of food safety emergencies.
3. Supporting science-based food safety governance and decisions by providing sound scientific advice to underpin food safety standards at national, regional and international levels.
4. Enhancing food safety management along food chains to prevent diseases and trade disruptions by supporting developing countries to apply risk-based food safety management along food chains that are appropriate for national and local production systems and in compliance with Codex texts.
5. Providing food safety platforms, databases and mechanisms which support networking, dialogue and global access to information and facilitating effective communication internationally on key food safety issues.
6. Evaluating new technologies to improve food safety and protect public health.

7. The Food Safety and Quality Programme supports an integrated and multidisciplinary approach to food safety management and holistic and feasible "food chain" solutions to specific food safety problems as laid out in FAO's Strategy for Improving Food Safety Globally.
8. FAO provides the technical information through its various publications such as A Training Manual on Food Hygiene and the HACCP System, Manual of Food Quality Control, Manual on Meat Inspection for Developing Countries, Manual on Simple Methods of Meat Preservation, Manual on Application of the HACCP System in Mycotoxin Prevention and Control, Manual on Risk Based Imported Food Control, Manual on Risk Based Food Inspection, Risk Communication Applied to Food Safety Handbook *etc.*

World Health Organization (WHO)

The WHO is mainly concerned with the health of the consumers and formulation of various specifications for hygienic food production. Role of WHO associated with food safety includes the following:

1. The WHO develops and promotes guidelines to minimize the public health impact of antimicrobial resistance associated with the use of antimicrobials in food of animal origin. WHO also provides technical support in the monitoring of antimicrobial use.
2. The WHO develops scientific risk assessments to define safe exposure levels which form the basis for the development of national and international food safety standards to protect the health of the consumers and ensure fair trade practices.
3. The WHO assists Member States in building capacity to prevent, detect and manage foodborne risks.
4. The WHO assists Member States in promoting safe food handling through systematic disease prevention and health educations programmes directed to food handlers, including the consumers.
5. The WHO together with FAO provides Member States with sound scientific advice on the assessments of foods derived from new technologies such as genetic modification and nanotechnology.
6. The WHO assists (through International Network of Food Safety Authorities: INFOSAN) Member States in managing food safety risks, ensuring rapid sharing of information during food safety emergencies to stop the spread of contaminated food from one country to another.
7. The WHO and FAO jointly developed International food standards (FAO/WHO Codex Alimentarius) which work on the provision of independent international scientific advice on microbiological and chemical hazards.
8. The WHO develops scientific risk assessments, guidelines for risk management, including risk communication messages for all stakeholders, including the end users to assist Member States to improve their capacity to prevent and control foodborne diseases.

9. The WHO works to promote the availability of safe, healthy and wholesome food for everyone to improve food safety and nutrition security.
10. In an attempt to promote safe food handling, the WHO has developed "Five keys to safer food", *viz.*, keep clean, separate raw and cooked, cook thoroughly, keep food at safe temperatures, and use safe water and raw materials.
11. The WHO provides the technical information through its various publications such as Risk Communication Applied to Food Safety Handbook, Antimicrobial Resistance, Chemical Risks, Foodborne Diseases, Food Hygiene, Food Technology, International Network of Food Safety Authorities (INFOSAN), Microbiological risks, Nutrition and Food Security *etc.*

Codex Alimentarius Commission (CAC)

The CAC is an intergovernmental body with 187 members. The CAC was established in 1963within the framework of the Joint Food Standards Programme by the FAO and the WHO. The CAC is commonly referred to as Codex. The CAC has its headquarters in Rome, Italy. The CAC meets in regular session once a year alternating between Geneva and Rome. The CAC coordinates inputs from member countries to develop and endorse the standards that comprise the international food code. The Codex Alimentarius are Latin words, meaning "Food Code" which is collection of standards, guidelines and code of practice adopted by the CAC. The Codex Alimentarius maintains food safety and quality by food hygiene, providing standards for maximum limits for food additives, pesticide residues and veterinary drugs, and maximum limits and codes for prevention of chemical and microbiological contamination. The CAC has the following aims:

1. To develop international and regional food standards and publish them in a food code.
2. To protect the health of consumers and ensure fair practices in the international food trade.
3. To promote coordination of all food standards work undertaken by international governmental and non-governmental organizations.
4. To develop harmonized standards, guidelines and codes of practice to protect the health of consumers.
5. To coordinate the food standards and testing methods internationally for food hygiene, food additives, residues of pesticides and veterinary drugs, contaminants, labeling, methods of sampling and analysis, import and export inspection and certification.

Office International des Epizooties (OIE)

The OIE is also known as World Organization for Animal Health. The OIE has its headquarters in Paris, France. It was established in 1924. At present, it has 182 Member States (member countries). The OIE is one of the international standards-setting organizations recognized in the Sanitary and Phytosanitary (SPS) Agreement, alongside the CAC and International Plant Protection Convention (IPPC). The SPS Agreement recognizes the OIE as reference international standard-setting organization for animal health and zoonoses.

The OIE is working with relevant organizations to reduce food-borne risks to human health arising from animal production. The OIE has established a permanent working group of internationally recognized experts from different organizations such as FAO, WHO and CAC. The OIE's International Food Safety Standards focus on eliminating potential hazards from consumers arising prior to the slaughter of animal or primary processing of their products (milk, meat, eggs *etc.*). The OIE provides guidance to the Member States to assists them in meeting the food objectives based on OIE and Codex Alimentarius standards. The OIE has developed standards on use of antimicrobial agents in terrestrial and aquatic animals. The OIE also publishes a list of antimicrobials of veterinary importance.

The OIE prescribes International Animal Health Code, which consists of six parts. Part I deals with definitions, epidemiological information/notification, veterinary ethics and import and export procedures. Part II mentions 15 animal diseases. Part III contains 68 diseases important in international trade. Part IV includes diagnostic tests for the purpose of international trade and general requirements of health and hygiene. Part V prescribes ten model international certificates approved by OIE. Part VI details diseases notifiable to OIE, sanitary certificate of meat, identification of meat and attestation of wholesomeness by veterinarians.

International Commission of Microbiological Specifications for Foods (ICMSF)

The ICMSF was founded in 1962 through the action of the International Committee on Food Microbiology and Hygiene, a Committee of the International Union of Microbiological Societies and to the WHO. The ICMSF has suggested many microbiological specifications and several HACCP systems for milk and milk products, meat and meat products, poultry, eggs and seafoods. The ICMSF provides basic scientific information through extensive study and makes recommendations without prejudice based on that information. Results of the studies published as books, discussion documents or refereed papers.

The ICMSF functions as a small "working group" of elected members and invited consultants, not a forum for the reading of papers or political debate. The commission has performed in-depth studies on all essential aspects relating to microbiological specifications for foods such as method of analysis, sampling plans and microbiological limits. In recent years, the commission has worked to improve the microbiological safety of foods through its books and other scientific publications, by participation in symposia and workshops, recommending sampling plans and microbiological criteria, and defining and promoting the use of GHP and HACCP and other systems of managing food safety. The commission meets annually since 1962.

International Occupational Hygiene Association (IOHA)

The IOSA was formally established, during a meeting in Montreal, on June 2, 1987. At present IOHA has the participation of 19 national occupational hygiene associations, with over nineteen thousand members from seventeen countries. The primary objective of IOHA is to promote and develop occupational hygiene throughout the world, at a high level of professional competence, through means that include the exchange of information among organizations and individuals, the further development of human resources and the promotion of a high standard of ethical practice. IOHA activities include scientific meetings and publication of a newsletter.

The World Food Safety Organization (WFSO)

The WFSO supports the EU, ISO and the corresponding accreditation schemes initiatives for the implementation of food safety management (HACCP) systems. The WFSO is an independent forum dedicated to the assurance of food safety. The WFSO is committed to assists the members in ensuring food safety from field to plate. Achieving a certification for food safety in line with the FAO, WHO, FDA and Codex Alimentarius supported by local legislation is of the essence. The WFSO approach is to provide the tools for professional consultants and auditors to verify an effective food safety management system through a process of assessments and continuous improvements. The WFSO has the following objectives:

1. To interpret and communicate food laws, regulations and standards in a language and through channels accessible to lay persons.
2. To provide information, education, and networking for food practitioners.
3. To provide accredited training programme.
4. To provide technical and legal help lines.
5. To provide a register of accredited consultants and auditors.
6. To provide the criteria for measuring an effective Food Safety Management system.
7. To provide the platform for all Food Agencies to jointly develop an International Food Safety Management Standard.

International Organization for Standardization (ISO)

The ISO is an independent, non-governmental international organization with 164 national standards bodies. The ISO has its headquarters in Geneva, Switzerland. The ISO was established in 1947. The ISO formulates international standards on food hygiene, food microbiology and food assurance of many foods including foods of animal origin. The BIS is operating Quality Management Systems on Certification Scheme (QMSCS) according to IS/ISO 9001:2008 as one of the schemes under Management Systems Certification. This standard is identical to ISO 9001: 2000. Most of the meat exporting plants has 9001-2000 or 2008 certification.

World Trade Organization (WTO)

The WTO was established in 1995. The WTO has its headquarters in Geneva, Switzerland. At present, the WTO has 164 member countries. The WTO is the only international organization dealing with rules of trade between nations. The WTO is a forum for governments to negotiate trade agreements. The WTO is run by its member governments. The goal of the WTO is to help producers of goods and services, exporters and importers conduct their business. The purpose of the WTO is to open the trade for the benefits of all and reduction of tariffs and other trade barriers.

The WTO itself does not develop standards related to SPS agreement. The standards are developed by leading scientists in the field and governmental experts on health protection. International standards are often higher than the national requirements of many countries including developed countries, but the SPS agreement explicitly permits governments to choose not to use the international standards. The WTO agreements cover the goods, services and intellectual properties. They spell out the principles of liberalization and the permitted exceptions. They include individual countries' commitments to lower customs tariffs and other trade barriers, and to open and keep open service markets. They require governments to make their trade policies transparent by notifying the WTO about laws in forces and measures adopted and through regular reports by the secretariat on countries trade policies. These agreements are often called the WTO' trade rules are actually agreements that governments negotiated.

Sanitary and Phytosanitary (SPS) Measures in Relation to Foods of Animal Origin

The agreement on application of SPS measures (the "SPS Agreement") entered into the force with the establishment of WTO on 1 January, 1995. It concerns the application of food safety and animal and plant health regulations. SPS measures are quarantine and biosecurity measures which applied to human, animal and plant life or health from risks arising from the introduction of establishment and spread of pests and diseases, and from risks arising from additives, toxins and contaminants in food and feed. SPS measure can take many forms such as requiring products to come from a disease-free area, inspection of products, specific treatment or processing of products, setting of maximum permissible limits of pesticides residues or permitted use of only certain additives in food.

The SPS agreement encourages governments to establish national SPS measures consist with international standards, guidelines and recommendations. This is known as "harmonization". SPS measures are governed by the WTO Agreement on the application of SPS measures and its committee of SPS measures (the SPS committee). The SPS agreement provides a framework of rules to guide WTO members in the development, adoption and enforcement of sanitary (human or animal life or health) and phytosanitary (plant life or health) measures which may affect trade. The objectives of SPS Agreements include the following:

1. To protect human or animal health from food-borne risks.
2. To protect human health from animal or plant carried disease.
3. To protect animals and plants from pests and diseases.

Food Safety Regulations

Almost all countries of the world have their own food safety regulations.

Food Safety Regulations in India

The Food Safety Regulations in India is being enforced through the following regulatory mechanisms:

1. The Food Safety and Standards Act 2006 (Food Safety and Standards Authority of India: FSSAI).
2. Food Safety and Standards Authority of India: FSSAI - Draft Safety Norms
3. Meat Food Product Order, 1973
4. Export of Raw Meat (Chilled/Frozen) Rules, 1992
5. Milk and Milk Product Order, 1992
6. Water (Prevention and Control of Pollution) Act, 1974
7. Air (Prevention and Control of Pollution) Act, 1981
8. Prevention of Food Adulteration Act, 1954
9. The Prevention of Cruelty to Animals Act, 1960
10. Export (Quality Control and Inspection) Act, 1963
11. Export (Quality Control and Inspection) Rules, 1964

Microbial Food Safety Regulations in India

The food safety concept is that the food will not cause any harm to the consumer when it is prepared and/or eaten according to its intended use. In India, microbial food safety is being enforced through the following regulatory mechanisms:

1. Prevention of Food Adulteration (PFA) Act, 1954
2. Quality control orders issued under Essential Commodities Act, 1955 such as Fruit Product Order (FPO), 1955; Meat Food Products Order (MFPO), 1973; and Milk and Milk Product Order (MMPO), 1992.
3. Agricultural Produce (Grading and Marking) Act, 1937.

Prevention of Food Adulteration Rules, 1956

Prevention of Food Adulteration (PFA) Rules, 1956 specifies microbiological requirements for pathogens such as *E. coli*, *Staphylococcus aureus*, *Salmonella* spp., *Shigella* spp., *Vibrio cholerae*, *V. parahaemolyticus*, *Clostridium perfringens*, *C. botulinum* and *Listeria monocytogenes* in foods commonly involved in food-borne diseases, such foods include seafood, milk and milk products, meat and meat products, fruit and vegetable products, packaged drinking water, mineral water *etc.*

Table 29.2. Prevention of Food Adulteration Rules, 1956: Microbiological Standards of Meat and Meat Products (Corned Beef, Luncheon Meat, Cooked Ham, Chopped Meat, Canned Chicken, Canned Mutton and Goat Meat)

Sl.No.	Microbiological Parameter	Standards
1	Total plate count	1000/g maximum
2	*E. coli*	Absent in 25 g
3	*Salmonella*	Absent in 25 g
4	*Staphylococcus aureus*	Absent in 25 g
5	*Clostridium perfringens*	Absent in 25 g
6	*Clostridium botulinum*	Absent in 25 g

Table 29.3. Prevention of Food Adulteration Rules, 1956: Microbiological Standards of Meat and Meat Products (Frozen Mutton, Goat, Beef and Buffalo Meat)

Sl.No.	Microbiological Parameter	Standards
1	Total plate count	10000/g maximum
2	*E. coli*	100/g maximum
3	*Salmonella*	Absent in 25 g
4	*Staphylococcus aureus*	100/g maximum
5	*Clostridium perfringens*	30/g maximum
6	*Clostridium Botulinum*	30/g maximum
7	*Listeria monocytogenes*	Absent in 25 g
8	Yeast and mould count	1000/g maximum

Essential Commodities Act, 1955

A number of quality control orders issued under Essential Commodities Act, 1955 which includes Fruit Product Order (FPO), 1955; Meat Food Products Order (MFPO), 1973; and Milk and Milk Product Order (MMPO), 1992. These aim at regulating sanitary and hygienic conditions at all levels of supply chain and lay down sanitary and hygienic conditions of premises, surrounding environment and personnel. Though FPO, 1955 also lays down product specifications, they do not address issues related to microbiological characteristics adequately.

Food Safety and Standards Act, 2006

Food Safety and Standards Act, 2006 (FSSA, 2006) was constituted to consolidate the laws relating to food and to establish the FSSAI for laying down science based standards for articles of food and to regulate their manufacture, storage, distribution, sale and import, to ensure availability of safe and wholesome food for human consumption and for matters connected therewith or incidental thereto. FSSA, 2006 bring the different pieces of legislation pertaining to food safety under one umbrella which overrides the PFA, 1955 and various Quality Control Orders under Essential Commodities Act,1955. The aim is to better coordinate and integrate food safety controls across India to give highest level of health protection.

Table 29.4. Prevention of Food Adulteration Rules,1956: (Microbiological Requirements for Sea Food)

Sl. No.	*Name of Product*	*Total Plate Count*	*E. coli*	*Staphylococcus aureus*	*Salmonella and Shigella*	*Vibrio Cholerae*	*Vibrio para-haemolyticus*	*Clostridium perfringens*
1	Frozen shrimps or Prawns (Raw)	Not more than 5 lakhs/g	Not more than 20/g	Not more than 100/g	Absent in 25 g	Absent in 25 g	Absent in 25 g	
	Frozen shrimps or Prawns (Cooked)	Not more than 1 lakhs/g	Absent in 25 g	Absent in 25g	Absent in 25 g	Absent in 25 g	Absent in 25 g	
2	Frozen Lobsters (Raw)	Not more than 5 lakhs/g	Not more than 20/g	Not more than 100/g	Absent in 25 g	Absent in 25 g	Absent in 25 g	
	Frozen Lobsters (Cooked)	Not more than 1 lakhs/g.	Absent in 25 g	Absent in 25 g	Absent in 25 g	Absent in 25 g	Absent in 25 g	
3	Frozen Squid	Not more than 5 lakhs/g	Not more than 20/g	Not more than100/g	Absent in 25 g	Absent in 25 g	Absent in 25 g	-
4	Frozen Fin Fish	Not more than 5 lakhs/g	Not more than 20/g	Not more than 100/g	Absent in 25 g	Absent in 25 g	Absent in 25 g	-
5	Frozen fish fillets or minced fish flesh or mixtures thereof	Not more than 5 lakhs/g	Not more than 20/g	Not more than 100/g	Absent in 25 g	Absent in 25 g	Absent in 25 g	-
6	Dried shark fins	Not more than 5 lakhs/g	Not more than 20/g	Not more than 100/g	Absent in 25 g	Absent in 25 g	Absent in 25 g	-
7	Salted fish/dried salted fish	Not more than 5 lakhs/g	Absent in 25 g	Absent in 25 g	Absent in 25 g	Absent in 25 g	Absent in 25 g	-
8	Canned finfish	Nil	Absent in 25 g	Absent in 25 g	Absent in 25 g	Absent in 25 g	Absent in 25 g	Absent in 25 g
9	Canned shrimp	Nil	Absent in 25 g	Absent in 25 g	Absent in 25 g	Absent in 25 g	Absent in 25 g	Absent in 25 g
10	Canned Sardines or sardine type products	Nil	Absent in 25 g	Absent in 25g	Absent in 25 g	Absent in 25 g	Absent in 25 g	Absent in 25 g
11	Canned salmon	Nil	Absent in 25 g	Absent in 25g	Absent in 25 g	Absent in 25 g	Absent in 25 g	Absent in 25 g
12	Canned crab meat	Nil	Absent in 25 g	Absent in 25g	Absent in 25 g	Absent in 25 g	Absent in 25 g	Absent in 25 g
13	Canned Tuna and Bonito	Nil	Absent in 25 g	Absent in 25g	Absent in 25 g	Absent in 25 g	Absent in 25 g	Absent in 25 g

30

Microbiological Criteria, Standards and Code of Hygienic Conditions

Foodstuffs should not contain microorganisms or their toxins or metabolites in quantities that present an unacceptable risk for human health. The safety of foodstuffs is mainly ensured by a preventive approach such as implementation of good hygiene practice (GHP) and application of procedures based on HACCP principles. As per the ICMSF any microbiological criterion for the food should contain the following information:

1. The sampling plan.
2. The microbiological limits.
3. The number of samples required for confirming microbiological limits.
4. A statement of the microorganism and/or concerned toxin.
5. The laboratory methods for detection and quantification of microorganism and/or concerned toxin.

Regulation contains microbiological criteria for carcases of cattle, sheep, goats and horses. Testing against the criteria set down in the Regulation should be undertaken when validating or verifying the correct functioning of systems in place. Various authorities/organizations such as World Health Organization, Codex Alimentarius Commission and European Union have attempted to lay down acceptance criteria for different types of meats. Although, as yet, there are no universally accepted criteria for the interpretation of bacteriological findings, the following criteria are frequently used:

1. **Microbiological Standards:** These are mandatory criteria with legal backing. This refers to "a criterion incorporated in a law or regulation, controlling foods produced, processed or stored in, or imported into the area of jurisdiction of the regulatory agency". Microbiological standards have been proposed from time to time, depending upon the necessity. Generally the decisions on the acceptability of food lots based on the examination of few samples are taken from the batch.
2. **Microbiological Specifications**: These are generally contractual agreements between a manufacturer and a purchaser to check whether the foods are of the required quality. HACCP and microbiological specifications under 2-3 class sampling (attribute) schemes are two excellent means of ensuring food quality. Microbiological specification refers to "the microbiological criterion that forms the conditions of acceptance of a specific food or food ingredient by a food manufacturer or private or public purchasing agency".

   Two types of microbiological specifications have been recommended for ensuring food safety and keeping quality. They are pathogenic and indicator organisms. Standards for specific pathogenic organisms have been proposed from time to time either by fixing maximum permissible limits per unit (g, ml, or cm^2) of food or by testing presence or absence of the organism in a given unit of food, *e.g.*, *Staphylococcus aureus* or *Salmonella* spp., respectively. Indicator organisms are organisms or group of organisms that indicate likely presence of hazard that may occur in food or water but they themselves may or may not be responsible for the hazard, *e.g.*, coliforms indicate the likely presence of other enteric pathogens in food.
3. **Microbiological Guidelines**: These are non-mandatory criteria usually intended as a guide to good manufacturing practice.

Attribute Sampling Schemes

The schemes state the criteria for acceptance of a food lot on the basis of examination of certain required number of sample units by defined methods that is the scheme prescribe sampling procedure and decision criteria. There are two types of sampling schemes.

1. **Two Class Plan:** This plan denotes two attributes that is presence or absence of an organism in a given sample unit. This scheme is applied for more hazardous type of organism like *Clostridium botulinum* and *Salmonella* spp.
2. **Three Class Plan:** This plan has three attributes which can be divided a lot into 3 categories that are acceptable (n, m); unacceptable (more than m) and marginally acceptable (c). The higher limit (M) is safety limit and values beyond this may be hazardous. "m" represents the lowest possible count that may occur under good manufacturing practices. "n" is the number of sample units to be examined.

Criteria for Devising Sampling Schemes

There are three most important factors that should be considered for devising attribute sampling schemes.

1. **Microbial Incidence:** It is essential to know the pattern of distribution of microorganisms in foods.
2. **Hazard:** If a product poses a potentially severe hazard, a two attribute scheme (presence or absence) has to be devised. In case of moderate or low hazard, a three class plan can serve the purpose.
3. **Processing Technique:** Milk, meat, eggs, poultry and other foods are subjected to a variety of processing techniques. In an abattoir or meat processing plant, the carcase may be subjected to cutting, deboning, mincing, chilling, freezing *etc.* Other may require pasteurization, ultra-heat treatment, scalding, radiation and incorporation of additives, chemical preservatives and vacuum or modified atmospheric packaging and canning. Hence, it is essential to investigate the survival or growth of microorganisms before finalizing the sampling scheme.

Table 30.1 ICMSF's Recommended Permissible Limits of Bacteriological Counts in Meat

Product	*Bacteriological Count*	*Permissible Limit*
Chilled meats	Total viable count at 35 °C (or 20 °C in case of chilled meats)	10^7/g
Chilled meats	*Salmonella*	Should be detected in not more than one of five 25 g samples
Frozen poultry	Count at 20 °C by using rinse method	10^7/ml of the rinsing solution
Poultry meat	*Salmonella*	Should be detected in not more than one of five 25 g samples

Microbiological Standards

Limits of microbiological parameters for processed food products are specified in respective Indian Standards. As far as microbiological criteria are concerned Indian Standards are elaborate. BIS has formulated standards on test methods for detection and enumeration of pathogenic microorganisms in food and specifications for ingredients used in media for microbiological work. BIS has also formulated code of hygienic conditions for various food industries. BIS has adopted Codex HACCP and Food Hygiene guidelines and ISO 22000 as Indian Standards.

Table 30.2. Microbiological Methods of Test

Standards	*Microbiological Methods of Test*
IS 14843: 2000/ ISO 13720: 1995	Meat and Meat products- Enumeration of *Pseudomonas* spp.
IS 14844: 2000/ ISO 13721: 1995	Meat and Meat products- Enumeration of lactic acid bacteria- Colony count technique at 30°C

Standards	Microbiological Methods of Test
IS 14920: 2001/ ISO 13681: 1995	Meat and Meat products – Enumeration of yeasts and mould – Colony count technique
IS 15463: 2004/ ISO 6391: 1997	Meat and Meat products – Enumeration of *Escherichia coli*– Colony count technique at 44 °C using membranes.
IS 5403: 1999/ ISO 7954: 1987	Methods for yeast and mould count of food stuffs and animal feeds (first revision)
IS 5887(Part 1): 1976	Methods for detection of bacteria responsible for food poisoning: Part 1 Isolation, identification and enumeration of *Escherichia coli* (first revision)
IS 5887(Part 2): 1976	Methods for detection of bacteria responsible for food poisoning: Part 2 Isolation, identification and enumeration of *Staphylococcus aureus* and faecal streptococci (first revision)
IS 5887(Part 3): 1999/ ISO 6579: 1993	Methods for detection of bacteria responsible for food poisoning: Part 3 General guidance on methods of detection of *Salmonella* spp. (second revision)
IS 5887(Part 4): 1999	Methods for detection of bacteria responsible for food poisoning: Part 4 Isolation and identification of *Clostridium perfringens, Cl. botulinum* and enumeration of *Cl. perfringens* (second revision)
IS 5887(Part 5): 1976	Methods for detection of bacteria responsible for food poisoning: Part 5 Isolation, identification and enumeration of *Vibrio cholerae and Vibrio parahaemolyticus* (first revision)
IS 5887(Part 6): 1999/ ISO 7932: 1993	Methods for detection of bacteria responsible for food poisoning: Part 6 Identification, enumeration and confirmation of *Bacillus cereus*
IS 5887(Part 7): 1999	Methods for detection of bacteria responsible for food poisoning: Part 7 General guidance on methods of detection of *Shigella* spp.
IS 5887(Part 8/Sec1): 2002/ ISO 6888-1: 1999	Methods for detection of bacteria responsible for food poisoning: Part 8 Horizontal method for enumeration of coagulase positive staphylococci (*Staphylococcus aureus* and other species) Section 1 Technique using Baird-Parker Agar Medium
IS 5887(Part 8/Sec2): 2002/ISO 6888-2: 1999	Methods for detection of bacteria responsible for food poisoning: Part 8 Horizontal method for enumeration of coagulase positive staphylococci (*Staphylococcus aureus* and other species) Section 2 Technique using rabbit plasma fibrinigen Agar Medium
IS 7402: 1996	Microbiology – General guidance for the enumeration of Enterobacteriaceae without resuscitation – MPN technique and colony count technique (first revision)
IS 14397: 1996	Detection, isolation and identification of pathogenic *Escherichia coli* in food
IS 14987: 2001/ ISO 8914: 1990	Microbiology – General guidance for detection of *Vibrio parahaemolyticus*
IS 14988(Part 1): 2001/ ISO 11290-1: 1996	Microbiology of food and animal feeding stuffs –Horizontal method for detection and enumeration of *Listeria monocytogenes*: Part 1 Detection method
IS 14988(Part 2): 2001/ ISO 11290-2: 1996	Microbiology of food and animal feeding stuffs –Horizontal method for detection and enumeration of *Listeria monocytogenes*: Part 1 Enumeration method

Table 30.3. Code for Hygienic Conditions

IS 4303(Part 1):1975	Code for hygienic conditions in fish industry: Part 1 Pre-processing stage (first revision)
IS 4303(Part 2):1975	Code for hygienic conditions in fish industry: Part 2 canning stage (first revision)
IS 8182:1976	Code for hygienic conditions for processed meat products

Table 30.4. Food Hygiene, HACCP and FSMS

IS 2491:1998	Food Hygiene – General Principles –Code of Practice (second revision)
IS 14595:1998	Food Hygiene – Microbiological Criteria Principles for establishment and application
IS 15000:1998	Food Hygiene – Hazard Analysis and Critical Control Point (HACCP) –Systems and guidelines for its application
IS 22000:2005 ISO 22000:2005	Food safety management systems –Requirements for any organization in the food chain

31

Cleaning and Sanitation of Slaughterhouse

Sanitation

Sanitation is the act or process of providing adequately hygienic conditions to ensure a safe, sound, wholesome product fit for human consumption and covers hygienic precautions regarding personal hygiene, process hygiene and cleaning and disinfection. Sanitation inactivates the microorganisms through disinfectants, and also combats pests such as insects and rodents through chemical substances using insecticides and rodenticides, respectively. It is important to remove all food products from the area when starting cleaning and disinfection/sanitation measures. This is because physical cleaning with pressurized water may stir up dirt or produce contaminated water droplets (aerosol), which could contaminate meat, present in such rooms. Moreover, chemical cleaning/disinfection may produce toxic residues when in contact with remaining meat or meat products. The same precaution measures are also applied while using insecticides and rodenticides for pest control.

A well-planned, well-executed and controlled cleaning and sanitation programme for rooms, machines and equipment is very important to achieve a hygienic standard. Effective sanitation measures in meat industry can prevent the contamination and thereby meat-borne diseases. Therefore, it is important for all food processing employees, regardless of their assigned duties, to realize the importance of their actions when producing food products. Their understanding of the basic concepts of sanitation and bacterial growth is vital in the production of a safe food supply. Efficient meat plant cleaning and sanitation is often neglected as it requires extra work and the positive effects are not immediately visible. However, failures in meat plant hygiene can cause high financial losses in the long run.

Unhygienic conditions in a meat plant result in unattractive and tasteless products, spoilage of valuable food and/or food-borne diseases.

Proper cleaning and sanitation is becoming increasingly important in modern meat processing as more perishable and hygienically sensitive meat products come on the market, particularly convenience foods such as pre-packed portioned chilled meat, vacuum-packed sliced sausage and ham products, meat products in controlled atmosphere packages *etc.* The microbial load of such products must be low to guarantee adequate shelflife and to avoid spoilage during distribution. Periodic cleaning and sanitation includes disinfection of meat plant premises and equipment which is an integral part of Good Hygienic Practice (GHP). Cleaning and sanitation is considered as one of the most important activities in the meat plant, as these measures provide the necessary environment for proper handling and processing of meat.

Sanitary Operations in Meat Plant

1. **Pre-operational Sanitation:** All floors, walls and equipment should be visually inspected for any contamination. If needed, the general cleaning and sanitizing procedures can be implemented. Boot dip mats filled with hot water and bleach can be placed in front of doors in order to prevent the entry of contamination in the meat production and processing areas.
2. **Operational Sanitation:** Carcase dressing should be performed under strict sanitary conditions and in such a manner to prevent contamination of the carcase. Cleaning of the hands, arms, aprons, boots and protective head gear is necessary during slaughter and dressing of carcase. Employees should clean and sanitize the knives, hand saws, hooks, and other equipment with hot water during slaughter and dressing of carcase to prevent contamination. The carcase splitting saws can be sanitized between carcases using hot water. Employees involving evisceration process should clean their hands, arms, clothes, aprons, boots and knives during the evisceration process.
3. **Post-Operational Sanitation:** After the end of each slaughter, all floors, walls and equipment are properly cleaned and sanitized.

Factors Influencing Sanitary Operations in Meat Plant

1. **Equipment Design:** Equipment of the meat plant should be designed in such a way and of such material and workmanship so that these can be adequately cleaned and properly maintained. Each piece of equipment should be installed and maintained as to facilitate the cleaning of the equipment and adjacent space. Food-contact surfaces should be maintained to protect food from being contaminated by any sources.
2. **Personnel Health and Hygiene:** Only healthy employees should be allowed to work in food contact areas. Any diseased person or person having cuts, open lesion, including boils, sores, or infected wounds, or any other possible source of microbial contamination should not be allowed to

work in meat processing operations. Personnel hygiene is also important in sanitary operations. Therefore, only personnel having good personal hygiene should be allowed to perform sanitary operations especially in meat work areas.

3. **Meat Plant and Its Surroundings:** Sanitary operations also depend on the meat plant and its surroundings conditions. Therefore, the meat plant and its surroundings should be kept in a sanitary manner to protect against possible contamination due to rodent infestation. Roads, yards, and parking lots should be maintained so that they do not contaminate food. Unloading facilities and holding pens for animals should be constructed of such materials so that they do not harbour the microorganisms. The slaughterhouse wastes must be properly collected and efficiently treated and disposed in such a manner so that they do not constitute a source of contamination of meat.

Cleaning and Disinfection

The main objective of a cleaning programme is to control microbial activity. Although an adequate cleaning programme will eliminate nearly all the soil present, it will not destroy or remove all the microorganisms. Therefore, this requires a second step that is disinfection.

Cleaning

Cleaning is a soil-removing process which removes the microorganisms. It includes removal of all organic substances such as dirt, manure, fat, protein particles, meat scraps *etc.* from surfaces of walls, floors, tools and equipment. Cleaning is the necessary prerequisite for sterilization and disinfection. A large number of microorganisms (90 per cent and more) present on these objects are removed through the cleaning procedures. However, many microorganisms stick very firmly to surfaces, in particular in tiny almost invisible layers of organic materials, so called bio-films, and cannot entirely be removed even by profound cleaning but persist and continue multiplying. Inactivation of microorganisms in organic materials or bio-film requires antimicrobial treatments. This is carried out in food industries through hot water or steam or through the application of disinfectants.

During the routine slaughter operations in an abattoir, the floor should be clean as possible by continuous removing of refuse and other wastes from the floor. If continuous washing of the floor is practiced, care should be taken not to splash and soil the carcases. The first step in floor and equipment cleaning is physical removal of scrap, *i.e.* coarse solid particles, with a dry brush or broom and shovel. This is usually referred to as "dry cleaning". This is useful in reducing the amount of water to remove these materials and to avoid the clogging of drains and to reduce the load of effluent treatment plant. More profound clean-up procedures require large amount of water.

Manual cleaning using brushes or scrapers is widely applied in small-scale operations although labour and time-intensive. High pressure cleaning method is commonly used in the meat industries. The pressurized water is applied by high pressure units and special spraying lances. The pressure should be between 30-70 bars. If hot water is used, the temperature should be 55°C at the nozzle in order to achieve sufficiently high temperatures on the surfaces, particularly for removal of fat.

Principles of Cleaning

Before planning a cleaning programme one of the essentials is to know which types of soil are present in the area to be cleaned. Fat and protein are the most common types of soil in the meat industry.

Cleaning Agents

Traditional cleaning substances for manual use are alkalis, such as sodium carbonates or washing soda. Alkalis are efficient in dissolving proteins and fats, but their pH above 11 may cause corrosion in tools and equipment. Commercially available cleaning agents in modern cleaning practices are complex compositions of alkaline, acid or neutral chemical substances. In order to improve their dirt loosening properties, surface-active agents (surfactants) or detergents are also added. Detergents decrease the superficial tension of water. Water can then penetrate into the small spaces between dirt particles and surfaces, where those particles are attached, thus facilitating their removal. Cleaning detergents dissolve the fat and prevents its settling after decreasing of water temperature, therefore, these are important in removal of fat by pressurized hot water.

1. **Alkaline Cleaning Agents:** These are generally suitable for removing organic dirt, protein residues and fat.
2. **Acidic Cleaning Agents:** These are used particularly for removal of encrusted residues of dirt or protein or of inorganic deposits ("scaling") such as water-stone, milk-stone, lime *etc.*
3. **Neutral Cleaning Agents:** These agents are less effective than alkaline or acid cleaning agents, but have mild impact on skin and materials and are useful for manual cleaning of smooth surfaces without encrusted dirt.

In practice, alkaline and acid cleaning substances should be used alternatively. The alkaline agent should be used for routine cleaning, but every few days an acid substance should be employed instead in order to remove encrusted residues, scaling *etc.* Cleaning substances together with the suspended dirt particles and fat must be rinsed off using potable water.

Cleaning Methods

Cleaning reduces a substantial amount of microorganisms but it does not have the potential to completely eliminate all surface contamination. Persistent microorganisms continue to grow in number by using remaining protein as nutrients and pose a further risk to the foods to be processed. The elimination of microorganisms is achieved through disinfection, either by using hot water or steam or chemical disinfectants. Chemical disinfectants are preferred for most applications in the meat industries as these are easy to use and do not involve the risk of accidents or damage to equipment by generating high humidity or water condensation, which may occur when using steam.

Cleaning and disinfection procedures in the meat industries are complex processes depending on type of surfaces and type of contamination. Selection of suitable chemicals for cleaning or for disinfection is also important. All these factors can make correct cleaning and disinfection a difficult task for the personnel involved. Cleaning and disinfection staff must be made aware that efficient cleaning and disinfection is of utmost importance for safety and quality of product.

High pressure water is efficient for surface cleaning after dry-cleaning of scrap. It serves for the removal of remaining small solid parts, blood and dirt from the entire floors and walls of processing sections as well as for the removal of meat and fat particles and layers of protein from tools and equipment. Hot water has better cleaning effect than cold water. Cleaning with equipment producing a pressurized steam/water mixture is even more efficient as impact temperatures of approximately 100°C can be achieved. The disadvantage of this method is the intense fog and aerosol formation, which cause spreading of unwanted microorganisms by water droplets (aerosol) and also affect installations and equipment through high humidity and excessive condensation. For these reasons a steam/water-mix is not suitable for meat processing facilities, therefore, cold or hot pressurized water cleaning is preferred.

The removal of loose dirt and meat/fat residues by water does not ensure complete cleaning. Sticky or encrusted layers of fat or protein may exist which can be effectively removed by using chemical cleaning solutions. The solutions can be applied manually by using brushes or scrapers for dismantled equipment or in general for smaller surfaces. Mechanical cleaning with high pressure equipment together with cleaning solutions is used for larger floor and wall areas as well as working tables, containers, equipment *etc.*

Best disinfection results are achieved when chemical disinfection is preceded by intensive dry/wet cleaning. Disinfection without pre-cleaning is not fully efficient as many microorganisms remain embedded in encrusted dirt, protein and fat, which cannot be properly dissolved by disinfection chemicals. Therefore, microorganisms remain protected against the disinfection chemicals. Moreover, remaining protein may inactivate chemical disinfectants.

Adequate rinsing with water after cleaning and prior to disinfection is also indispensable as chemical disinfectants may be neutralized by remaining cleaning substances. These should be taken into account to ensure efficient disinfection and prevent the wastage of money. Foam cleaning is a relatively new cleaning method for the food industry, in particular the larger-scale plants. Water foam containing detergents and other cleaning agents is sprayed on wetted walls, floors and surfaces of equipment. The foam does not immediately run off but clings to the surfaces. This allows a longer term contact on the surfaces to be cleaned. After a sufficient impact period (minimum 15 minutes) the foam is washed down with water using water hose or low-pressure water spray. High pressure water spraying for washing off the foam should be avoided in order to prevent the spread of water droplets (aerosol) in the room.

Disinfection

Disinfection is the process of destruction of microorganisms, but not usually bacterial spores. Disinfection does not necessarily kill all microorganisms but reduce them to a level acceptable for a defined purpose.

Principle of Disinfection

Complete removal of wastes, soil, dirt *etc.* from the surfaces of floor, walls and equipment is essential before disinfection to achieve effective disinfection.

Disinfectants for the Meat Industry

Disinfectants should be effective and rapidly acting in killing microorganisms. The chemical composition of disinfectants varies depending on the specific target (slaughterhouse, meat processing, easily accessible open processing lines or closed food pipeline systems) and on chemical formulations by the individual disinfectant manufacturer. Modern disinfectants are mostly mixtures of different chemical substances. Combinations of disinfection chemicals achieve a synergistic effect and result in the elimination of various types of microorganisms. Different types of compounds are used as disinfectants in meat industry as follows:

1. **Aldehydes, Phenols, Alcohols, Alkalis and Acids:** These groups include important compounds such as formaldehydes, sodium hydroxide, organic compounds *etc.*
2. **Quaternary Ammonium Compounds (QUATs):** These compounds are used in food industries due to their non-corrosive nature, odourless, additional cleaning properties (surfactant) and effectiveness on cell wall of bacteria. These compounds are not fully effective on spores.
3. **Oxygen Releasing Substances and Peroxides:** These include hydrogen peroxide and per acetic acid. These compounds can penetrate into cells and effective on all microorganisms including spores and virus. These compounds are odourless but may be corrosive in concentrations more than 1 per cent.
4. **Chlorine Containing Compounds:** These include sodium or calcium hypochlorite and gaseous chlorine. These compounds are effective against a wide range of bacteria, penetrate cell walls, but has a corroding effect on equipment.

Commercially available efficient disinfectant is a combination of components such as organic acids, surfactants (surface active agents) and peroxide compounds. The organic acids have sanitizing effect and also decrease the pH as some disinfectants are more efficient at lower pH. The surfactants help to penetrate the organic materials. The peroxide compounds kill the viruses by coagulation and denaturation of proteins and destruction of bacteria by penetration through cell walls.

Different types of chemical disinfectants act differently on certain groups of bacteria and under certain pH-ranges. Chemical disinfectants should be applied alternate periodically to achieve a maximum disinfection effect. Application of suitable alternative substances inactivates those bacteria, which were possibly surviving the previous sanitation process. Application of alternative substances also helps to counteract the development of resistant bacteria in the meat plant.

Combined Cleaning and Disinfection Agents

A combined cleaning agent and disinfectant are used to remove the problems of separately using the cleaning agent and disinfectant. They are made on the basis of quaternary ammonium compounds, which have surfactant and disinfectant properties. The combined method should be considered only in cases of very little dirt contamination. Adequate cleansing and disinfection facilities in the slaughterhouse are essential for disease control and biosecurity.

Frequency of Disinfection

The frequency of cleaning (daily or several times a day) is decided according to type and product lines or activity of each individual meat plant and requirements.

1. **Several Daily Disinfections:** These are necessary for hand tools, meat saws and cutting boards. These are carried out by using hot water or chemicals.
2. **Daily Disinfection:** It is useful for dismantled equipment such as parts of grinders, fillers, stuffers *etc.*
3. **Weekly Disinfection**: It is recommended for other equipment and floors and walls of processing and chilling rooms.

Cleaning and Disinfection Schemes

The personnel of the meat industry must be made fully aware of the need for proper cleaning. Cleaning should be treated as an integral part of the production process. Cleaning should be done carefully and not just superficially or in a rush at the end of the production process. Periodic cleaning and sanitation includes disinfection of meat plant premises and equipment which is an integral part of Good Hygienic Practice (GHP). Cleaning and sanitation is considered as one of the most important activities in the meat plant, as these measures provide the necessary environment for proper handling and processing of meat.

Sequence of Sanitation in a Meat Plant

The correct sequence of sanitation in a meat plant is discussed as follows:

1. **Gross Cleaning:** This is first and important step which involves removal of all larger pieces of food product, meat *etc.* before application of detergent. These objects can be removed by hand-picking, scraping and shoveling. All waste should be collected in bins and removed entirely from the area.
2. **Pre-Rinsing:** This is carried out for removal of deposits (*e.g.*, blood, small meat pieces, particles *etc.*) which cannot easily be removed by hand-picking, scraping and shoveling. Pre-rinsing can be carried out using warm water (below 50°C to avoid coagulation) especially on the surface

having thick deposits of grease, *e.g.*, cutting board. After pre-rinsing, the water left on the flat surface must be removed to avoid loss of efficiency of disinfectant due to its dilution with the water.

3. **Detergent Application:** The detergent is applied to remove the thin tenacious layers of protein, grease *etc.* that are still on the surface which may be already look fairly clean. Detergent foam should be applied to ensure that all surfaces have been covered in the foam, both top and bottom.
4. **Post-Rinsing:** This is an important step to remove the detergent applied on the surfaces. Care should be taken to minimize the amount of splash and generation of aerosol, which may contaminate previously cleaned surfaces. After post-rinsing, the surface should be free of all visible particles, soil and detergent residues. After rinsing, any pool of water should be removed from surfaces and vessels.
5. **Disinfection:** This should only be carried out on a visually clean, well-rinsed surface, with minimal amounts of surface water. The concentration of disinfectant should be optimum to achieve efficient disinfection. Direct food contact surfaces should be disinfected daily, while other surfaces such as doors, walls *etc.* disinfected on a regular basis.
6. **Final Rinsing:** Most of the disinfectants are safe to leave on the surfaces without final rinsing. Disinfectant residues often help to maintain a low microbial count for considerable time after the cleaning sequence is finished. However, in some sections of food industry, European Union Directives and national regulations require rinsing of food-contact surfaces with potable water after the disinfectant has acted unless the disinfectant is specifically designed to remain unrinsed.

Verification of Sanitation Activities

Verification is the use of methods, procedures, or tests in addition to those used in monitoring to determine if the operation is in compliance with the written plan (sanitation standard operating procedures). It indicates whether the sanitation programme is addressing the needs of the plant or it needs modification. Verification is long term and does not always call for immediate changes; however, it may necessitate some modification of the existing sanitation programme to ensure a safe finished product.

32

Disposal of Unsound Meat and Slaughterhouse Wastes

The meat which has any physical, chemical, biological and microbiological agents or factors that can cause health hazards in the consumers is known as unsound meat. Unsound meat, offal and condemned materials may harbour several types of pathogenic organisms such as bacteria, viruses, fungi and parasites. Moreover, the residues of some chemical substances such as dugs, antibiotics, hormones *etc.* may be found beyond permissible limit. Unsound meat is unfit for consumption and may cause environmental pollution. Careless disposal of unsound meat may produce offensive odour in the vicinity and attract flies, dogs, cats, vultures and crows. Therefore, its proper disposal is important.

Pre-Disposal Care

Condemned meat or carcase should be immediately removed from the slaughterhouse and covered with a sack soaked in a disinfectant to discourage flies, dogs and other predators. Slaughterhouse floor, equipment, instruments and other objects should also be disinfected after scrubbing and washing.

Method of Disposal of Unsound Meat/Carcase

There are several methods of disposal of unsound meat/carcase. Animal carcases are usually disposed of by traditional methods like burial, incineration, rendering and composting. Each method has some advantages as well as disadvantages. Burial of unsound meat in a pit can lead to ground water contamination. Incineration is expensive and can potentially pollute the air. Rendering of carcase is associated with transportation cost. Therefore, disposal method should be safe, practical and economical. Some methods of disposal of unsound meat/carcase are discussed as follows:

1. **Burial:** Burial is the most common method of disposal of condemned meat and offal. For burial of condemned meat, offal, other materials *etc.*, a 2-3 meter deep trench is dug and these are disposed in it. Quicklime or other appropriate disinfectant is layered on the disposed materials and then covered with at least 1 meter thick layer of soil. The burial area should be protected from stray and wild animals by using barbed fencing wire or strong smelling disinfectant like formalin. Environmental constraints associated with this disposal method may be contamination of air, soil and water. Social concerns with this disposal method are odour and fly menace. Therefore, great care must be taken to dispose the carcase away from the water sources in order to avoid its contamination.
2. **Incineration:** Incineration is the most efficient method of disposal of unsound meat and offal particularly in case of condemnation of these due to highly infectious diseases. Wood, sawdust, coal, oil, gas *etc.* can be used for incineration of unsound meat, offal and other contaminated materials. The economic constraints associated with this disposal method are increase in the costs of burning materials like kerosene, diesel, coal, wood *etc.* for burning of carcase. Other issues related to this method are the labour cost, availability of land *etc.*
3. **Rendering into Meat By-Products:** Rendering is a process of heating of slaughterhouse by-products through which fats is separated from water and protein residues for the production of edible lards and dried protein residues. Rendering is the most economical method of disposal of unsound meat and offal affected with non-infectious diseases. In this method, the condemned carcase and offal are converted into useful by-products such as meat meal, bone meal, blood meal *etc.* which can be used as feed for livestock and poultries. Other by-products such as tallow and lard can also be obtained from rendering of these condemned materials. Rendering of unsound meat and offal can be done either by wet rendering system or by dry rendering system.

   In wet rendering method, the condemned meat along with bone and offal are finely chopped and autoclaved at 15 lb pressure for half an hour. After autoclaving the wet dough of the autoclaved material can be dried in the sunlight or by mechanical driers. The obtained material may be meat meal or meat-cum-bone meal which can be used as feed for livestock and poultries or as a fertilizer.

   In dry rendering method, the condemned meat and offal are finely chopped and cooked in a double-walled vessel using steam at 15 lb pressure for half an hour. Excess amount of fat is drained out using outlet at the bottom and this fat is obtained as tallow or lard. Cooked material can be converted into meat meal or meat-cum-bone meal. These meals can be used as feed for livestock or can be enriched with phosphate and used as fertilizer. This method is more useful as it gives better yield.

Treatment and Disposal of Slaughterhouse Effluent

Effluents of slaughterhouses constitute one of the most serious causes of environmental pollution, bad odours and health hazards. Major contribution to the total waste load includes blood and paunch manure. Blood is the most important source of pollution. Of all waste products, the waste in the form of blood has the highest polluting value. Blood has a high BOD value (150,000 - 200,000 mg/l). In order to reduce the waste load, attempts should be made to collect and process the blood (*e.g.*, drying).

Drying of blood can be done by direct or indirect heating. Direct heating of blood produces large quantities of blood-water (corresponding waste load approximately 1.3 kg BOD per ton of LWK). Indirect (external) heating of blood is preferred as it produces less waste load (corresponding waste load approximately 0.3 kg BOD per ton of LWK). Paunch manure is the second most important source of pollution. It may substantially contribute to the total waste load if not properly handled. Dumping (sewering) of the entire paunch content gives a BOD of 2.5 kg per ton of LWK. Minor contribution to the total waste load includes stockyards and pens, slaughtering wastes, meatpacking, rendering *etc.*

Cutting and deboning operations produce trimmings, blood, bones and bone dust. The total of raw waste loads for meat processing plants (including cutting and deboning) has been estimated at 5.7 - 6.7 kg BOD per ton of product. Meat processing operations produce a raw waste load from blood, tissues and fat that reaches the sewer during cleaning activities. The wastes generated from curing process include sugar and salt. Pickling can cause a high chloride waste (only 25 per cent of the curing brine remains in the product). Both wet-rendering and continuous rendering at low temperatures produce polluted tank water containing residues of fat and protein (2 kg BOD per ton of LWK).

Raw waste effluent from food-processing industry has extremely high level of soluble organic matter. The strength of wastewater varies considerably according to plant, specific process and raw product characteristics. From an economic point of view, it is normally less costly to treat a high-strength, low-volume waste than a large-volume, diluted waste. The liquid portion of food-processing waste is normally screened or separated after settling, as a preliminary step in any treatment process, so that these wastes can be disposed of as garbage or combined with other solids in a by-products recovery programme.

The treatment of wastewater involves a variety of physical, chemical and biological methods. Primary treatment includes sedimentation, filtration, flocculation, flotation, centrifugation, ion exchange, reverse osmosis, carbon absorption and chemical precipitation. Secondary treatment is involves biological processes that are necessity to reduce the BOD value and reach wastewater effluent standards. The BOD value can be reduced by mixing higher concentrations of organisms and oxygen in the waste stream to provide rapid oxidation and stabilization of the waste stream prior to their discharge into the water bodies or on the land. Other secondary methods that are widely used include the activated sludge process, aerobic trickling filters, spray irrigation and the use of a variety of ponds and lagoons. Odour nuisances have been associated with ponds of inadequate depth and improperly managed. Odours from anaerobic processes can be removed by the use of soil filters that can oxidize objectionable polar gases.

Prevention of Waste Production from Slaughterhouse

The quality of overall cleaning-up practices determines to a large extent of the total waste load produced. It has been established that the waste load decreases with a decrease of the water being used. As much blood as possible, should be collected and processed. Indirect heating can reduce the amount of wastewater (and thus the waste load), compared with direct heating from 1.3 to 0.3 kg BOD per ton of LWK.

Paunch can be handled in different ways. Dry dumping (dumping for subsequent rendering or disposal as solid waste without needless water flushing) can reduce the waste load. Dry dumping has a waste load of 0.2 kg BOD per ton of LWK; while dumping in full into the sewer has a waste load of 2.5 kg BOD per ton of LWK and wet dumping (washing out and screening the wet slurry on gross solids) has a waste load of 1.5 - 2 kg BOD per ton of LWK.

Dry animal pen clean-up reduces the amount of wastewater. If the pens are covered, no rain or snow water can enter, which reduces the amount of wastewater. Hog hair, recovered from the dehairing process, can be disposed as solid waste, washed and baled for direct marketing, or hydrolyzed by pressure cooking for marketing as a feed supplement.

Glossary of Terms

Abattoir (slaughterhouse): A facility where animals are killed and processed into meat products.

Abrasion: A type of open wound that's caused by the skin rubbing against a rough surface. Abrasion is sustained through knocks from equipment, falling gambrels *etc.* It may also occur during transportation, loading and unloading of animals, and fighting of animals in the lairage.

Abscess: A localised, encapsulated collection of pus in a cavity formed by disintegrating tissues.

Acne: A pustular infection of hair follicles and sebaceous glands with a surrounding acute inflammation. Acne is also known as folliculitis or pyoderma.

Activated sludge process: A modern method of sewage treatment which is carried out in either aeration tank or trickling filter.

Adulteration: A fraudulent practice involving substitution or mixing of flesh of cheaper variety which is objectionable for the reasons of health, religion and economy.

Aerosol: A suspension of fine solid or liquid particles in a gas, usually air.

Anaplasia: An occurrence of pleomorphism in the cells with abnormal constituents.

Ante-mortem inspection: The clinical inspection (before slaughter) of the live animal by a qualified veterinarian.

Antibiotic: A substance, produced by bacteria or fungi that destroys or prevents the growth of other microorganisms (bacteria and fungi).

Antiseptic: A substance which prevents the growth of organisms either by inhibiting their activity or destroying them. It is suitable for application to living tissues such as a wound.

Antiserum: The blood serum containing antibodies.

Antitoxin: A substance that specifically neutralizes the toxin.

Anton test: A test used for assessing the virulence of *Listeria* spp. Conjunctivitis is produced when about 10^6 cells of *Listeria monocytogenes* are administered into the eye of a rabbit or guinea pig.

APEDA: A government agency of India that plays important role in the export of agricultural and processed food products. It is responsible for the development and promotion of the export of agricultural and processed food products.

Aplasia: The absence of an organ in the body that may be due to genetic factors, intrauterine infections and intake of teratogenic drugs.

Atrophy: The shrinking and reduction in size of an already fully developed organ that may occur due to decrease in size and number of its cells as a result of necrosis and autolysis.

Baby beef: The meat of calf of 12-18 months of age.

Back bleeding: A type of contamination that occurs on the perital pleura, particularly along the posterior edges of the ribs due to puncture of pleura during sticking that leads to aspiration of blood into thoracic cavity. This contamination is known a back bleeding or over-sticking.

Bacon: The meat from a pig that is treated with smoke or salt, and is often cooked in rashers (thin pieces).

Bacteraemia: The occurrence of bacteria transiently in the bloodstream.

Barley beef: The beef production from cattle by feeding them a diet which largely composed of barley.

Basso Macello: A system for the sale of meat of inferior quality in Italy. The meat requires some treatment before sale. The meat is sold at low price under strict governmental or municipal supervision. The system is similar to Freibank system of Germany and Switzerland.

Beef: A technical term which is used for the meat of cattle species.

Biltong: A type of preserved meat which is prepared mainly from beef (sometimes flesh of game animals, zebra or ostrich) by cutting into strips followed by curing in salt, sometimes immerged in vinegar, followed by drying. This method of meat preservation is used in South America.

Biological hazards: The harmful or toxic substances, chiefly microorganisms or infectious agents. Biological agents include microorganisms which have been genetically modified, cell cultures and human endoparasites which may be able to provide any infection, toxicity or allergy.

Biosafety: The safety measures that are adopted in industries or laboratories dealing with biological materials or organisms.

Biosecurity: The measures that are taken to prevent the spread or introduction of harmful organisms to humans, animals and plant life.

Black beef: A troublesome condition in which the colour of the musculature of freshly slaughtered animal appreciably darker than normal. This is also known as dark cutting beef. The condition develops in cattle when they are exposed to extreme weather conditions, injury, excitement or fatigue during detention in a cold, damp lairage, particularly in autumn when the days are warm and nights are cold.

Black disease: An infectious disease which is caused by *Clostridium oedematiens* and characterized by necrotic hepatitis and black colouration of the inner surface of the skin due to marked and widespread engorgement of the minute blood vessels. The carcase of affected animal is condemned.

Black pith: A type of melanin pigmentation (melanosis) involving lungs, liver, brain and spinal cord. This condition is commonly termed as "black pith" by the butchers.

Black spot: A type of meat spoilage which is caused by a mould "*Cladosporium herbarum*".

Bleeding: A process of cutting of part of the neck and removal of blood from the body of animal during its slaughter.

Blood splashing: The muscular hemorrhages in the slaughtered animal in various locations of the body. Delay between stunning and sticking; and increased blood pressure arising violent muscular contraction at stunning are the main causes of blood splashing.

Bloom of carcase/meat: The development of bright colour associated with the formation of oxymyoglobin on the lean surface of a cut of meat when exposed to oxygen.

Blower can: A can with bulged ends by internal pressure due to production of gas by microbial activities or chemical reactions.

Boar odor: An unpleasant odour of boar found in pork from a proportion of uncastrated male pigs due to the presence of 5α-androst-16-en-3-one, indole and skatole (3-methyl indole). These compounds are mainly located in adipose tissue and have been associated with odours such as 'perspiration-like' or 'urine-like' in cooked pork.

Boar: A male pig after puberty, intended for breeding.

Bobby calves: The calves which are slaughtered at the age of 1-7 days, with the object to prevent the loss of milk due to consumption by calves. There is little muscular development and the flesh is fit only for use in manufactured food products.

Boiling test: A test applied for physical evaluation of quality of meat by boiling the meat and observing the smell from cut surface of the boiled meat.

Bone discolouration: A type of pigmentation of bone which is attributed to an iron containing pigment "porphyrin".

Bone taint: The growth of putrefactive bacteria (mostly anaerobic spore-forming bacteria) in both the musculature around the bone and in the bone marrow, due to slow cooling of deep seated musculature.

Booth system: A system of slaughter of animal in which the slaughter, bleeding and dressing operations are carried out on the same booth.

Bovine spongiform encephalopathy (BSE): A viral zoonotic disease which is caused by prions. It is a highly contagious viral disease of cattle which affects mainly the nervous system.

Brine staining: A type of contamination of meat with calcium chloride which is commonly used in the brine circulating system for chilling or freezing of meat.

Broiler meat: The meat obtained from young male or female chicken about 1.5 kg in weight and 8 weeks of age.

Broiler: The chicken that is bred and raised specifically for meat production. The term "broiler" is mostly used for a young chicken, 6 to 10 weeks old and weighing 2.5-3.5 pounds, and is interchangeable and sometimes in conjunction with the term "fryer", *e.g.*, "broiler-fryer".

Browning: A defect in cured meat which occurs due change of pigment nitrosomyoglobin to brown metmyoglobin due to dehydration caused by low humidity, high temperature and oxidation caused by prolong exposure to air, excessive nitrate and poor packaging.

Bruise: A condition which involves release of blood from ruptured blood vessels into surrounding tissues due to traumatic injury. Bruise affects the quality of meat which requires local condemnation of carcase.

Bunging: A technique of sealing the rectum in order to prevent faecal contamination of carcase. In bunging, a cut is made around the rectum to free it from the carcase, and then it is tied off to prevent spillage of fecal material.

Calcification: The deposition of calcium salts, small amount of iron, magnesium and other minerals in the body tissues.

Canning: A method of food preservation involves heating of food above boiling point in sealed cans. Canning virtually kills all microorganisms.

Canpak system: A continuous conveyorized method in which the carcases are suspended by heavy beef trolleys or runners from the overhead rail. Spreader or gambrel is not used in this method.

Capon: A castrated male chicken that weigh 6-8 pounds.

Carabeef: The technical term which is used for the meat of buffalo species.

Carbuncle: A deep-seated skin abscess, caused commonly by staphylococci.

Carcase yield: The weight of carcase in respect to the weight of live animal. This is also known as killing out percentage or dressing out percentage.

Carcase: The body of an animal after its slaughter and dressing. It is the whole body of the animal after slaughter, bleeding, removal of skin, head, extremities of the limb at the carpus and tarsus, udder and tail and evisceration. The skin and udder are included in the pig carcase.

Casualty Slaughter: A category of animals on ante-mortem inspection in which the animals are not in acute pain or immediate danger of death but affected with some more chronic conditions such as obturator paralysis, post-partum paraplegia following milk fever in cow and benign superficial tumor.

Chalef: A razor sharp knife which is used in Jewish method of slaughter of animals.

Chemical stunning: A method of stunning by using chemical substances like carbon dioxide, nitrous oxide or mixture of argon and carbon dioxide or oxygen.

Cheviot flesh: The meat obtained from Cheviot sheep.

Cheviot sheep: The Cheviot originated in the Cheviot Hills, on the border of England and Scotland. The Cheviot is a distinctive white-faced sheep, with wool-free head and legs, pricked ears, black muzzle and black feet. It is a very alert, active sheep, with a stylish, lively carriage.

Chevon: A technical term which is used for goat meat.

Chicken: A technical term which is used for poultry meat.

Chill room/hall: A room/hall which is used for the treatment of the carcase immediately after dressing for its rapid cooling in a dry atmosphere.

Chilling: A method of food preservation using low temperature storage (between 1°C and 4°C). It reduces the spoilage by slowing down the action of microorganisms and their enzymes.

Chitterlings: The cooked food products prepared from stomach and large intestine of pig.

Cholangitis: The inflammation of bile ducts. It may be caused by many bacteria and parasites.

Cholecystitis: An inflammatory condition of the gall bladder.

Cholelithiasis: The formation of stones in the gall bladder (gallstones) and bile ducts (biliary calculi).

Cirrhosis: The chronic variety of hepatitis.

Cleaning: A soil-removing process which removes the microorganisms. It involves the removal of soil, food residue, dirt, grease or other objectionable matter. Cleaning is the necessary prerequisite for sterilization and disinfection.

Cleaning-in-place (CIP): A system of cleaning in which cleaning compounds are circulated by a pump through a series of pipes to the components to be cleaned. This can be applied in meat industry for the internal cleaning of mixers, choppers and other equipment that necessitate the use of tanks.

Cold shortening: A condition develops due rapid cooling of meat/carcase (below 10°C) before falling of muscle pH below 6.0. This condition is characterized by contraction of muscle fibres and toughening of meat.

Communicable disease: The disease capable of being directly or indirectly transmitted from human to human or animal to animal or from environment to human or animal.

Condemnation: A decision making process to declare the carcase/meat unfit for human consumption.

Condemned meat room: The room where condemned carcase/meat is kept for a short period of time before its disposal.

Condemned: The carcase and parts of slaughtered animal which has been examined and judged by a competent authority, as being unsafe or unsuitable for human consumption and requiring appropriate disposal.

Contagious disease: An infectious disease in which the causative agent(s) are spread by direct animal-to-animal (also human-to-human or animal-to-human) contact.

Contaminant: Any physical (foreign matter, or other substance), chemical or biological agent which has not been intentionally added to food but may affect the food safety or suitability.

Contamination: The presence of an infectious agent on a body surface, also on or in clothes, beddings, surgical instruments or other inanimate articles or substances including water, milk and food.

Continuous powered system: A line dressing system in which the dressing line is in continuous motion. The system has mechanical hide puller, moving inspection table *etc.*

Contusion: A region of injured tissue or skin in which blood capillaries have been ruptured.

Copper sulphate test: A test applied for chemical evaluation of quality of meat to determine the staleness of meat.

Critical control points (CCPs): A location, practice, procedure or process at which control can be exercised over one or more factors which if controlled, could minimize or prevent a hazard.

Critical limit: A criterion which separates acceptability from unacceptability.

Cross-contamination: The physical movement or transfer of microorganisms from one person, object or place to another. Food and kitchen tools and surfaces may become contaminated from raw food products such as meat and poultry. Microorganisms can be transferred from one food to another by using the same knife, cutting board or other utensil without washing the surface or utensil in between uses.

Curing: A method of meat preservation which involves the addition of salt, sugar, nitrite and/or nitrate *etc.* to the meat which results in a conversion of the meat pigments, predominantly myoglobin, to the nitroso or cured form. High salt brine solution of curing ingredients inhibits the growth of microorganisms.

Cutting room: The room where carcases are cut into different meat pieces.

Cyst: A close cavity containing clear fluid, pus or semisolid material, which is usually 1-5 cm in diameter.

Cystitis: The inflammation of the urinary bladder.

D Value (Decimal reduction value): The dose (time) required to destroy 90 per cent of the initial population of microorganisms. When the does required to reduce the population from 1000 000 to 100 000 is the same as the dose required to reduce the population from 100 000 to 10 000, *i.e.* the D value remains constant over the full range of the survivor curve. D value reflects the resistance of an organism to a specific temperature.

Dark cutting meat: An abnormal muscle condition in which it remains dark purplish-red on exposure to air instead of a bright red colour. This condition develops due to long-term stress before slaughter or starvation uses up the glycogen so that less lactic acid is formed after slaughter. Dark cutting meat develops in the carcases of cattle.

Dark firm dry (DFD) meat: The meat which is unpleasant purplish-red colour, direr and darker than the normal meat, lower shelflife, less desirable colour and flavour but fit for human consumption. DFD meat develops in the carcases of pigs.

Decapitation: A process of removal of head of an animal during its slaughter. In Jhatka method of slaughter, the head of the animal is decapitated in a single blow by using a sword.

Defeathering: A process of removal of feathers of birds during their slaughter.

Degeneration: The change of tissue to a lower or less functionally active form or deterioration (impairment) of an organ or cell due to changes in its size.

Dehiding: The process of removal of hide of the slaughtered or dead animals.

Delayed Slaughter: A category of animals on ante-mortem inspection that may be fatigued, inadequately rested, excited or affected with transit sickness, and such animals require rest or treatment.

Detained meat room: A separate room where suspected carcases are detained for detailed post-mortem inspection, laboratory testing *etc.*

Detergent: A surface cleaning agent which has power to remove soil or dirt from surfaces. It works by lowering the surface tension.

Disinfectant: A chemical agent that is capable of disinfection. Disinfectant may kill vegetative bacteria, viruses, fungi and occasionally spores by the destruction of proteins, lipids or nucleic acids in the cell or its cytoplasmic membrane. Disinfectant is used on inanimate objects and surfaces.

Disinfection: The process of destruction of microorganisms, but not usually bacterial spores. Disinfection does not necessarily kill all microorganisms but reduce them to a level acceptable for a defined purpose.

Disinfestations: The process of physical or chemical removal or destruction of the parasites from the body of the host, inanimate objects *etc.*

Dressing percentage: The percentage of carcase weight with respect to the live weight of slaughtered animal.

Dressing: The process of progressive separation of the body of a slaughtered animal into a carcase and other edible and inedible parts.

Drip: The presence of a watery, blood stained fluid which escapes from frozen meat when thawed. This occurs due to rupture of muscle fibres by ice crystals and irreversible changes in muscle plasm. Drip consists mainly of water, salts, proteins, damaged blood cells and extractives.

Drying: A method of meat preservation by sun drying or mechanical drying of meat.

Dysplasia: The alteration in the size, shape and orientation of adult cells. Usually epithelial cells are affected, though sometimes mesenchymal cells may also be involved. This is caused by chronic irritation and inflammation. This is seen in the epithelium of skin, oesophagus and cervix.

Edible red offal: The offal includes liver, heart, kidney, tongue, head and tail.

Effluent treatment plant: A system for the treatment of effluent released from factories in order to safe disposal of treated effluent.

Effluent: The solid-liquid wastes released from factories containing large amounts of organic and inorganic matters.

Electrical stunning: A method of stunning which involves passing of alternating current in the body of animal to render it unconscious. It causes massive polarization of neurons in the brain, resulting in an epileptic seizure.

Emaciation: A pathological condition of an animal developed either due to lower food intake or some metabolic disturbances, usually during the course of chronic disease.

Emergency Slaughter: A category of animals on ante-mortem inspection in which the animals are in acute pain or suffering from a condition where a delay in slaughter would be contrary to animal welfare and may lead to economic losses due to death of animal.

Endotoxin: A component of the outer membrane of Gram-negative bacteria, released from the bacterial surface following natural lysis of the bacterium or by disintegration of the organism in vitro.

Enterotoxin: A toxin which acts on the intestinal tract to cause excessive exudation of fluid.

Evernazione: A method of slaughter in which a sharp double edged knife is plunged into occipito-atlantal space at the nape of the neck of the animal to sever the medulla oblongata.

Evisceration: A process of the removal of the viscera, including edible offal, heart, intestines, paunch, liver, lungs, spleen and pancreas.

Exhaustion of animal: A situation in which an animal losses its energy due to excessive physical activity like walking or traveling for a long distance, fighting *etc.*

Exotoxin: The diffusible proteins secreted into the external medium by the pathogen. Exotoxins are produced by Gram-negative and Gram-positive bacteria. Exotoxins are very toxic, highly antigenic and used for toxoid preparation.

Extract release volume (ERV): A test applied for physical evaluation of quality of meat by observing the extent of extract release volume from homogenized meat. Spoilt meat releases less volume than the fresh meat.

Exudative diathesis: A condition of young domestic fowls and turkeys involving breakdown of the blood capillaries, and also leakage of plasma due to increased capillary permeability. The condition is characterized by accumulation of gelatinous yellow fluid beneath the skin.

F value: The equivalent time in minutes at 250°F, of all heat considered, with respect to its capacity to destroy spores or vegetative cells of a particular organism. This represents a measure of the capacity of a heat processes to reduce the number of spores or vegetative cells of a given organism per container.

Falsification: A fraud practice in selling of meat of different animals or mixing the inferior or undesirable meat with superior or costly meat.

FAO: An international organization of the United Nations "Food and Agriculture Organization" which is associated with the food production and food safety.

Farmer's lungs: An allergic condition in persons working in meat plants which occurs due to contact with infected animals or inhalation of mouldy hay and straw containing large amount of spores of a bacterium *Micropolysporum faeni* and characterized by rhinitis, asthma, dyspnoea, cough, fatigue and chills.

Fasting: A process of withholding the feed from animals for a certain time period (*e.g.*, 6 hours for cattle) before their slaughter with a view to minimize the emigration of bacteria from intestinal tract during digestion.

Fatal syncope: A condition in exhausted and fatty pigs which is characterized by fever, skin blanching, erythema and dyspnoea followed by sudden death. This is also known as porcine stress syndrome (PSS), shock heart failure syndrome and herztod.

Felt: A textile material that is produced by matting, condensing and pressing fibers together. Dried hair of cattle has insulating property which can be used as felt for placing beneath carpets.

Fever: The abnormal elevation of central body temperature. It is one of the most common and well known manifestations of disease.

Fiery carcase: A condition of carcase in which blood is trapped in the capillaries located in the fat or connective tissue layers over muscle, resulting in a pink tinged carcase. The condition occurs when an animal has been excited shortly before slaughter.

Fillet: A fleshy boneless piece of meat from near the loins or the ribs of an animal.

Final inspection: A more thorough inspection of carcase and viscera which are found diseased and abnormal on routine post-mortem inspection, to determine whether the carcase and visceral organs are fit for human consumption or not.

Flaying: The technique of removal of skin from a slaughtered animal.

Flipper: A condition of early spoilage in the canned foods which is characterized by normal appearance of the can but flipping out one end of the can when strike against a solid object and regaining its normal position after applying light pressure.

Flowdiagram: A systematic representation of the sequence of steps or operations.

Fomites: The inanimate or non-living objects that carry viable organisms.

Food animals: The animals which are used mainly for the purpose of milk, meat *etc.*

Food hygiene: The conditions and measures necessary to ensure the safety of food from production to consumption.

Food infection: A type of food-borne illness which involves ingestion of viable pathogenic bacteria along the food leads to their lodgment and establishment in organs of consumers.

Food intoxication: A type of food-borne illness which involves ingestion of toxins already produced by microorganisms in the food and brings about poisoning syndrome in the consumers.

Food poisoning organisms: The organisms causing food poisoning due consumption of food contaminated with them or their toxins.

Food poisoning: The poisoning caused due to ingestion of contaminated foodstuffs.

Food safety: Limiting the presence of those hazards whether chronic or acute, that may make food injurious to the health of consumers.

Food security: The availability, accessibility and affordability of food to all people at all times.

Food suitability: The assurance that food is acceptable for human consumption according to its intended use.

Food-borne disease: Any disease of an infectious or toxic nature caused by consumption of food.

Food-borne outbreak: An incident in which two or more persons experience a similar illness, usually gastrointestinal after ingestion of a common food, and epidemiological analysis implicates the food as the source of the illness.

Frankfurter: A cured, cooked and smoked sausage of beef or beef with pork, turkey, *etc.*, made in cylindrical links a few inches long, now usually without a casing; wiener. The frankfurter originated some 350 years ago in Frankfurt, Germany and German immigrants brought the technology to the USA. The terms frankfurter, wiener or hot dog are practically interchangeable today. When the term "beef frankfurter" is used, the sausage is made of pure beef only.

Freeze drying: The process of removal of water content from the frozen foods. In order to ensure the removal of water content the meat should be sliced or diced and packaged in completely moisture proof packaging material.

Freezer burn: The damage of the surface of imported frozen offal usually liver and kidney due to loss of moisture from the outer tissues. The condition is characterized by development of soft and spongy with withered patches on the surface of the offal.

Freezing: A process of food preservation involving the refrigeration of carcase/meat or other foods at temperatures below 0°C. Freezing diminishes the dangers of most pathogenic organisms and slows the growth of spoilage organisms and destroys certain parasites.

Freibank system: A system for the sale of meat of inferior quality in Germany and Switzerland. This meat requires some treatment before sale. The meat is sold at low price under strict governmental or municipal supervision.

Fresh Meat: The meat that apart from refrigeration has not been treated for the purpose of preservation other than through protective packaging and which retains its natural characteristics.

Frozen meat: The meat which has been subjected to refrigeration below 0°C.

Fryer or young rabbit: A rabbit weighing not less than 1.5 pounds and rarely more than 3.5 pounds, and less than 12 weeks of age. The flesh is tender, fine grained, and a bright pearly pink colour.

Fryer chicken: The chicken between 7 and 10 weeks old and weighing between 2.5 and 4.5 pounds when processed.

Furuncle: The large size of abscess in the dermis.

Gangrene: The death of body tissue (necrosis), generally in considerable mass, usually associated with loss of vascular supply and followed by invasion and putrefaction by saprophytic bacteria.

Germicide: A substance which destroys vegetative bacterial cells but not necessarily bacterial spores and thus akin to a disinfectant. It is usually applied in relation to disease producing bacteria.

Giblet: A culinary term for the edible offal of a fowl, typically including the heart, gizzard, liver, and other visceral organs.

Goads: The device which is used to encourage animals to move.

Good hygienic practices (GHPs): The set of requirements to prevent contamination of food in order to provide safe food to the consumers.

Good manufacturing practices (GMPs): The part of quality assurance which ensures that products are consistently produced and controlled to the quality standards appropriate to their intended use and as required by the marketing authorization.

Gravity rail system: A system of carcase dressing in which the carcases are suspended from a spreader and single wheel trolley or runner and are gravitated to each station and stopped by a manually operated stop on overhead rail. The system is most compact and economical but slaughter rates are lower.

Green offal: The offal including stomach and intestine of any animal, other than poultry or fish, and their contents.

Green rot: A condition associated with marked changes in colour and odour of hen's eggs due to growth of *Proteus* or *Pseudomonas*. The changes in colour can be black or red, therefore also known as "black rot" or "red rot".

Green struck: A condition similar to "heated beef" or 'sour side" is known in the trade which occurs in rabbits, hares and game which are packed in hampers or baskets while still warm.

Greening: A type of defect in cured meat which is caused by using excessive nitrate and due to bacterial contamination. The greening caused by hydrogen peroxide occurs commonly on frankfurters as well as on other cured and vacuum packaged meats due to implication of *Lactobacillus viridescens*. The greening caused by hydrogen sulphide production by *Pseudomonas mephitica* occurs commonly on fresh red meat that are held at 1-5°C and stored in gas-impermeable or vacuum-packaged containers.

Gut and tripe room: The room where the initial separation and emptying of stomachs and intestines is carried out.

Guttery: A place for removing the guts or cleaning the guts of slaughtered animals.

HACCP plan: A document prepared in accordance with the principles of HACCP to ensure control of hazards which are significant for food safety in the segment of the food chain under consideration.

HACCP: A system which identifies, evaluates, and controls hazards which are significant for food safety.

Haematoma: A lump formed from a blood clot in tissues or organs that may be associated with trauma or a clotting defect.

Haemorrhage: The escape of blood from an artery, vein or capillary to the outside (external haemorrhage) or into the body cavity or into tissues (internal haemorrhage).

Halal: A method of religious slaughter which is carried out in Muslim method of slaughter of animals.

Hazard analysis: The process of identification or recognition of hazard which may be present in the food chain.

Hazard: A physical, chemical or biological agent in, or condition of, food with the potential to cause an adverse health effect.

Health education: A heath programme which is executed to create awareness among the occupational groups and public in general about source of infection and contamination, mode of spread of disease and its severity and method of prevention and control of occupationally important diseases.

Health: A state of complete physical, mental and social well-being and not merely the absence of disease or infirmity.

Heated beef: A condition known in Britain to freshly-killed carcase which occurs due to inability to dissipate heat rapidly when carcases are hung too close to each other, thus preventing a proper circulation of air around the sides.

Hepatitis: An inflammatory condition of liver involving various degenerative processes. Hepatitis may be either infectious or non-infectious or toxic.

Herztod: A condition in exhausted and fatty pigs which is characterized by fever, skin blanching, erythema and dyspnoea followed by sudden death. This is also known as fatal syncope or shock heart failure syndrome.

Hind: A female deer, also known as doe.

Hippophagia: A technical term which is used for eating of horse flesh.

Host: An organism (human, animal or plant) that harbours a parasite. Host may be of different types such as intermediate, definitive (final), dead-end, accidental (incidental) and amplifier hosts.

Hot boning: The process of cutting up meat without proper cooling of the freshly-killed carcase in an attempt to make an integration of slaughtering and cutting operations, with vacuum packaging of the primal cuts. The process has been used in Germany and US for many years.

Hot meat: The fresh meat immediately after slaughter of animal, prior to setting of rigor mortis in the carcase.

Humane slaughter: The slaughter of animal with its previous stunning.

Hydrogen swell: A type of chemical spoilage in the canned foods which occurs due to internal corrosion in the acidic medium (pH 3.5-4.5).

Hygiene: The conditions and practices that help to maintain health and prevent the spread of diseases.

Hyperplasia: An increase in the overall number of cells in an organ or tissues.

Hypertrophy: An increase in size of the cells and consequently size of an organ. Hypertrophy and hyperplasia usually coexist in the same organ of the body and occur due to increased functional demand by the organ.

Hypoplasia: The failure of an organ to develop to its normal size due to decrease in size and number of cell.

ICMSF: An international commission that provides timely, science-based guidance to government and industry on appraising and controlling the microbiological safety of foods. It provides scientific basis for microbiological criteria and to promote principles for their establishment and application.

Icterus or jaundice: A clinical sign of a faulty liver or bile duct malfunction that lead to an abnormal accumulation of bile pigments in the blood.

Immaturity: The status of an animal which has not attained its age for optimum time of slaughter. The meat obtained from immature animals is not suitable for consumption due to reasons of taste, health and economy. The flesh of immature animals is regarded as unwholesome. The flesh of immature animals also has high proportion of water and bone.

Imperfect bleeding: A reduced degree of bleeding in slaughtered animal due to stress, nervousness, tiredness, and excitation.

Incubation period: The time period between infection and occurrence of clinical symptoms.

Indicator organisms: The organisms or group of organisms indicate likely presence of hazards that may occur in the food or water but they themselves may or may not be responsible for the hazards, *e.g.*, coliforms indicate likely presence of other enteric pathogens in foods.

Infarct: An area of coagulative necrosis that results due to the sudden blocking of an artery which has no collateral circulation.

Infarction: The coagulative necrosis of tissues due to the sudden blocking of an artery which has no collateral circulation.

Infection: The invasion of the body by infectious agents and their successful propagation in tissues and body fluid. There are several levels of infections such as colonization, subclinical or inapparent infection, latent infection and manifest or clinical infection.

Infectious disease: The disease caused by an infectious agent.

Infestation: The lodgment and development of parasites in or on the body of a host.

Intermediate host: The host in the course of heterogenous development in which asexual development (sometimes multiplication) takes place.

Intermittent powered system: A type of on-the-rail carcase dressing systems which involves mechanical moving of carcases suspended on a spreader (gambrel) and trolley along a rail at intervals by means of a variable timing device.

Intoxication: A type of food poisoning which arises due to consumption of food already containing microbial toxins in it (intradietetic or pre-formed toxins in the food).

Invasion: The entry of a pathogen across the epithelial cell surface (local invasion) or enter across the epithelial cell surface and spread systemically through the body.

Irradiation: A process of destruction of microorganisms and parasites by using radioactive substances like infrared rays, ultraviolet rays, gamma rays *etc.*

ISO: An International Organization for Standardization which formulates international standards on food hygiene, food microbiology and food assurance of many foods including foods of animal origin.

Isolation block: A miniature abattoir, with a lairage for up to four cattle, a slaughterhall and hanging room which is situated near to the suspect meat detention room.

Isolation pen: The pen for keeping of diseased/abnormal animals after their ante-mortem inspection in the lairage. Diseased/abnormal animals are kept in the isolation pen for their rest/treatment.

Isolation: The separation for the period of communicability of infected person or animals from others in such places and under such conditions, as to prevent or limit the direct or indirect transmission of the infectious agent from those infected to those who are susceptible or who may spread the agent to others. Isolation is the oldest communicable disease preventive measure.

Jelly pockets: A defect in cured meat which occurs due to injection of brine into connective tissues. It causes denaturation of connective tissues.

Jerked beef: A type of preserved meat which is prepared from beef by cutting into strips and dried rapidly in the air. This method of meat preservation is used in South America.

Jewish slaughter: A method of slaughter of animal as per the Jewish law.

Jhatka: A method of slaughter of animals by Sikh community. This is also known as Sikh method of slaughter.

Jowl: The cheek of a pig used as meat.

Judgement: A decision making process regarding acceptance or rejection of carcase/meat during post-mortem inspection.

Killing-out percentage: The per cent weight of carcase in respect to the weight of live animals.

Kinophagia: A technical term which is used for eating of dog flesh.

Knocking: The process of rendering the animal unconscious or insensible before its slaughter. It is also known as stunning.

Kosher slaughter: A method of slaughter in which the animal is slaughtered as per the Jewish Board of Shechita for production of kosher meat.

Kosher: The carcase/meat which fulfills the requirements of Jewish law and fit for Jews consumption. For production of kosher meat, the animal must be alive and healthy, have suffered no injury and not stunned at slaughter.

Labeling: A process of putting the label on the packaged meat regarding quantity, date of package, period of safe consumption *etc.*

Lairage: An enclosure in the slaughterhouse intended for providing rest, water and feed to the animals arriving at slaughterhouse. In order to conduct efficient ante-mortem inspection of animals arrived at the slaughterhouse; a well-constructed lairage with all essential facilities should be available.

Latent infection: The infection in which the host does not shed the infectious agent which lies dormant within the host without symptoms and often without demonstrable presence in blood, tissue or body secretions of the host.

Law or Act: The regulatory instrument approved by a legislative body in accordance with the law-making procedures provided in the constitution. Food safety laws include all legislation that has an impact, directly or indirectly, on food safety.

Leaker: A condition of microbial spoilage in canned foods which is characterized by hole in the can, which may be contaminated by air and escaping of food content through it.

Lean meat: The meat having little or no fat in it.

Legislation: The law or set of laws suggested by a government and made official by a parliament.

Local body: The local self-government.

Lymphadenitis: The inflammation of lymph nodes. This may be local, general or non-specific.

Lymphangitis: The inflammation of lymphatic vessels.

Malachite green test: A test applied for chemical evaluation of quality of meat to determine the efficiency of bleeding of carcase.

Maximum residue limit (MRL): The maximum concentration of residue per kg fresh weight of food.

Meat food product order (MFPO): A regulatory order associated with meat food product that shall contain no poisonous chemicals in excess of permitted quantity.

Meat hygiene: The expert supervision of all meat products with the object of providing wholesome meat for human consumption and preventing danger to public health.

Meat inspection: The supervision of the food animals and the meat derived from them. It includes ante-mortem and post-mortem inspection, veterinary supervision of municipal abattoirs, export oriented abattoirs, retail meat markets and butcher shops. This is important to ensure the safety of consumers.

Meat swelling capacity (MSC): A test applied for physical evaluation of quality of meat by observing the extent of swelling of meat. The MSC of spoilt meat is higher than the fresh meat.

Meat technology: The application of science and engineering methods for production, processing, packaging, preparation and utilization of meat and meat products.

Meat: The flesh or other edible parts of animals (usually domesticated cattle, swine, and sheep) used for food, including not only the muscles and fat but also the tendons and ligaments.

Melt: A synonym of spleen.

Metaplasia: The transformation of one type of cell into another type. It is occasionally encountered in meat inspection.

Microbiological contamination: The non-intended or accidental introduction of microbes such as bacteria, viruses, fungi, parasites and their toxins and byproducts in the food, water *etc.*

Microbiological guidelines: The non-mandatory criteria usually intended as a guide to good manufacturing practice.

Microbiological specification: Contractual agreements between a manufacturer and a purchaser to check whether the foods are of the required quality.

Microbiological standards: The mandatory criteria with legal backing.

Minced meat: Boneless meat which has been reduced into fragments.

Misrepresentation: A type of fraud in meat selling which involves replacing/adding a particular part of the carcase/animal body part (head, tail *etc.*)/organ (testes).

Missed shock: A condition produced due to application poorly positioned electrical electrodes on the body of animal during electrical stunning. The condition is characterized by paralysis in the animal but it is fully conscious.

Mixing: A process which involves pooling of meat from different animal species or good quality meat with poor quality/infected meat.

Monitoring: The act of conducting a planned sequence of observations or measurements of control parameters to assess whether a CCP is under control.

Morbidity: The amount of disease in a population. It is commonly defined in terms of incidence or prevalence.

Mortality: The measure of the number of deaths in a population. It represents the lot of animals dying or dead in a population.

Mulberry heart disease: The degeneration of heart muscle in pigs, caused by deficiency of selenium or vitamin E. The disease is sporadic and acute with rapid onset and affects fast growing pigs. The disease is characterized by a distended pericardial sac containing fluid, and extensive haemorrhages in the myocardium. Death occurs suddenly and may be precipitated by exercise.

Muslim method of slaughter (Halal): A method of slaughter of animals by Muslim community.

Mutton: A technical term which is used for sheep flesh.

Myoglobin: A type of pigment in the meat which is responsible for its redness. Myoglobin is the deciding component of meat for its broad classification into "red" and "white" meat.

Necrosis: The local death of cells or tissue in a living body.

Nitrosamine: A substance which is formed from nitrites. Nitrite is used in curing salts which can react with amines commonly present in food, to form nitrosamines, which has been shown to be carcinogenic in all species of animals.

Nitrosohaemochrome: An expected heat-stable pink pigment found in cured meat products in which nitrite is purposefully added at levels to assure the formation of this pigment. Nitrite can be reduced to nitric oxide, which reacts with the heme ring of myoglobin and forms nitrosohemochrome upon cooking, thereby promoting the pinking.

Nitrosomyoglobin: A bright red colour pigment is formed from metmyoglobin due to action of nitric oxide.

Notifiable disease: The disease which requires immediate reporting to the higher health authority to take necessary action for preventing further spread.

Occupation hygiene: The science of the anticipation, recognition, evaluation and control of hazards arising in or from the workplace, and which could impair the health and well-being of workers, also taking into account the possible impact on the surrounding communities and the general environment.

Occupational disease: The disease that occurs in persons involved in certain occupation such as animal handling and slaughtering, meat processing *etc.* Slaughterhouse and meat processing occupational groups may be exposed to zoonotic pathogens and affected with several zoonotic diseases.

Occupational injuries: The injuries that occur in persons involved in certain occupation such as animal handling and slaughtering, meat processing *etc.* They get injured during handling and slaughtering of animals and meat processing.

Oedema (dropsy): An excessive accumulation of fluids in the intercellular spaces and body cavities.

Offal room: The room where offal such as lung, liver, kidneys *etc.* are trimmed and then placed in a chilling or freezing room.

Offal: The fresh meat other than that of a carcase whether or not naturally attached to the carcase, *e.g.*, liver, kidney, heart *etc.*

Official veterinarian: The veterinarian appointed by the competent authority.

On-the-rail dressing system: A system of carcase dressing, where the carcases are conveyed by gravity or power along an overhead rail after stunning and bleeding. The system is useful for hygienic dressing of carcase as well as hygienic meat production.

Opportunistic infection: An infection by an organism that takes the opportunity provided by a defect in host immune system to infect the host and hence cause disease.

Opportunistic pathogens: The pathogens that rarely cause disease in individuals with intact immunological and anatomical defenses. They cause disease only when the host defenses are impaired or compromised, as a result of congenital or acquired disease or by the use of immunosuppressive therapy or surgical techniques.

Organoleptic inspection: The inspection using the senses of sight, touch, taste and smell for identification of diseases and defects.

Osteohaematochromatosis: A congenital condition which occurs in calves, young cattle and pigs affecting all the bones of the skeleton. It is attributed to an iron-containing porphyrin pigment.

Outbreak: Foodborne disease outbreak is an incident in which two or more persons experience a similar illness, usually gastrointestinal after ingestion of a common food, and epidemiological analysis implicates the food as the source of the illness.

Oxymyoglobin: An oxidized product which is formed due to exposure of myoglobin to oxygen. Oxymyoglobin imparts red colour to the meat.

Ozonization: A method of meat preservation which restricts the growth of moulds and yeasts on the surface of meat. This is also effective against bacteria, when moisture content on the superficial tissues is low. Ozonization is valuable in the treatment of various types of taints such as oil taint, fruit taint *etc.* in the frozen imported meat.

Pale soft exudative (PSE) meat: An abnormal condition of pork which develops due to stress in pigs just prior to their slaughter.

Papule: A small, circumscribed, raised solid elevated skin lesion.

Parasite: An organism that lives and subsists in or on the body of a host.

Paratenic host: The host in which infective stages may accumulate but do not develop further.

Pathogen: An organism that parasitizes an animal, plant or human and produces a disease.

Pathogenesis: The process of developing a disease.

Pelts: The skin of a beast with the hair.

Pemmican: A type of preserved meat which is prepared from beef, venison or fish by smoking followed by drying and pulverization and addition of equal amount of fat. This method of meat preservation is used in North America.

Pen: A component of the lairage, where animals are kept for resting, watering and feeding. A suitable lairage has a series of pens with wall or tubular partitions.

Percussive stunning: A type of physical or mechanical stunning which is carried out either by penetrative or non-penetrative object.

Peritonitis: An inflammatory condition of peritoneum which may be acute or chronic.

Peroxide test: A test applied for chemical evaluation of quality of meat to determine the freshness of meat.

Phosphorescence: A type of microbial spoilage which is caused by *Pseudomonas phosphorescens.* The surface of the affected meat shows scattered luminous areas on when seen in a dark room.

Picric acid test (PAT): A test applied for chemical evaluation of quality of meat to determine the spoilage/freshness of meat.

Pigmentation: Abnormal colouration in the skin and other organs that may occur due to deposition of certain type of pigments. Abnormal deposition of melanin is encountered mainly in lungs, liver and on the meninges and spinal cord of calf and ox.

Pithing: A process of inserting a long thin rod or closely coiled wire into the hole (made by the penetrating bolt of the pistol during stunning) in the skull to destroys the motor centres of the brain, so that reflex muscular action does not occur. The purpose of pithing is to avoid injury to operatives and speeding carcase dressings.

Plant sanitation: The process of removal of waste materials from meat plant in order to prevent or minimize contamination.

Plate count: A process of determining the load of microorganisms in food or surface of equipment.

Pleurisy (Pleuritis): The inflammation of pleura.

Pluck: The internal organs comprising larynx, trachea, lungs, heart and liver in sheep, calf and pig.

Pneumonia (Pneumonitis): An inflammatory condition of the lungs in which exudate is produced in the alveolar spaces that causes consolidation of the lung tissues.

Poisoning: A condition which is caused due to infection and/or intoxication by organisms that results into severe health hazards and economic losses.

Porcine stress syndrome (PSS): A condition which is observed in exhausted and fatty pigs and characterized by fever, skin blanching, erythema and dyspnoea followed by sudden death. Porcine stress syndrome is also known as herztod or fatal syncope.

Porging: A process of removal of larger blood vessels from meat/carcase particularly from forelimb and hind limb in Jewish method of slaughter.

Pork rinds: A type of edible product which is prepared from pork skin. In the United States, Latin America, Europe and some Asian countries, pork skin is immersed, boiled, dried and then fried to make a snack food.

Pork scratching: Pork rinds in UK are known as pork scratching.

Pork: A technical term which is used for pig meat.

Post-mortem inspection: Any procedure or test conducted by a competent authority on all relevant parts of slaughtered/killed animals for the purpose of judgement of safety and suitability and disposition.

Preservation: A method of storage of foods for long duration without deterioration of their conditions.

Pre-slaughter animals handling: A practice of gentle handling of animals before their slaughter in order to avoid stress in them and prevent the deterioration in the quality of meat obtained from them.

Pre-slaughter check: A process of determining the animal's unconsciousness after stunning to ensure its humane slaughter.

Pre-slaughter stress: The stress in animals before their slaughter that may occur due to inadequate rest to them, their rough handling, fighting among them, stampede *etc.* Pre-slaughter stress in animals causes poor quality meat production.

Primary production: The steps in the food chain such as harvesting, slaughter, milking, fishing *etc.*

Prolapse: Downward or outward displacement of an organ or part from its normal position. It may be due to weakness in the supporting tissues or follow usually heavy loading or stress.

Protein quality: The quality of a protein is a measure of its ability to satisfy human requirements for the amino acids. The quality of dietary protein can be measured as the ratio of the available amino acids in the food or diet compared with needs.

Puntilla: A short double-edged knife which is used for slaughter of animals in neck-stab (Evernazione) method of slaughter.

Purple staining: A Type of chemical spoilage which involves breakdown of sulphur containing proteins by *Clostridium nigrificans.* In this process, hydrogen sulphide is liberated and a thin layer of tin sulphide is formed on the inside of the can.

Pyaemia: The spreading of pus forming bacteria through the blood stream resulting in metastatic abscessation in other parts of the body.

Quality assurance (QA): The planned and systematic activities implemented within the quality system and demonstrated as needed, to provide adequate confidence that an entity will fulfill requirements for quality.

Quarantine: A restraint, placed upon the movement of the animals, humans, plants or goods which are suspected of being carriers or vehicle of infection or of having been exposed to infection. In a general sense refers to detention of animals suspected of disease at ports or land borders.

Quick chilling: The rapid lowering of the carcase temperature within an hour after slaughter.

Raceways: The necessary lanes for animals to walk or be led on/off vehicles and platforms into holding pens or slaughter facilities *etc.*

Ready-to-eat (RTE) products: Products that are intended to be consumed without any further biocidal steps.

Rearing: The growing of animals, particularly for milk and meat production purpose. The rearing of animals under hygienic condition is an important step in the production of quality meat.

Red offal: The offal including heart, liver, kidneys, spleen, tongue, lung, pancreas *etc.*

Red rot: A type of microbial defect associated with marked changes in colour and odour of hen's eggs due to growth of *Proteus* or *Pseudomonas.*

Refrigeration: A process of food preservation at low temperature. It is the most commonly and widely used method of meat preservation at commercial as well as domestic levels.

Regulation: The set of rules prescribed by the competent authority, as prescribed in the constitution and/or in primary legislation, generally by the executive body.

Re-inspection: A process of repeat inspection of carcase/meat to ensure the handling of product in clean environment, the product free from harmful ingredients and the product have not been mislabeled.

Rejection: A process of condemnation of unsound carcase/meat in order to prevent health hazards in consumers.

Religious method of slaughter: The slaughter of animals according to the laws of different religions.

Rendering: A process of heating of slaughterhouse by-products through which fats are separated from water and protein residues for the production of edible lards and dried protein residues.

Resazurin dye reduction test: A test which is applied for microbiological evaluation of quality of meat.

Reservoir: An animate or inanimate object on or in which an infectious agent usually lives and serve as a source of infection. Reservoirs are usually domestic or wild animals, birds and human beings that may or may not show clinical disease but help to maintain the infection for long periods.

Residues: These are "pharmacologically active substances (whether active principles, recipients or degradation products) and their metabolites which remain in foodstuffs obtained from animals to which the veterinary medicinal products (VMPs) has been administered".

Resting: The process of providing an adequate period of rest (before slaughter) to the food animals arrived at slaughterhouses.

Rigor mortis: A phenomenon that occurs in a slaughtered or dead animal, which is characterized by the hardening and contraction of all the voluntary muscles, loss of transparency of the surface of the muscles (dullness), stiffness of joints and slight rise in temperature of the carcase to 1.5°C or more above normal (in case of beef carcases).

Rimmington and Fowrie test: A test applied for chemical evaluation of quality of meat for differentiation between jaundiced carcase and carotene pigmented carcase.

Risk assessment: The process that is used to estimate and characterize risk, quantitatively or qualitatively.

Risk communication: The exchange of information and opinions concerning risk and risk-related factors among risk assessors, risk managers, consumers and other interested parties.

Risk management: The weighing and selecting of options, and implementing controls as necessary to ensure an appropriate level of protection.

Risk: A function of the probability of an adverse event and the magnitude of that event, consequential to a hazard(s) in food.

Ritual slaughter: The slaughter of animals according to the religious beliefs. This is also known as religious slaughter.

Roaster or mature rabbit: A mature rabbit of any weight, but usually over 4 pounds and over 8 months of age. The flesh is firm and coarse grained, and the muscle fibre is slightly darker in colour and less tender. The fat may be more creamy in colour than that of a fryer or young rabbit.

Roaster chicken: An older chicken about 3 to 5 months old and weighing between 5 and 7 pounds. Roaster chicken yields more meat per pound than a fryer and is usually roasted whole, but it can also be used in other preparations, like chicken cacciatore.

Rodding: A process of separation of oesophagus from trachea and surrounding tissues using a metal rod and tying of oesophagus with a thread in order to prevent spillage of gastrointestinal contents and thereby contamination of interior and exterior surface of the carcase.

Rothera's test: A test applied for chemical evaluation of quality of meat to determine the presence of ketone bodies (acetone) in the carcase/meat. The test is also known as sodium nitroprusside test.

Safe and wholesome meat: The meat which is fit for human consumption, free from contamination and risk of food-borne infection or intoxication and does not contain residues in excess of permissible limits.

Safe for human consumption: The food that has been produced by applying all food safety requirements appropriate to its intended end-use; meets risk-based performance and process criteria for specified hazards; and does not contain hazards at levels that are harmful to human health.

Safety: A state in which or a place where a person is safe and not in danger or at risk.

Salmonellosis: A bacterial disease which is caused by various *Salmonella* serovars and characterized by sudden onset of nausea, vomiting, foul-smelling diarrhoea and fever. Complications may lead to systemic infections and cause arthritis, osteomyelitis, pyelonephritis, peritonitis, cholecystitis, meningitis and endocarditis.

Sanitary indices: The indices that is associated with sanitary condition of a product such as standard plate count, coliform count and other microbial counts in foods or on equipment surfaces.

Sanitation standard operating procedures (SSOPs): A documented system for assuring that personnel, facilities, equipment and utensils are clean and where necessary, sanitized to specified levels prior to and during operations.

Sanitation: The provision of facilities and services for the safe disposal of human urine and faeces. It also refers to the maintenance of hygienic conditions, through services such as garbage collection and wastewater disposal.

Sanitization: The process of destruction of all pathogenic and almost all non-pathogenic microorganisms from equipment surface, wounds or clothing, through the use of heat or chemicals.

Sanitizer: A substance which reduces the number of microorganisms on the skin and surfaces of equipment and other objects.

Sarcocystosis: A type of parasitic infestation of carcase/meat with *Sarcocystis* spp. Sarcocysts are frequently seen in the skeletal muscles. Carcases severely affected with sarcocysts are needed to be condemned.

Scalding: A process of immerging the slaughtered animal (usually pigs and poultries) into the hot water.

Sealer: The person who assists the shochet is known as sealer. Sealer is also known as shomer. Sealer is responsible for putting the kosher mark on the brisket and on edible offal.

Searching: A type of post-mortem examination, carried out by a Jewish butcher (shochet) by making an incision posterior to the xephoid process and inserting the arm to detect any adhesions in the thoracic cavity.

Seedy cut: A type of melanosis in which the melanin pigment is accumulated in various organs such as lymph nodes, pig skin and belly fat or mammary tissue in female pigs. This condition is called "seedy cut" or "seedy belly" since the black colour in the mammary tissue resembles round, black seeds.

Septicaemia: An overwhelming invasion of the bloodstream from a focus of infection.

Sereny test: A test used for assessing the virulence of viable bacterial cultures (*e.g., E. coli* and shigellae). The test is carried out by administering a drop of cell suspension containing 1.5 x 10^{10} to 2.3 x 10^{10}/ml in phosphate buffered saline into the conjunctiva of guinea pig and examined daily for 5 days for the development of keratoconjuntivitis.

Shackling: Securing of birds to reduce wing flapping during their slaughter.

Shechita: The Jewish method of slaughter. This is controlled in Britain by the Jewish Board of Shechita. In this method a swift cut is made across the neck with a very sharp knife.

Shelflife: The length of time in which a food can be kept safe for its use/consumption.

Shipping fever: A condition which is induced due to transportation of food animals, caused by *Pasteurella haemolytica* and characterized by an acute, toxaemic bronchopneumonia with high fever. It mainly affects stored cattle in a poor condition that have become fatigued due to a long journey by road, rail or ship without providing sufficient rest to them.

Shochet: The butcher who performs Jewish slaughter.

Shock heart failure syndrome: A condition which is observed in exhausted and fatty pigs and characterized by fever, skin blanching, erythema and dyspnoea followed by sudden death. This is also known as herztod or fatal syncope.

Shock: A common grave medical emergency, characterized basically by a reduction in the effective circulating blood volume and in the blood pressure.

Shomer: A slaughterhouse personnel who assists the shochet in Jewish method of slaughter. Shomer is also known as sealer. Shomer is responsible for putting the kosher mark on the brisket and edible offal.

Sikh method of slaughter: A method of slaughter in Sikh community in which sheep or goat is decapitated by one stroke of a special sword.

Singeing: The process of removing hair on a hog carcase by burning it off which may be done with a hand-held gas torch or automated systems. Singeing removes any remaining hairs, shrinks and sets the skin, decreases the number of adhering microorganisms and leaves an attractive clean appearance.

Skinning: A process of removal of skin of a slaughtered or dead animal.

Slaughter authorized under special condition: A category of animals on ante-mortem inspection that shows localized conditions such as fracture, injuries, abscess, tumors *etc.* which will show the post-mortem lesions, need to be segregated and subjected to a detailed examination.

Slaughterhall: An important component of a slaughterhouse with sufficient floor space, good ventilation and adequate light, where food animal are slaughtered.

Slaughterhouse: A slaughterhouse also called an "abattoir, is facility where animals are killed and processed into meat products.

Slimy spoilage: A microbial spoilage caused by yeasts and lactic acid bacteria such as *Lactobacillus* spp., *Enterococcus* spp. and *Bacillus thermosphacta*. It occurs on the outside of casings, especially of frankfurters, and may be seen in early stage as discrete colonies which may later coalesce to form a uniform layer of gray slime.

Smoking: A method of meat preservation using smoke from wood sawdust in a generator. Smoke contains chemicals which inhibit the growth of microorganisms.

Soil (deposits): The substances in the meat plant, deposited from animals themselves and other components used in manufacturing processes. These product-derived soils include fats, oils, greases, proteins, carbohydrates, starches *etc.*

Sour side: A condition known in North America to a freshly-killed carcase which is similar to "heated beef" known in Britain.

Source of infection: Any host organism, object or substance from which an infectious agent passes to a susceptible individual. A source of infection thus may be patient, carrier, reservoir, vector, vehicle *etc.*

Souring: A type of microbial spoilage that takes place underneath the casing of frankfurters, bologna, sausage and luncheon meat. It is caused by *Lactobacillus* spp., *Enterococcus* spp. and other organisms.

Splashing: A condition which involves the release of blood in the muscles due to rupture of blood vessels. It is also known as blood splashing or muscular hemorrhage. It is an undesirable phenomenon which affects the keeping quality of meat.

Spoilage: The process of deterioration of food items due to action of microorganisms.

Springer: A type of spoilage in canned foods which is characterized by bulging of one end of the can when struck against a solid object and regaining of its normal position on applying pressure but bulging of the other end. The condition develops due to improper canning operations such as inadequate exhaustion of air before sealing and overfilling.

SPS measures: The quarantine and biosecurity measures which are applied to human, animal and plant life or health from risks arising from the introduction, establishment and spread of pests and diseases, and from risks arising from additives, toxins and contaminants in food and feed.

Standard plate count: A test applied for microbiological evaluation of quality of meat to assess the microbial load in meat.

Sterilization: The process used to achieve sterility, an absolute term meaning the absence of all viable microorganisms.

Stewing chicken: A chicken (usually hen) over 10 months old and weighing 5-7 pounds.

Sticking: The process of cutting the throat of slaughter animal using a very sharp knife to sever the major blood vessels of the neck to ensure the rapid blood loss and therefore death.

Stress: Any factor that threatens the health of an organism or has an adverse effect on its functioning.

Stunning: The process of rendering the slaughter animals unconscious/insensible.

Substitution: A type of adulteration which involves replacement of some parts of carcase or its organs of one species of animal with other animal species. This practice is objectionable for the reasons of health, religion and economy.

Sulphiding: A purple staining on the inner surface of the can in which sulphur containing foods (fish and meat products especially liver, kidneys and tongue) are packed.

Surveillance: An epidemiological investigation involves gathering, recording and analysis of data; and dissemination of information to interested parties, so that action can be taken to control disease.

Survey: An epidemiological investigation involves collection of information in which a causal hypothesis usually is not tested.

Survivor curve: The logarithm of the number of survivors plotted against the received lethal dose. This is independent of the size of the original population of microorganisms and is approximately liner.

Sweetbread: A variety meat that is composed of the thymus from a young beef animal. Thymus constitutes true sweetbread, while, pancreas is termed as gut sweetbread.

Swell: This is same as blower.

Taint in ham: A condition in the ham develops due to production of amino acids and ammonia from breakdown of proteins due contamination with proteolytic organisms like *Clostridium sporogenes, Clostridium putrefaciens* and *Clostridium putrificum*. The condition is also known as souring.

Tanning: The process of treatment of raw hide with chemicals like lime, acids *etc.* to make finished leather.

TBT Agreement: A WTO Agreement on Technical Barriers to Trade (the "TBT Agreement") which establishes rules and procedures regarding the development, adoption, and application of voluntary product standards, mandatory technical regulations, and the procedures (such as testing or certification) used to determine whether a particular product meets such standards or regulations.

Telangiectasis: A pathological condition of liver which is characterized by bluish black and irregular with depressed surfaces and dilated blood filled hepatic sinusoids. This is also known as "plum pudding".

Terefa: The carcase/meat which does not fulfill the requirements of Jewish law or the carcase showing any disease lesion *i.e.* unfit for Jews consumption.

Thaw rigor: A condition that occurs when muscle is frozen pre-rigor and kept for a short time in cold storage, it is still able to contract and go into rigor after thawing.

Thermal death point: The temperature necessary to kill a given number of microorganisms in a fixed time, usually 10 minutes.

Thermal death time (TDT): The time necessary to kill a given number of organisms at a specific temperature. By this method, the temperature is kept constant and the time necessary to kill all cells is determined.

Toxaemia: A condition of the presence of microbial or endogenous toxins in the blood circulation.

Toxico-infection: A type of food-borne illness which involves infection of intestine with a certain group of organisms ingested along with the food and produce toxins in situ to bring about symptoms of poisoning, *e.g.*, toxico-infection with *Clostridium perfringens.*

Toxigenicity: The toxin producing property of the organism.

Toxin: A poisonous substance, usually of biological origin and especially produced by a bacterium. Toxin is protein in nature and capable of inducing antibody production.

Transit fever: A catarrhal and often fatal disease caused by *Pasteurella haemolytica* and characterized by an acute, toxaemic bronchopneumonia with high fever. It mainly affects stored cattle in a poor condition that have become fatigued due to a long journey by road, rail or ship without providing sufficient rest to them. Transit fever is also known as shipping fever.

Transit tetany: A condition which is characterized by violent muscle contraction that may develop due to long and stressful journey. This condition is observed mainly in cows particularly in advance pregnancy. Warm weather acts as a predisposing factor for development of transit tetany.

Transmission: The passing of a pathogen causing communicable disease from an infected host or group of a particular individual or group, regardless of whether the other individual was previously infected.

Transport mortality: The mortality in the animals during transport may occur due to long distance transportation without providing sufficient rest to the animals during transport, transportation stress, extreme weather conditions, overloading of animals in transport vehicle, inadequate supply of food and water to animals, poor design of transport vehicle *etc.*

Travel oedema: A condition analogous to PSE meat, which may develop in lambs slaughtered after an arduous journey in which the carcase particularly the shoulders is unduly watery. This problem can be alleviated by providing proper rest to the animals before their slaughter and immediate chilling of carcase especially in summer months.

Trichinellosis: A parasitic infestation of carcase/meat with a parasite *"Trichinella* spp.". It is also known as trichinosis or trichiniasis.

Tripe: The intestine that has been emptied after slaughter of animal.

Tripery: A place where tripe is prepared.

Ulcer: The localized destruction of an epithelial surface with exposure of underlying tissues.

Unfit: A category of animals on ante-mortem inspection that may contaminate the premises and meat and may pose a risk for slaughterhouse personnel, include dead animals and animals affected with any condition that would entail condemnation of the carcase at post-mortem inspection.

Veal: A technical term which is used for the meat of calf.

Vector: An invertebrate animal (frequently an arthropod) responsible for transmission of an infectious agent from an infected individual or its excreta to susceptible individual (or to some immediate source of infection such as food or water).

Vehicle: Any nonliving thing or substance including food, milk, meat, water, pus, serum *etc.* by which or upon which an infectious agent passes from an infected individual to a susceptible one.

Venison: The meats from the deer species as well as elk, moose, caribou. Sometimes the meats of antelope are also known as venison.

Vesicle: A small sac (usually less than 5 mm in diameter) on the skin, containing liquid.

Virulence: The relative capacity of a microbe to cause damage in a host.

Viscera: The offal found in the thoracic, abdominal and pelvic cavities including the trachea and oesophagus.

Waste disposal: A process of removal of waste (that may contain hazardous organisms) from slaughterhouse and rendering it harmless before its release in the water bodies in case of liquid waste, and burial or incineration in case of solid waste.

Waste processing plant: The plant having facilities to render the harmful wastes into harmless wastes that do not cause health hazards and environmental pollution.

Watering: The process of providing water to the transported animals arrived at slaughterhouse in order to fulfill the need of water for metabolic activities and survival of the animal. It is also important in reducing the microbial load in the intestine and facilitates the removal of hide or pelts during dressing of the slaughtered animal and also facilitates efficient electrical stunning.

Watery pork: A condition that develops in pork due to severe, short-term stress in pigs just prior to slaughter. This occurs due to rapid breakdown of muscle glycogen and the meat becoming very pale with pronounced acidity (pH values of 5.4- 5.6 immediately after slaughter) and poor flavour. Watery pork is also known as pale soft exudative (PSE) meat.

Weasand: The meat surrounding the esophagus.

Weeping: The presence of a watery, blood stained fluid which escapes from frozen meat when thawed. Weeping is also known as drip. Drip consists mainly of water, salts, proteins, damaged blood cells and extractives. Weeping occurs after thawing of frozen meat which is an undesirable feature.

Wet carcase syndrome: A pathological condition in lambs in South Africa which is characterized by lesions in the form of layers of slimy translucent jelly-like encapsulated material on the buttocks, briskets and flanks.

Whiskers: A type of fungal spoilage in meat which is caused by *Thamnidium* spp. and *Mucour* spp., and characterized by hair like growth of fungi.

White offal: The offal including brain, spine, bone marrow, testicle, teats *etc.*

White spot: A type of spoilage which occurs on the surface of imported meat. This is caused by *Sporotrichum* spp. and *Chrysosporium* spp. The spores of these organisms can develop at -8°C and become profuse at temperature above 0°C.

Wiener: A meat product which is also known as frankfurter.

Withdrawal period: The interval necessary between the last administration to the animals of the drug under normal condition of used and the time when treated animal can be slaughtered for the production of safe foodstuffs. It is also known as the depletion or clearance period which may be few hours to several days or even weeks.

Yellow discolouration: A microbial spoilage occurs in vacuum packaged luncheon-style meat which is caused by *Enterococcus casseliflavus*. The discolouration appears as small spot on products stored at 4.4°C.

Yoke gall: A localized acute inflammation of the skin and subcutis on the dorsum of the neck of working cattle and buffalo bullock due to constant friction caused by yoke.

Z value: The degrees Fahrenheit required for the thermal destruction curve to traverse one log cycle. Mathematically, this value is equal to the reciprocal of the slope of the thermal death time (TDT) curve. Z value provides information on the relative resistance of an organism to different destructive temperatures.

References

BIS. (2007). IS 14904: 2007, Indian Standard: Transport of Livestock- Code of practice (First Revision), ICS 03.220.01; 65.020.30. Bureau of Indian Standards, Manak Bhavan, 9 Bahadur Shah Zafar Marg, New Delhi 110002

CAC. (1993). Joint FAO/WHO Food Standards Programme. Codex Alimentarius Commission, 20th Session Geneva, Switzerland, 28 June – 7 July 1993. Report of the 7th Session of the Codex Committee on Meat Hygiene Rome, 29 March – 2 April 1993. ALINORM 93/16A.

CAC. (2001). Codex Alimentarius Food Hygiene Basic Texts. Food and Agriculture Organization of the United Nations, World Health Organization, Rome

CAC. (2003). CAC/GL 50 (2003) adopted by the 26th Session of the Codex Alimentarius Commission (July 2003).

CAC. (2005). Code of hygienic practices for meat. CAC/RCP 58-2005.

CAC. (2014). Principles and guidelines for the conduct of microbiological risk assessment. CAC/GL 30-1999.

FAO. (2001). Guidelines for Humane Handling, Transport and Slaughter of Livestock. Food and Agriculture Organization of the United Nations, Regional Office for Asia and the Pacific. RAP Publication 2001/4.

FAO. (2019). Technical guidance principles of risk-based meat inspection and their application. Rome, FAO

FAO/WHO. (2006). Updating the principles and methods of risk assessment: MRLs for pesticides and veterinary drugs. Food and Agriculture Organization of the United Nations and World Health Organization, Rome and Geneva 2006.

FAO/WHO. (2015). Codex Alimentarius Commission. Procedural Manual, 24th edition. ftp://ftp.fao. org/codex/Publications/Proc Manuals/Manual_24e.pdf

Gracy, J. F., Collins, D. S. and Huey, R. J. (1999). Meat Hygiene, 10th edn., W. B. Saunders Company Ltd. London, Edinburgh, New York, Philadelphia, Sydney, Toronto.

Jay, J. M. (2005). Modern Food Microbiology. CBS Publishers and Distributers, 4596/1-A, 11 Darya Ganj, New Delhi-110002. (India).

OIE. (2009). Codex Code of Hygienic Practice for Meat. OIE Regional Representation for Eastern Europe Workshop on Food Safety, Sofia, 22-24 April, 2009.

WHO. (2018). Pesticide residues in food. Fact sheet, Revised January 2018.

Index

A

D

E

I

J

K

L

M

N

O

P

Q

R

S